Artemisia annua

Prospects, Applications and
Therapeutic Uses

Artemisia annua
Prospects, Applications and Therapeutic Uses

Edited by
Tariq Aftab
M. Naeem
M. Masroor A. Khan

CRC Press
Taylor & Francis Group
Boca Raton London New York

CRC Press is an imprint of the
Taylor & Francis Group, an **informa** business

CRC Press
Taylor & Francis Group
6000 Broken Sound Parkway NW, Suite 300
Boca Raton, FL 33487-2742

First issued in paperback 2021

ISBN-13: 978-1-03-209589-9 (pbk)
ISBN-13: 978-1-138-63210-3 (hbk)

Library of Congress Cataloging-in-Publication Data

Names: Aftab, Tariq, author.
Title: Artemisia annua : prospects, applications and therapeutic uses /
Tariq Aftab, M. Naeem, and M. Masroor A. Khan.
Description: Boca Raton : Taylor & Francis, 2018. | "A CRC title, part of the
Taylor & Francis imprint, a member of the Taylor & Francis Group, the
academic division of T&F Informa plc." | Includes bibliographical
references.
Identifiers: LCCN 2017034216 | ISBN 9781138632103 (hardback : alk. paper)
Subjects: LCSH: Artemisinin--Therapeutic use. | Artemisia annua.
Classification: LCC RC159.A7 A38 2018 | DDC 616.9/362061--dc23
LC record available at https://lccn.loc.gov/2017034216

Visit the Taylor & Francis Web site at
http://www.taylorandfrancis.com

and the CRC Press Web site at
http://www.crcpress.com

Contents

Preface

Internationally, scientists are making unstinted efforts to improve the understanding of malarial biology and to develop more effective malaria treatments. Malaria remained the major scourge of mankind until the Chinese introduced artemisinin to the world as a remedy. The antimalarial drug artemisinin was discovered by Tu Youyou, a Chinese scientist, who was awarded half of the 2015 Nobel Prize in Medicine for her discovery. Since the discovery of artemisinin, treatments containing artemisinin derivatives (artemisinin combination therapies, or ACTs), have been standardized for the treatment of *Plasmodium falciparum* malaria worldwide. Artemisinin is certainly one of the most promising natural products investigated in the past couple of decades. The plant has potent therapeutic potential beyond its antimalarial activity, including anticancer, immunosuppressive, anti-inflammatory, antihypertensive, antioxidative, antimicrobial, antiparasitic, and antiviral activities. However, artemisinin-derived drugs are not available to millions of the world's poorest people because of the low yield (0.1%–0.5% of dry weight) of artemisinin in naturally grown *Artemisia* plants. The present demand for artemisinin far outstrips supply; therefore, researchers around the world are working toward improving the artemisinin content of the plant by various means.

The editors' efforts, in the form of this comprehensive volume, detail recent updates to the applications, current research, and future prospects of *Artemisia annua*. Since the intact plant contains artemisinin in very low concentrations, its commercial extraction requires huge amounts of plant biomass to be processed. Massive demand and low yield of artemisinin from the plant has led to exploration of alternative means of production, including the cultivation of *A. annua* on scientific lines. Considering the significant benefits of various properties of the plant to human health, we present this exclusive volume entitled *Artemisia annua*: Prospects, Applications and Therapeutic Uses. As per the rationale, this volume focuses on various scientific approaches, namely, agricultural, pharmacological, and pharmaceutical aspects, *in vitro* technology, and nutrient management strategies, as well as omics technologies for the regulation of artemisinin biosynthesis in *A. annua*. The book also contains a plethora of information about various scientific approaches to the cultivation of this medicinal plant. Also, it includes information about the plant's survival under conditions of environmental stress.

The book comprises 14 chapters, most of them being reviewed articles written by experts from around the globe. We are hopeful that this volume will meet the needs of all researchers who are working or have interest in this particular field. Undoubtedly, this book will be helpful to research students, teachers, ethnobotanists, oncologists, pharmacologists, herbal growers, and anyone else with an interest in this plant of paramount importance.

We are greatly thankful to the CRC Press, Taylor & Francis Group, USA, for their expeditious acceptance and compilation of this scientific work. Sincere thanks are expressed to the team members of the Taylor & Francis Group for their dedication, sincerity, and friendly cooperation in producing this volume. With great pleasure, we extend our sincere thanks to all the contributors for their timely response, outstanding and up-to-date research contributions, support, and consistent patience.

Lastly, thanks are also due to the well-wishers, research students, and authors' family members for their moral support, blessings, and inspiration in the compilation of this book.

Tariq Aftab
M. Naeem
M. Masroor A. Khan
Aligarh Muslim University, Aligarh, India

About the Editors

Tariq Aftab received his PhD from the Department of Botany at Aligarh Muslim University, India, and is currently an assistant professor there. He is the recipient of the prestigious Leibniz-DAAD Fellowship from Germany, Raman Fellowship from the Government of India, and Young Scientist Awards from the State Government of Uttar Pradesh (India) and Government of India. He has worked as a visiting scientist at IPK, Gaterleben, Germany, and in the Department of Plant Biology at Michigan State University, United States. He is also a member of various scientific associations in India and abroad. He has published research articles in several peer-reviewed national and international journals and is the lead editor of the book *Artemisia annua: Pharmacology and Biotechnology*. His research interests include physiological, proteomic, and molecular studies on medicinal and aromatic plants.

M. Naeem is an assistant professor in the Department of Botany at Aligarh Muslim University, India. For more than a decade, he has devoted his research to improving the yield and quality of commercially important medicinal and aromatic plants (MAPs). His research focuses on escalating the production of MAPs and their active principles using a novel and safe technique involving radiation-processed polysaccharides (RPPs) as well as the application of potent plant growth regulators (PGRs). To date, he has successfully run three major research projects as principal investigator, two of which were sanctioned by the Department of Science & Technology, New Delhi, while another was awarded by the Council of Science and Technology, Uttar Pradesh, Lucknow. Dr. Naeem has published 7 books and more than 80 research papers in reputable national and international journals. He has also participated in various national and international conferences and acquired life memberships to various scientific bodies in India and abroad. Based on his research contributions, Dr. Naeem has been awarded a Research Associateship from the Council of Scientific & Industrial Research, New Delhi; a Young Scientist Award (2011) from the State Government of Uttar Pradesh; a Fast Track Young Scientist Award from the Department of Science & Technology, India; a Young Scientist of the Year Award (2015) from the Scientific and Environmental Research Institute, Kolkata; a Rashtriya Gaurav Award (2016) from the International Friendship Society, New Delhi; and a Distinguished Young Scientist of the Year Award (2016) from the International Foundation for Environment and Ecology, Kolkata.

 **M. Masroor A. Khan** is a professor in the Department of Botany at Aligarh Muslim University, India. After completing his PhD, he worked as a postdoctoral fellow at the Ohio State University, USA (1987–1988). He then worked as pool scientist for the Council of Scientific and Industrial Research and as a research associate for the Council for Scientific and Industrial Research and the University Grants Commission before he joined Aligarh Muslim University as an assistant professor. In his 30 years of teaching experience at university level, he has guided 8 PhDs, 2 MPhils, and 30 MSc students. For the convenience of his students, he has also developed study materials and launched more than 100 class notes and PowerPoint presentations on his website. Professor Khan has run 6 research projects sponsored by national and international funding agencies and has published about 120 research papers in reputed journals and books. Prof. Khan has presented his research work in various conferences held in Australia, Canada, Egypt, Finland, Greece, New Zealand, Saudi Arabia, South Africa, Turkey, USA, etc. In his research, Professor Khan is working to promote the productivity and active ingredients of medicinal and aromatic plants using different strategies such as the application of mineral nutrition, PGRs, and nanoparticles. Professor Khan has contributed greatly toward the establishment of RPPs as plant growth promoters. In collaboration with the Bhabha Atomic Research Centre, Mumbai, he is researching the appropriate doses of RPPs that would help farmers increase the productivity of medicinal and aromatic plants and their active constituents. Another field of his research is nanoparticles, some of which he has found can enhance the essential oil production of some aromatic plants. He has also filed a patent in this regard.

Contributors

Malik Zainul Abdin
Centre for Transgenic Plant Development
Department of Biotechnology
Hamdard University
New Delhi, India

Tariq Aftab
Plant Physiology Section
Department of Botany
Aligarh Muslim University
Aligarh, India

Athar Ali
Centre for Transgenic Plant Development
Department of Biotechnology
Hamdard University
New Delhi, India

Anna Rita Bilia
Department of Chemistry
University of Florence
Florence, Italy

Ashish Bharillya
Green Technology Department
Ipca Laboratories Ltd.
Sejavta, India

Ebiamadon Andi Brisibe
Bio-products Development and
    Plant Cell and Tissue Culture
    Research Laboratory
Department of Genetics and Biotechnology
University of Calabar
Calabar, Nigeria

Nina Cedergreen
Københavns Universitet
Det Natur- og Biovidenskabelige
    Fakultet
Sektion for Miljøkemi og Fysik
Copenhagen, Denmark

Andreia Corciova
Department of Drug Analysis
University of Medicine
    and Pharmacy
"Grigore T. Popa" Iasi Universitatii
Iasi, Romania

Elisabeth Hsu
Institute of Social and Cultural
    Anthropology
University of Oxford
Oxford, United Kingdom

Bianca Ivanescu
Department of Pharmaceutical Botany
University of Medicine and
    Pharmacy
"Grigore T. Popa" Iasi
Iasi, Romania

Dharam Chand Jain
Green Technology Department
Ipca Laboratories Ltd.
Sejavta, India

M. Masroor A. Khan
Plant Physiology Section
Department of Botany
Aligarh Muslim University
Aligarh, India

Mather Ali Khan
Bond Life Sciences Center
University of Missouri-Columbia
Columbia, Missouri

Bushra Hafeez Kiani
Department of Bioinformatics and
    Biotechnology
International Islamic University
Islamabad, Pakistan

Anjana Kumari
Laboratory of Morphogenesis
Department of Botany
Banaras Hindu University
Varanasi, India

Bhupendra Kumar Mehta
Natural Products Research Laboratory
School of Studies in Chemistry and
    Biochemistry
Vikram University
Ujjain, India

Darshana Mehta
Natural Products Research Laboratory
School of Studies in Chemistry and
    Biochemistry
Vikram University
Ujjain, India

Caroline Meier zu Biesen
Centre National de la Recherche
    Scientifique (CNRS)
Centre de recherche médecine,
    sciences, santé, santé mentale,
    société (Cermes3)
Paris, France

Himanshu Misra
Natural Products Research Laboratory
School of Studies in Chemistry and
    Biochemistry
Vikram University
Ujjain, India
and
Green Technology Department
Ipca Laboratories Ltd.
Sejavta, India

M. Naeem
Plant Physiology Section
Department of Botany
Aligarh Muslim University
Aligarh, India

Neha Pandey
Laboratory of Morphogenesis
Department of Botany
Banaras Hindu University
Varanasi, India
and
Central Institute of Medicinal and
    Aromatic Plant Sciences (CIMAP)
Lucknow, India

Shilpi Paul
G.B. Pant Institute of Himalayan
    Environment and Development
Kosi Katarmal
Almora, India

Abdul Qadir
Quality Control, Hamdard
    Laboratories
New Dehli, India

Shashi Pandey-Rai
Laboratory of Morphogenesis
Department of Botany
Banaras Hindu University
Varanasi, India

Sanjay Kumar Rai
Horticulture Department
Dr. Rajendra Prasad Central
    Agricultural University
Pusa, India

Mauji Ram
Green Technology Department
Ipca Laboratories Ltd.
Sejavta, India

Parul Saxena
Centre for Transgenic Plant
    Development
Department of Biotechnology
Hamdard University
New Delhi, India

Asfia Shabbir
Plant Physiology Section
Department of Botany
Aligarh Muslim University
Aligarh, India

Karina Knudsmark Sjøholm (nee Jessing)
Københavns Universitet
Det Natur- og Biovidenskabelige
 Fakultet
Sektion for Miljøkemi og Fysik
Copenhagen, Denmark

Bjarne W. Strobel
Københavns Universitet
Det Natur- og Biovidenskabelige Fakultet
Sektion for Miljøkemi og Fysik
Copenhagen, Denmark

Salisu Muhammad Tahir
Department of Biological Sciences
Kaduna State University
Kaduna, Nigeria

Deepika Tripathi
Laboratory of Morphogenesis
Department of Botany
Banaras Hindu University
Varanasi, India

CHAPTER 1

Diverse Biologies and Experiential Continuities
A Physiognomic Reading of the Many Faces of Malaria in the Chinese Materia Medica

Elisabeth Hsu

CONTENTS

1.1 INTRODUCTION

This chapter explores whether, and how, the enormously rich and rewarding biomedical research into the antimalarial efficacy of artemisinin, contained in *A. annua* plant materials, might be useful for textual scholarship.* Admittedly, such a project is fraught with problems, as social historians working with pre-twentieth century medical texts are apprehensive of any attempt to identify the referential meanings of the terms translated. *Malaria*, for instance, is a modern scientific nosological term for which there is no equivalent in the

---

* An earlier version of this article was published in Wallis (2009).

premodern Chinese medical texts. Modern scientists have translated malaria into the Chinese *nüeji* 瘧疾, which derives from a term that occurs in premodern Chinese texts—just as malaria is derived from premodern terminology, *mal'aria* ([caused by] bad air). However, neither *nüeji* nor *mal'aria* referred to malaria as a *disease* category in these texts. The premodern Chinese had notions of *bing* 病 (disorder), *hou* 候 (conditions, "syndromes"), *zheng* 證 (evidence, patterns; patterned evidence), and the like, as perceived through the prisms of morality, *adhoc* (magical) intervention, and legal practice, among others. The premodern Chinese term *nüeji* was a *bing* or a *hou* and not a "disease" in the modern scientific sense. Yet, today, it is used as the standard term into which the biomedical disease category "malaria" is translated.

Furthermore, regarding plant identification, ethnobiologists have demonstrated that the modern species-concept is just as historically evolved and socially constructed as modern biomedical terms of human pathology. However, since the mid-20th century, ethnobiologists have attenuated their cultural relativist claims by demonstrating that the plant world is cross-culturally considered to be marked by discontinuities, some of which can be reproduced with great constancy; this is in stark contrast to the worlds of sickness and disease (Hsu, 2010).

Textual scholars do not profess to have deep medical understanding, and they often rely on their colleagues in medical schools, and commonsense biomedical understandings of disease and the body. The Cartesian view of body and mind, which provides the foundations for the biomedical understanding of the body, has given rise to a prevailing assumption in textual scholarship that plots an underlying "nature," which is real and of the body, against "culture," which is constructed and of the mind. Currently, recipes in premodern *materia medica* texts tend to be read either in an almost naïve realist way, where premodern terminology is imbued with contemporary scientific meanings (*nüeji* means *malaria*), or in a cultural constructivist way, where they are read as consisting of a somewhat random assemblage of information on how to treat rather arbitrary, culturally constructed, states of misfortune and bodily dysfunction. More recently, however, some medical anthropologists have been inspired by Merleau-Ponty's ([1945]1962) *Phenomenology of Perception* to demonstrate that these culturally specific terms need not be entirely arbitrary and incommensurable with others in cross-cultural comparisons. The key concept that will be mobilized here is Merleau-Ponty's notion of *physiognomy*, which draws on and further develops the Gestalt psychologists' notion of *Gestalt*.

As argued here, it is particularly research applied to *individual* cases, with *practical* implications and easily perceived *immediate effects* that enable a textual scholar to undertake critical comparisons across time and space. Recipes or formulas (*fangji* 方劑) and recipe texts are meant to have *practical* effects and can be tested on *individual* patients, as can the application of *materia medica* (in the sense of "herbs" or "medicinal drugs" 藥) and *materia medica* literature (*ben cao* 本草). Since these texts often present therapeutic procedures with perceived *immediate effects*, they qualify as a genre worth investigating here. Their practical significance makes it possible for us to test a physiognomic reading.

1.2 PHYSIOGNOMIC READING OF A RECIPE TO TREAT INTERMITTENT FEVERS WITH QING HAO

Let us start by reading the famous physician Ge Hong's 葛洪 (284–363 CE) prescription against "intermittent fevers" (*nüe* 瘧) in his *Zhou hou bei ji fang* 肘後備急方 (emergency prescriptions kept in one's sleeve). Let us first ask whether he considered these fevers malarial, which would make his prescription an herbal antimalarial, and second, query what might make the reading of his formula (or recipe) physiognomic.

> Another recipe: *qing hao*, one bunch, take two *sheng* [2×0.2 liters] of water for soaking it, wring it out, take the juice, ingest it in its entirety (*you fang: qing hao yi wo yi shui er sheng zi, jiao qu zhi, jin fu zhi* 又方 青蒿一握 以水二升漬 絞取汁 盡服之).
> (*Zhou hou bei ji fang, juan 3,* "Zhi han re zhu nüe fang" 治寒熱諸瘧方16: 734–407)

If we ask whether he recognized the intermittent fevers as a sign or symptom caused by what today is malaria, the answer has to be "no." Although he used the term *qing hao*, we cannot be certain that Ge Hong used what today is considered the Chinese herb or drug, or more aptly, the Chinese *materia medica*, called *qinghao*; that is, plant materials of the species *A. annua* L. As the practice of *zhongyi* 中醫, Chinese medicine, is a living tradition, there is no guarantee that any term in use today designates the same taxon as it did in Ge Hong's time. From a biomedical perspective, we also know that intermittent fevers are a symptom not only of malaria but also of other diseases. We can be quite certain that intermittent fevers, read as a symptom or sign of a biomedical disease category, occur in many more conditions than those caused by malarial parasites.

Now, if we assume that *qing hao* in Ge Hong's recipe and *qinghao* today are constituted of plant materials of the same species,* *A. annua*, we may deduce that Ge Hong and other premodern Chinese physicians sometimes got it right (when the intermittent fevers were malarial), but not always (not all intermittent fevers are malarial). In line with our progressivist view of humankind, the suggested reading of the recipe reaffirms our conviction that the ancients engaged in science, but that our modern scientific knowledge is more precise and accurate than premodern knowledge.

However, did the Chinese physicians conceive of intermittent fevers as symptoms or signs of a biomedical disease category? As established in the preceding paragraphs, they did not. So how should a textual scholar relate to the term used for intermittent fevers, *nüe*? Here, Merleau-Ponty's (1945) insights become important on the theoretical and methodological levels, as this question can be reformulated: how might we read premodern terms that refer to a lived experience of the body, such as that of fevers that come and go intermittently? Merleau-Ponty posited that how we know the world depends on how we project our body into the world. In contrast to the assumption that the world and the self can be separated from one another, as posited by empiricist

---

* Writing convention in this chapter: the monosyllabic transcription refers to terms in premodern texts, for example *qing hao*, but not in modern ones, where *qinghao* is used.

science grounded in a Cartesian view of mind and body, Merleau-Ponty insisted that the body–self formed an inextricable part of the phenomenal field through which it moved. Accordingly, the researcher's body is part of the lived world he or she inhabits and aims to research. While a natural scientist, as the subject who does the research, is expected to investigate a research object in a detached manner (even if "objectivity" may be performed in different ways; Daston and Galison, 2007), Merleau-Ponty stressed that the body has a spatiality that is part of the spatial field around it. It cannot be disentangled from its surroundings, just like its parts cannot be considered a [random] "assemblage of organs juxtaposed in space." Rather, they form a whole, and are, in not entirely arbitrary ways, "enveloped in each other" (Merleau-Ponty, 1962, p. 98). Merleau-Ponty thereby provided a basis for critiquing objectivist disease categories. Accordingly, intermittent fevers are not a symptom or sign caused by the disease of malaria, because any biomedical disease category presupposes an objective description of the world, and its relation to the sign is grounded in cause–effect relations established through objective scientific study.

Following Merleau-Ponty, I suggest instead to read "intermittent fevers" as a physiognomy of the spatial field. As already said, Merleau-Ponty emphasizes that the body has a spatial dimension inextricably entwined with the phenomenal field of its surroundings. This spatial field has physiognomies that arise from a practical engagement of the self with its surroundings. Merleau-Ponty's philosophical concept, physiognomy, makes "intermittent fevers" an aspect of the spatial field with which the body–self is practically engaged. This practical engagement arises from the body–self experiencing perceived demands from specific configurations in the spatial field to "do" something. The demands affect the self on multiple levels and are responsible for prompting the body–self into action. Intermittent fevers thus become relevant for the patient and physician as physiognomies of a spatial field demanding a practical intervention from the body–self.

Physiognomies are perceived wholes. Merleau-Ponty points out that to a person for whom meanings are no longer embodied in the world, the world no longer has any physiognomy (ibid, p. 132). Much like the Gestalt psychologists emphasized that the whole is *other* than the sum of its parts (a saying that is often misquoted as "the whole is greater than the sum of its parts"), Merleau-Ponty emphasized that perception relies on an organism's ability to perceive wholes in a single instant. The Gestalt psychologists and the philosopher Merleau-Ponty argued against the behaviorists of the day, against the "empiricists," "intellectualists," and "sensationalists," who all posited that sensory perception is an additive process. Merleau-Ponty spoke of knowing "without thinking" (ibid, p. 129), and underlined immediacy in the perceiving of a whole. His concept of physiognomy is important for us here as it refers to an "immediate practical recognisability" (Morris, 2012, p. 25), where the practical is not merely opposed to the intellectual, but directly appeals to our practical capabilities. As the Gestalt psychologists emphasized, the perception of the practical significance of objects, entities, and events lies not merely in their functional characteristics, but also in so-called "demand characters" or "appeals," "attractions," "exigencies," and "solicitations" that arise from the perceiver's unmediated, often affective relation to them (Koffka and Gauillaume, cited in Morris, 2012, p. 40).

Focusing on a physiognomy of a malarial episode such as "intermittent fevers" as a lived experience emphasizes that its perception and recognition is inextricably related to the way in which it is enacted and acted upon. This appeal to practical intervention is very important when interpreting prescriptions, recipes, and formulas that arguably include an antimalarial ingredient, *qinghao*. Rather than assessing malaria as a disease in terms of objectively given structures and functions, we shall ask what is gained if we read recipes including *qing hao* as providing a practical response to the many different faces of malaria and their physiognomy.

This does not appear to circumvent the problem of reading a retrospective biomedical diagnosis into a premodern text, however. The skepticism of the self-critical textual scholar cautions us. So, might we frame and conceive of the problem differently? Might we be able to explore modes of countering our skepticism by means of an equally rigorous and critical way of thinking about, or rather of *doing*, science? A physiognomy has practical significance, which asks for, or demands, a practical intervention.

A natural scientifically minded realist will turn to experimentation. We know beyond reasonable doubt that the chemical substance artemisinin is a highly effective antimalarial and that *A. annua* plant materials contain artemisinin. We also know that today *A. annua* plant materials make up the Chinese medical *materia medica* called *qinghao*. As already noted, the *qing hao* that Ge Hong used some 1500 years ago may not have consisted of the same plant materials, nor can we be certain that they were identical in morphology and chemical composition.

The easiest way to practically assess the effects of Ge Hong's procedures, surely, is to re-enact them, much in the way current science historians have built and re-enacted significant nineteenth-century machinery to understand how nineteenth-century scientists arrived at their scientific concepts (e.g., Sibum, 1998). This is exactly what an interdisciplinary team of researchers did from January to June 2006; namely, a gardener, who with the permission of his patron grew a mini-plantation of *A. annua* at Hayley House in Oxfordshire; a pharmacognosist and his research team at the Department of Pharmacy of the University of Bradford; and a malariologist and his team at the Swiss Tropical Institute, with a medical anthropologist as tentative initiator, project designer, and go-between (Wright et al., 2010). Our research was written up as a project in which modern science was used to validate the effectiveness of a premodern recipe. Yet in this chapter a proposal is made to "read" and interpret the project differently.

Relevant for textual scholarship is the interpretation of the experiment as one where the physiognomy of intermittent fevers prompted the enactment of body techniques to make a *qinghao* juice that affected it instantly. In other words, our experiment can be framed as one where we treated intermittent fevers in mice (induced through the injection of *P. berghei* parasites into their bloodstream several days earlier) as a specific physiognomic phenomenon of practical significance in so far as it demanded a specific treatment, a *qinghao* juice. We produced this juice through a practical intervention, namely the body techniques of either wringing out plant materials soaked overnight or pounding fresh plant materials on the same day. Our practical procedure had immediate effects, which were observed in individuals (three mice in each cohort tested, and the juice of a certain concentration, injected twice and thrice, respectively, at six-hour intervals into the gut of infected mice effected 95% and 96% recoveries, respectively, in each cohort (based on plasmodial density in the blood a few days later).*

As noted elsewhere (Hsu, 2015), it was the juice obtained by pounding fresh plant materials that contained concentrations that were highly effective in mice infected with *P. berghei* and not the juice wrung out from plant materials that had been soaked overnight, as recommended by Ge Hong. In other words, it was in fact the virtuosity in textual interpretation of the Ming dynasty (1368–1644) physician Li Shizhen (1518–93) that made Ge Hong's recipe effective in our experiment:

> Recipe from the *Zhou hou fang*: use one bunch of *qing hao*, two *sheng* [2×0.2 liters] of water, pound with a pestle to a juice and ingest it (*yong qing hao yi wo, shui er sheng, dao zhi fu zhi* 用青蒿一握, 水二升, 搗汁服之).

Naturally, given the brevity of these instructions, it is difficult to know whether the techniques *jiao* 絞 and *dao* 搗 that the Chinese physicians Ge Hong and Li Shizhen recommended were indeed re-instantiated in our experiment. Techniques such as wringing out (*jiao*) or pounding (*dao*) are "body techniques" in the sense that they have a bodily component. Marcel Mauss ([1935]1973) made an important observation when, in his study on body techniques, he emphasized that they are culturally learned, and arbitrary in a sense. Yet for Merleau-Ponty, their cultural arbitrariness is not so much a problem as is the observation that there are intrinsic limitations to the body as a spatial configuration: "bodily space and external space form a practical system" (ibid, p. 102). Techniques enacted through the body are part of a space imbued with practical significance, and hence cannot be entirely arbitrary. Accordingly, even if a human being follows an instruction given over one thousand years ago, and even if his or her body does not have the skillfulness of a physician who did so daily, certain human movements that have specific practical significance with immediate effects are not entirely arbitrary, as the anatomy and physiology of the body limits the possibilities of how they can be enacted. This enhances the likelihood of them being re-enacted in a similar mode, even 1500 years later.[†]

In this way, by *doing* science and by interpreting the administration of a *qing-hao* juice as a practical response to the demand characteristics of the physiognomy of intermittent fevers, the mice's "malarial intermittent fevers" were perceived as a situationally given, instantly recognizable spatial configuration with a practical significance. Their existence was evidenced by them being radically affected by the

---

* Although sometimes characterized as such, ours was not actually an "ethno-archaeological" project because our research was not concerned with the study of contemporary people to explain an archaeological record of the past.

[†] Techniques of wringing out plant materials (like a house wife wrings out washed clothes) continue to be enacted today by healers in Africa (fieldwork observation, early 2000s) and in the Amazon (personal communication, F. Barbira-Freedman, late 2000s).

practical intervention of administering *qinghao* juice, such that they disappeared and the mice recovered.

In summary, Ge Hong's extraction technique of wringing out the whole plant, which resulted in an emulsion of water, flavonoids, aromatic oils, and artemisinin as contained in the leaves, in particular, is likely to have yielded artemisinin in larger quantities than earlier techniques of preparation recorded in the Chinese *materia medica,* and this is in all likelihood directly linked to the recommendation of using it for treating acute fever episodes of intermittent fevers (Hsu, 2014). Not just artemisinin, but also several flavonoids, found in stems and leaves alike, have antimalarial properties, and synergistic effects may or may not have played an additional role (Willcox et al., 2004). Furthermore, resistance to the antimalarial, artemisinin, is much less likely to occur in whole plant preparations (Elfawal et al., 2014).

1.3 PHYSIOGNOMY OF COMPLAINTS FOR WHICH THE CHINESE MATERIA MEDICA LITERATURE RECOMMENDS THE APPLICATION OF QING HAO

In the Chinese *materia medica* literature, wherein *qing hao* is mostly known as *cao hao* 草蒿, it is recommended for a variety of complaints other than intermittent fevers (for a chronically ordered, comprehensive translation of these recommendations, see Hsu, 2010). Medical historians have been quick in dismissing those as biologically unfounded "culture-bound syndromes," "illnesses," or "sicknesses." In what follows, we discuss the treatment recommended in the Chinese *materia medica* from the first century CE to 1596 for such culture-specific notions in light of the scientifically known biological variations of malaria. If the pathology of malaria in regions where it is endemic need not always manifest as fever bouts, its cultural perceptions may vary accordingly. In what follows, we explore what is gained by reading these recommendations as a response with practical significance to the solicitations of the physiognomy of the lived experience of malaria's diverse biologies.

The Chinese *materia medica* literature is a genre that consists of long lists of *materia medica* identified by name and synonyms; by flavor, quality, and other properties, such as whether the *materia medica* in question has potency/toxicity (*you/wu du* 有 / 無毒); by main indications; and sometimes also by "pharmaceutical" information on how to prepare and when to administer them. The first canonical *materia medica*, which is no longer extant in its original form but has been reconstructed from multiple citations in later works, is *Shennong's Canon of Materia Medica* (*Shennong ben cao jing* 神農本草經), presumably compiled in the first century CE. It has an entry on *cao hao*, the "herbaceous *hao*."

> The herbaceous *hao*. Its flavor is bitter, cold. It treats *jie* itches, *jia* itches,* and ugly wounds. It kills lice and lingering heat between bones and joints. It brightens the eyes.

---

* *jie sao jia yang* 疥瘙痂痒 is an itching that can affect toes and fingers. See, for instance, *jie jia* 疥痂 in *Ling shu* 靈樞10:307, or *jie chuang* 疥瘡 in *Zhu bing yuan hou lun* 諸病源候論 50:1411.

Another name is *qing hao*, another name is *fang kui*. It grows in river waste lands.* (*Shennong ben cao jing, juan* 4: 341)

At a first glance, it appears as though the authors of this text were completely unaware of the potential antimalarial use of *cao hao*. They primarily recommended it for treating different kinds of itches, ugly wounds, and lice.† However, joint aches are a common lived experience among patients in regions where malaria is endemic. Likewise, the "lingering heat between bones and joints" in *Shennong's Canon of Materia Medica* may have alluded to this face of endemic malaria.

Shennong's Canon of Materia Medica also recommends *cao hao* for "brightening the eyes" (*ming mu* 明目). Incidentally it is also mentioned in the *Materia Medica for Successful Dietary Therapy* (*Shi liao ben cao* 食療本草) of 721–739 among a list of terms indicating enhancement of one's vitality:

> They say *qing hao* is cold, enhances *qi* 氣, causes growth of head hair, can make the body feel light, supplement the interior and prevent ageing, brighten the eyes, and halt wind poison. (*Zheng lei ben cao, juan* 10:20b)

Chinese medical historians generally do not consider *materia medica* that can "brighten the eyes" to be antimalarial. However, if we take into account that endemic malaria causes anemia, which is experienced as lethargy and tiredness, we can see why a *materia medica* with antimalarial effects might be considered vitality- and longevity-enhancing.

The *Materia Medica for Successful Dietary Therapy* uses the raw plant, after soaking it in urine and making it into a powder and pill. It also recommends *qing hao* as a pickle, and Tao Hongjing (456–536) seems to have recommended it as an unprocessed food supplement in his *Notes to [Shennong's] Canon of the Materia Medica* ([Shennong] *Bencao jingji zhu* [神農]本草經集注) around 500 CE.‡

> It is everywhere, this one is today's *qing hao*, people even take it mixed with fragrant vegetables for eating it. ([*Shen nong*] *Ben cao jing ji zhu, juan* 5, p. 363)

Although a syntactic reading of the text is that the vegetables were "fragrant," there is little doubt that they were considered such due to the fragrance of *qing hao*, which in other texts is called "fragrant *hao*" (*xiang hao* 香蒿) or even "stinking *hao*" (*chou hao* 臭蒿). As food supplement, the presumably fresh and fragrant *qing hao* may have been seen as a preventive health measure.

---

* *ze* 澤 is here rendered as "wasteland," in accordance with Bodde ([1978]1981). This is the ecological niche of *Artemisa apiacea* and *Artemisia annua*, rather than interpreting *ze* as meaning swamps and wetlands.

† This is a recommendation much in line with the first extant text on the therapeutic use of *qing hao* in a manuscript unearthed from a tomb closed in 168 BCE near Mawangdui (Harper, 1998: 272–273), and it is one that prevailed in the *materia medica* literature for about one thousand years (Hsu, 2014: table 1).

‡ Notably, these three passages do not recommend heating the plant extract, which is in line with the modern scientific finding that heat changes the molecular structure of artemisinin to one without antimalarial effects (Hien and White, 1993).

Later *materia medica* texts recommend *cao hao* (a synonym of *qing hao*) for treating "bone steaming" (*gu zheng* 骨蒸) and conditions of "exhaustion arising due to heat/fevers" (*re lao* 熱). Although Chinese medical historians do not generally consider these two terms to describe malarial conditions, with hindsight, knowing that *qing hao* can be used as antimalarial, it is possible that they sometimes referred to malarial fevers. Furthermore, it is conceivable that the effective treatment of malarial cases of *gu zheng* and *re lao* by *cao hao* may have led to this recommendation.

> For treating bone steaming, take one *liang* 兩 [41.3 g.] of urine to soak it overnight, dry it, turn it into powder and make a pill. It entirely eliminates exhaustion arising due to heat/fevers. (*Shi liao ben cao* 食療本草 of 721–739, as quoted in the *Zheng lei ben cao, juan* 10:20b)

In this context, it is worth noting that a *hao* is also recommended for a wide range of convulsive disorders including "daemonic *qi*," "*rigor mortis* possession disorder,"* and *fu lian,*† as in the *Supplements to the Materia Medica* (*Ben cao shi yi* 本草拾遺) of the eighth century, which recommended, in line with Ge Hong, wringing out the juice from a presumably fresh plant:

> *Hao* controls daemonic *qi, rigor mortis* possession disorders*, fu lian*, the blood *qi* of women, fullness inside the abdomen and [perceptions of] intermittent cold and hot, and chronic diarrhea. In autumn and winter, use the seeds, in spring and summer, use the sprouts, together pound them with a pestle, wring out the juice, and ingest. Alternatively, dry it in the sun and make it into a powder, and apply it in urine. (*Zheng lei ben cao, juan* 10:20a)

Notably, cerebral malaria can also present as convulsions, but Chinese medical historians do not associate "daemonic *qi*," the "*rigor mortis* possession disorder," and *fu lian* with malaria. Rather, they tend to relegate these conditions into the domain of culture-bound possession behavior or mental illness, and these have rather distinctive physiognomies; for instance, when associated with pollution through contamination with the dead (e.g., Li, 1999). In this case, it is possible but unlikely that *qing hao*'s effectiveness against cerebral malaria motivated its recommendation. Rather, another observation comes to mind: the component *qing*, in the name *qing hao*, has a phonoaesthetic that alludes to lightness, transparency, and purity, and hence it may have been used for treating conditions of pollution. Although the term *qing* is not given in this quote, it likely was implied. In the highly medicalized formula literature, *qing hao* is the usual term, but not in the *materia medica* literature, where we have *cao hao* in its stead.

---

* For *zhu* 注 "possession disorders," see Chao Yuanfang 巢元方's *Zhu bing yuan hou lun* 諸病源候論 (Origins and Symptoms of Medical Disorders) of 610, *juan* 24: 690–715.
† For *fu lian* 伏連 lit. "to harbour and connect," see Wang Tao's 王燾 *Wai tai mi yao* 外台秘要 (Arcane Essentials from the Imperial Library) of 752, *juan* 13: 358.

1.4 SUMMARY

We started with the modern bioscientific finding that *qinghao*, a Chinese herbal *materia medica* that today is composed of *A. annua* whole plant materials, contains the chemical substance artemisinin, a molecule with a peroxide bridge that has been proven beyond reasonable doubt to be an effective antimalarial (WHO 2005, WHO 2006). Although skeptical textual scholars likely adhere to the cultural constructivist stance that renders this bioscientific finding irrelevant, we asked whether this incisive bioscientific research might have any relevance for textual scholarship nevertheless. Our aim was not to validate premodern texts through modern scientific research, but rather to work out a method to make sense of *materia medica* texts and propose a reading of them that is sound, theoretical, and rigorous.

It is hoped that this study has demonstrated that *materia medica* texts have relevance if one accords the lived body a theoretical importance, not as an object of study but as starting point for making sense of the world, as did Merleau-Ponty. To overcome some obstacles causing the current impasse between naïve realism and cultural constructivism, we proposed to work with Merleau-Ponty's notion of "physiognomy": a physiognomic reading accords the text a practical significance and in this way, reduces its culturally constructed arbitrariness (which can obstruct meaningful cross-cultural comparison).

We still do not know how to read texts containing premodern scientific knowledge. So, what is gained by reading some select *materia medica* excerpts in a physiognomic way; that is, in a way that according to Merleau-Ponty attended their practical significance? A physiognomic reading of the selected *materia medica* passages made it possible to relate to them not merely as poetry about culturally constructed disorders for which there is no biological basis, but as recommendations that have a practical significance to individual cases. To be sure, the proposed physiognomic reading was not applied to all passages (some passages clearly recommended *qing hao* because of its wound-healing properties; others, likely, alluded to its general purificatory powers). Rather, we asked what is learned from specific felicitous text passages if *qing hao* is read as having a practical significance due to its antimalarial properties? A physiognomic reading offers advantages on the theoretical and methodological levels. It makes it possible to grant premodern physicians and scientists the capacity to accurately assess specific physiognomies of the lived experience of the diverse biologies of acute, recurrent, and endemic malaria, and of cerebral *falciparum* malaria, and it imbues the responses to them with equally accurate practical significance.

The preparation methods presented here varied from using fresh *qing hao* as a food supplement (to enhance vitality and longevity) to wringing out the whole plant after soaking it in water or pounding it when fresh (for treating acute fever bouts). The experiment with plants grown in Oxfordshire (Wright et al., 2010) was just as much a scientific legitimation project for validating the efficaciousness of a specific premodern recipe as it has become an explorative project here for better understanding the problems that textual scholarship is confronted with.

1.5 OUTLOOK

Many retrospective biomedical diagnoses are grounded in the assumption that biological processes give rise to "diseases" that are universally the same, while "illness" is culturally specific. The prime example of such a concept of disease and illness is what scientists eventually identified as Creutzfeldt–Jakob disease, which the Fore people of Papua New Guinea considered a phenomenon of *kuru* sorcery (Lindenbaum, 1979). In this particular case, both the Fore and the scientists referred to people suffering from the very same morbid condition, which was very distinctive and could not be confused with others. However, in general, local and biomedical terms rarely refer to exactly the same phenomena. Sometimes, they are entirely unrelated, but often they overlap in respect of certain easily identifiable phenomena.

Paul Unschuld (2002) has argued that malaria has a biological reality, which cross-culturally can be recognized in the local terminology of *intermittent fevers*; that is, *han re* (寒熱; intermittent coldness and heat) and *nüe* (瘧; intermittent fevers) in Chinese medicine. Some intermittent fevers are malarial, others not. In full awareness that other culture-specific terms cannot be mapped onto "empirical reality" as straightforwardly, Unschuld argued against an extremist cultural relativist interpretation of different illness categories.

Unschuld was following the lead of medical anthropologists who adhere to the notion that a biological substratum of disease is universal, even if the culture-specific way in which it is expressed differs in degrees. For instance, the use of the term *somatization* in psychiatry and applied medical anthropology made it possible to conceive of depression as a universal condition, experienced in primarily somatic ways in cultures other than those of Europe and North America (Kleinman, 1980). However, the usefulness of the disease/illness dichotomy has been abandoned since (also by Arthur Kleinman himself), and the concept of disease as a universal biological substratum onto which are grafted culturally specific illnesses has proven untenable. Margaret Lock's (1993) concept of "local biologies," in particular, underlined that women during menopause who suffered from hot flashes were not merely complaining about a cultural construct (affecting their minds), but possessed locality-specific physiologies that were bio-scientifically validated yet not universal, that is not identical cross-culturally.

Thomas Csordas (1994), by combining Merleau-Ponty (1945) and Bourdieu ([1977]1980), took the discussion in a direction that is even more applicable to this study. In *The Sacred Self*, Csordas underlined in a similar vein the complex interplay of physiology, bodily posture, psychology, faith, rhetoric, and the socio-cultural, but he gave the bodily processes primacy when he claimed that the body is the foundation of one's existence, even if one's experience of it is never pre-cultural. Following in his footsteps, I argue here that it is not only "mental" conditions, but also biologies that can be very diverse, as in the case of what is often conceived as the "physical" condition *par excellence*, the very real, and often deadly, disease malaria.

Although malaria tends to be used as a prime example of a disease with a biologically universal pathology (Unschuld, 2002), its biology can be locally specific. It can be expressed very distinctively in different populations, depending on the genetics of both the parasites and the human beings.* As bioscientific research has now established, malaria presents many different faces, and non-biomedical terms, like the Chinese medical ones, may well be based on the experience of practically engaging with these manifestations and their physiognomy.

In areas of Africa where malaria is endemic, it is rarely perceived as a life-threatening sickness, whereas missionaries, soldiers, tourists, and other foreign visitors experienced malaria as a horrendous, if not deadly, disease. Locals, due to their acquired immunity, complained of pain in the joints, headaches, flu-like fevers, a general unease, and "low energy" (Marsland, 2005). This is not a matter of over-diagnosing malaria in patients suffering from other conditions such as, for instance, depression.† Nor is it a matter of mind and culture only, but rather it is intrinsic to the biology of malaria and its frequent co-morbidities (e.g., Kelly, 2014). The perception of this lived experience, as a holistically comprehended physiognomy, is dealt with in practical interventions that are often much more to the point than a cultural relativist reading tends to give them.

Malaria may also present yet another face, that of convulsions. *P. falciparum* can cause the, often lethal, cerebral malaria. Considering the peculiar manifestation of this condition, it is not surprising that locals relate to its physiognomy differently, and consider it a distinctive disorder, not *homa ya malaria* (e.g., Winch et al., 1996). In Tanzania, convulsions in infants and toddlers that *P. falciparum* may induce are attributed to *dege dege*.‡ In Swahili, *dege dege* means bird, which in such cases presumably indicated preeminent death and the ensuing flight of the soul.

Finally, and perhaps most importantly for this study, the physiognomic reading highlighted that diseases like malaria have different faces and physiognomies reflecting lived experience, to which premodern texts attend with specific and practical therapeutic recommendations. So, rather than adhering to a simplified model of a biomedically constructed "disease" and culturally constructed "illness," more attention should be paid to the *Gestalt* and physiognomy of peoples' lived experience of the diverse biological manifestations of a morbid condition. This approach will sharpen the comparativist lens.

REFERENCES

Bodde D. [1978] 1981. Marches in the *mencius* and elsewhere: A lexicographic note. In *Essays on Chinese Civilisation*. Princeton University Press, Princeton, 416–425.
Bourdieu [1977] 1980. Outline of a theory of practice. Translated by Nice, R. Cambridge University Press, Cambridge.

* Nor is it a matter of race, for malaria is just as deadly to the African infant as it is flu-like for the European missionary who has been able to build up immunity after a lifetime's exposure to it.
† Rachel Jenkins, personal communication; and own fieldwork experience (e.g., Pemba, January 2004).
‡ *Dege dege* is a term used for making sense of many symptoms other than convulsions, ranging from general irritability to complete apathy; it designates an age-specific sickness that only affects infants and toddlers (Makemba et al., 1996).

Csordas T. 1994. *The Sacred Self: A Cultural Phenomenology of Charismatic Healing.* University of California Press, Berkeley.

Daston L. and Galison P. 2007. *Objectivity.* Zone Books, New York.

Elfawal M.A., Towler M.J., Reich N.G., Weathers P.J., and Rich, S.M. 2014. Dried whole-plant *Artemisia annua* slows evolution of malaria drug resistance and overcomes resistance to artemisinin. *Proceedings of the National Academy of Sciences* 112(3): 821–826.

Harper, D. 1998. Early Chinese Medical Literature: The Mawangdui Medical Manuscripts. Translated by Harper, D. Kegan Paul, London.

Hien T. and White N. 1993. Qinghaosu. *Lancet* 341: 603–608.

Hsu E. 2006. The history of *Qing hao* 青蒿 in the Chinese materia medica. *Transactions of the Royal Society of Tropical Medicine and Hygiene* 100(6): 505–508.

Hsu E., in collaboration with F. Obringer. 2010. *Qing hao* 青蒿 (Herba *Artemisiae annuae*) in the Chinese materia medica. In E. Hsu and S. Harris (eds) *Plants, Health, and Healing: On the Interface of Medical Anthropology and Ethnobotany.* Berghahn, Oxford, 83–130.

Hsu E. 2014: How techniques of herbal drug preparation affect the therapeutic outcome: Reflections on *Qinghao* 青蒿 (Herba *Artemisiae annuae*) in the history of the Chinese materia medica. In T. Aftab, J.F.S. Ferreira, M.M.A. Khan, and M. Naeem (eds) *Artemisia annua – Pharmacology and Biotechnology.* Springer, Heidelberg, 1–8.

Hsu E. 2015. From social lives to playing fields: "The Chinese antimalarial" as artemisinin monotherapy, artemisinin combination therapy and *qinghao* juice. In L. Pordié and A. Hardon (eds) *Stories and Itineraries on the Making of Asian Industrial Medicines, Special Issue, Anthropology & Medicine* 22(1): 75–86.

Li J.M. 1999. Contagion and its consequences: The problem of death pollution in ancient China. In Y. Otsuka, S. Sakai, and S. Kuriyama (eds) *Medicine and the History of the Body.* Ishiyaku Euro America, Tokyo, 201–222.

Kelly T. 2014. Plants, power, possibility: Maneuvering the medical landscape in response to chronic illness and uncertainty. PhD Thesis, Institute of Social and Cultural Anthropology, University of Oxford.

Kleinman A. 1980. *Patients and Healers in the Context of Culture: An Exploration of the Borderland between Anthropology, Medicine, and Psychiatry.* University of California Press, Berkeley.

Lindenbaum S. 1979. *Kuru Sorcery.* Mayfield, Palo Alto.

Lock M.M. 1993. *Encounters with Aging: Mythologies of Menopause in Japan and North America.* University of California Press, Berkeley.

Makemba A., Winch P., Makame V., Mehl G., Premji Z., Minjas J., and Schiff C. 1996. Treatment practices for *degedege*, a locally recognized febrile illness and implications for strategies to decrease mortality from severe malaria in Bagamoyo district, Tanzania. *Tropical Medicine and International Health* 1(3): 305–313.

Marsland R. 2005. Ethnographic malaria: The uses of medical knowledge in southwest Tanzania. PhD thesis in Anthropology, School of Oriental and African Studies, University of London.

Mauss M. [1935] 1973. Techniques of the body. *Economy and Society* 2(1): 70–88.

Merleau-Ponty M. [1945] 1962. *Phenomenology of Perception,* tr. C. Smith. Routledge, London.

Morris K.J. 2012. *Starting with Merleau-Ponty.* Continuum, London.

Sibum O. 1998. Les gestes de la mesure. Joule, les pratiques de la brasserie et la science. *Annales. Histoire, Sciences Sociales* 53(4): 745–774.

Unschuld P. 2002. Diseases in the Huang Di Neijing Suwen: Facts and constructs. In A.K.L. Chan, G.K. Clancey, and H.C. Loy (eds) *Historical Perspectives on East Asian Science, Technology and Medicine.* Singapore University Press, Singapore, 182–197.

Wallis F. (ed) 2009. Medicine and the soul of science: Essays by and in memory of Don Bates. Special Issue. *Canadian Bulletin of Medical History* 26(1).

WHO (World Health Organization). 2005. Malaria control today: Current WHO recommendations. http://www.who.int/malaria/publications/mct_workingpaper.pdf, retrieved on 1 December 2014.

WHO (World Health Organization). 2006. Guidelines for the treatment of malaria. http://whqlibdoc.who.int/publications/2010/9789241547925_eng.pdf?ua=1, retrieved on 1 December 2014.

Willcox M., Bodeker G., Bourdy G., Dhingra V., Falquet J., Ferreira J.F.S., Graz B., et al. 2004. *Artemisia annua* as a Traditional Herbal Antimalarial, in M. Willcox, G. Bodeker, and P. Rasoanaivo (eds) *Traditional Medicinal Plants and Malaria.* CRC Press, Boca Raton, 43–59.

Winch, P., Makemba A., Kamazima S.R., Lurie M., Lwihula G.K., Premji Z., Minjas J.N., et al. 1996. Local terminology for febrile illness in Bagamoyo district, Tanzania and its impact on the design of a community-based malaria control programme. *Social Science and Medicine* 42(7): 1057–1067.

Wright C.W., Linley P.A., Brun R., Wittlin S., and Hsu E. 2010. Ancient Chinese methods are remarkably effective for the preparation of artemisinin-rich extracts of Qing Hao with potent antimalarial activity. *Molecules* 15(2): 804–812.

Zhongyao dacidian 中藥大辭典 *(Great Dictionary of the Chinese Materia Medica)* 1986. *Jiangsu xinyi xueyuan* 江蘇新醫學院 *(eds). Shanghai keji chubanshe,* Shanghai.

PREMODERN SOURCES

Ben cao gang mu 本草綱目 *(Classified Materia Medica). Ming,* 1596. *Li Shizhen* 李時珍. *4 vols. Renmin weisheng chubanshe,* Beijing, 1977–1981.

Huang di nei jing 黃帝內經 *(Yellow Emperor's Inner Canon). Zhou to Han, 3rd c. BCE to 1st c. CE. Anon. References to Huangdi neijing zhangju suoyin* 黃帝內經章句索, *edited by Ren Yingqiu* 任應秋. *Renmin weisheng chubanshe,* Beijing, 1986.

Ling shu 靈樞, *see Huang di nei jing* 黃帝內經.

[Shennong] Ben cao jing ji zhu 神農本草經集注 *(Notes to Shennong's Canon on Materia Medica). Liang, ca.* 500. *Tao Hongjing* 陶弘景. *Annotated by Shang Zhijun* 尚志鈞 *and Shang Yuansheng* 尚元勝. *Renmin weisheng chubanshe,* Beijing, 1994.

Shennong ben cao jing 神農本草經 *(Shennong's Canon on Materia Medica). Han, 1st c. CE. Anon. References to Shennong bencaojing jizhu* 神農本草經輯注, *annotated by Ma Jixing* 馬繼興. *Renmin weisheng chubanshe,* Beijing, 1995.

Wai tai mi yao 外台秘要 *(Arcane Essentials from the Imperial Library). Tang,* 752. *Wang Tao* 王燾. *Renmin weisheng chubanshe,* Beijing, 1955.

Zheng lei ben cao 證類本草 *(Materia Medica Corrected and Arranged in Categories). Song, ca.* 1082. *Tang Shenwei* 唐慎微. *References to Chong xiu Zheng he jing shi zheng lei bei yong ben cao* 重修正和經史證類備用本草. *Facsimile of the 1249 edition. Renmin weisheng chubanshe,* Beijing, 1957. *And to Siku yixue yeshu congshu* 四庫醫學叢書, *annotated by Cao Xiaozhong* 曹孝忠. *Shanghai guji chubanshe,* Shanghai, 1991.

Zhou hou bei ji fang 肘後備急方 *(Emergency Recipes Kept in One's Sleeve). Jin, 4th c. (340 CE?). Ge Hong* 葛洪. *Si ku quan shu* 四庫全書 *(Collection of the Works from the Four Storehouses). References to Wen yuan ge Si ku quan shu* 文淵閣 四庫全書. *Shangwu yinshuguan, Taibei,* 1983.

Zhu bing yuan hou lun 諸病源候論 *(Treatise on the Origins and Symptoms of Medical Disorders). Sui, 610. Chao Yuanfang* 巢元方. *References to Zhubing yuanhoulun jiaozhu, annotated by Ding Guangdi* 丁光迪. *Renmin weisheng chubanshe, Beijing,* 1991.

CHAPTER 2

Artemisia annua and Grassroots Responses to Health Crises in Rural Tanzania

Caroline Meier zu Biesen

CONTENTS

2.1 INTRODUCTION: BACK TO THE ETHNOBOTANICAL ROOTS

I met Adamu Bekeli* during my visit to the isolated village of Malya in the northwestern Mara Region of Tanzania. Malya, which lies roughly 80 km from Mwanza and numbers some 2000 inhabitants, has very little in the way of infrastructure and only sporadic access to electricity. However, it is the lack of adequate medical care from which villagers suffer most. Malya's proximity to Lake Victoria means that the village lies in a region of highly endemic malaria. Increases in malaria cases correlate with the monsoon, during which period most of the village's inhabitants farm in order to secure their livelihoods. Illnesses caused by malaria during the most economically important time of the year mean that many families face financial crises (Mboera et al., 2007). In addition to lower productivity through illness, endemic regions are burdened by higher expenditure on prophylactic measures, treatments, and antimalarial medicines. The first-line antimalarial treatment in Tanzania is

* All names are fictitious.

artemisinin-based combination therapy (ACT). ACTs have reduced the incidence of malaria worldwide. Yet, for the poorest patients in remote areas, they are often unaffordable unless available through a heavily subsidized scheme (Bhatt et al., 2015).

As a response to this, the international non-governmental organization (NGO) Action for Natural Medicine (ANAMED) introduced a standardized phytomedicine derived from the Chinese plant *Artemisia annua* L. (hereafter *Artemisia*), which is supposed to provide a fast-acting and accessible treatment that can be applied in most settings, especially in remote areas. Ethnobotanical practice with *Artemisia* involves the internal use of tea derived from the whole plant. ANAMED members have a progressive political commitment: they understand the need to share information on *Artemisia*'s therapeutic value via public relations, workshops, and seminars. Since 1998, ANAMED* has distributed over 1200 *Artemisia* "starter kits" (containing *Artemisia* seeds[†] and instructions for their therapeutic use) to partners in 75 countries. ANAMED key workers have also run week-long training seminars on natural medicine in 20 different countries, mostly in Africa.

In 2006, Adamu attended one such ANAMED seminar. Immediately after receiving training, he began growing a handful of *Artemisia* plants on a small plot of land. Conditions in Malya are not ideal for the cultivation of this sensitive plant species: poor soils and a chronic lack of water makes all crop farming very difficult. Against the odds, Adamu succeeded in cultivating runners from a parent plant. In order to attain the ideal of an affordable plant source, which in turn encourages agricultural and economic independence among farmers, ANAMED recommends vegetative propagation, as plants reproduced using this method demonstrate the same level of artemisinin yield as the parent plant. From his first harvest, Adamu was able to produce large amounts of *Artemisia* raw material, which he packaged cleanly and according to the doses stipulated by ANAMED and sold in his small shop (Swahili: *duka*). As a result of the regular presence of ANAMED workers in the Mara Region, knowledge of *Artemisia*'s therapeutic potential spread quickly and, in so doing, created increasing demand for the plant. Adamu soon came to be a key figure in local *Artemisia* production in Malya. He was able to establish a corresponding market culture for *Artemisia* in the village and determine the rules of competition (*ushindani kwa soko*). According to his own reckonings, Adamu earns up to 300,000 Tanzanian shillings (approximately US$135) per month—a large sum in a community in which the average monthly income is around US$50.

I use this scene in Malya as a starting point for raising the question of how people—especially those without financial means—raise resources to protect their health. In particular, I explore what networks of care emerge in response to uneven access to high-quality antimalarials. Traditional medicine has received renewed attention due to epidemics such as malaria, HIV/AIDS, and tuberculosis, and has

---

* ANAMED was founded in 1985 in Congo/Zaire by the German pharmacist Dr. Hans-Martin Hirt, the Congolese teacher Bindanda M'Pia, and various Congolese healers. Originally conceived as a regional movement, ANAMED sparked attention internationally and, with the founding of the association Anamed International e.V. with headquarters in Germany in 1994, was established as a legal entity.

† Seeds used by ANAMED are F1 hybrid seeds of the tropicalized *Artemisia* plant.

become an important element in the global governance of health (Gaudillière, 2014; Meier zu Biesen, 2017). Yet the matter of how, where, and to what end herbal medicine ought to be deployed in response to limited access to publicly funded medicine and/or medical emergencies remains under-debated within global health circles and under-represented in scholarship (Craig, 2015).

Current studies furnish evidence of the high value of medicinal plants; however, they also emphasize the necessity of investigating phytomedical knowledge using an improved methodological basis and driving forth innovative research in order to integrate these into malarial treatment provisions (Aftab et al., 2014; Etkin et al., 2004; Ginsburg and Deharo, 2011; Graz et al., 2011). Under the guise of a "reverse pharmacology" approach, which aims quite literally for a "back to the roots of medicine and hence to medicinal plants" ethos, there is currently an effort to develop a standardized antimalarial phytomedicine (Wells, 2011; Willcox et al., 2011c). The concept of reverse pharmacology was coined in India to develop pharmaceuticals from Ayurvedic medicines, and involves a classical pathway of isolating compounds for further development (see Pordié and Gaudillière, 2014). However, the primary objective of Willcox and others—in particular that of members of the Research Initiative on Traditional Antimalarial Methods (RITAM)—is not to develop new drugs, but to improve the utilization of medicinal plants already in use and to develop complementary antimalarial formulations from ethnobotanical leads. Researchers suggest that rigorous evaluation of traditional medicines through controlled clinical trials in parallel with agronomical development for more reproducible levels of active compounds could improve the availability of therapies at an acceptable cost and serve as a source of income.*

The "revolutionary potential" of this concept apparent to many relates explicitly to the *Artemisia* plant and its—historically proven—use as a synergetic remedy (cf. De Ridder et al., 2008; Mueller et al., 2000). ANAMED is the only organization worldwide that strongly advocates this approach—in particular in medical emergency settings—in the context of humanitarian intervention. Since 1997, inspired by traditional treatment methods from Chinese pharmacopoeias (see Hsu, 2006), ANAMED has been conducting research jointly with other scientists on how *Artemisia* in tea or decoction form can be deployed in the treatment of malaria. Research was carried out in "reverse order," with observational clinical studies as the starting point. This led to the conviction that *Artemisia* was a highly effective and inexpensive form of malaria treatment that fostered the independence of patients and the local community.

In this chapter, I will not focus on/review the controversies over the application of non-pharmaceutical forms of *Artemisia* on a global level (see Meier zu Biesen, 2010, 2013). Nor do I intend to depict folk healing as somehow superior to biomedical therapies (cf. Farmer, 1997). Artemisinins are the most effective and rapidly

---

* In order to develop a standardized phytomedicine, Willcox et al. (2011c) tested the reverse pharmacology approach, where clinical evaluation was prioritized from the start. They suggested developing a standard scoring system to prioritize plants used in folklore practices that includes meticulous analysis and scoring of ethnobotanical data for frequency of citation of a plant, efficacy *in vitro/vivo*, and safety and activity based on retrospective treatment outcome.

acting antimalarial drugs ever discovered (Li and Weina, 2010; White, 2008); therefore, it is not realistic to aim to develop an herbal treatment that can outperform ACTs (Willcox et al., 2011c: 7).

As a researcher in the field of critical medical anthropology, I wish to step back a little from binary epistemologies possessing little creative power and instead suggest that we rethink what traditional medicine is "good for" in global health. To do so, I engage in the complexities surrounding the process of seeking a treatment for malaria in rural Tanzania and take a closer look at how people have responded to a multitude of uncertainties; that is, shortages of ACTs, fake medicines, or high costs to cover antimalarial therapy. Rather than merely following instructions for treatment and acting in accordance with the knowledge mediated by biopolitical regimes (cf. Dilger, 2012), people are trying to act on threats to their health by drawing on alternative resources that have emerged in response to challenging political–economic conditions. ANAMED's grassroots movements' attempt to organize humanitarian intervention using alternative methods for delivering artemisinin to patients represents a unique case in terms of access to therapeutic care: special attention is directed to the economic independence of patients and to the strengthening of the community via a low-cost, local, effective, and reliably produced treatment that does not only decrease mortality rates of malaria in isolated regions, but also spurs on economic development.

2.2 GLOBAL CONCEPTS OF EFFICACY, LOCAL STRATEGIES OF ACTION

In 2008, ANAMED Tanzania (a local branch of the international NGO) recorded over 300 active members in the country's various regions, in particular those in which healthcare facilities were sparsely available.* The main administrative office and one ANAMED training center are located in Mwanza. A number of individual ANAMED village, city, and regional committees (*kikau cha anamed*), which meet regularly to exchange information and ideas about their projects, have since grown. The organizational structure of the groups is highly formalized: each group has an elected chair, a deputy chair, and a treasurer whose task it is to administer the annual dues (US$2). All meetings commence with a prayer.

---

* For 13 months (between 2006 and 2008), I was an intermittent resident of Musoma/Nyasho in the Mara Region, the center of ANAMED's activities. I conducted regular follow-up visits to Tanzania until 2015. My stay, together with the eight-year-long relationship I built up with people in Mara, helped me to understand the history of the introduction of *Artemisia* to Tanzania. By way of individual and group interviews with ANAMED members, I was able to document what channels had been utilized for the dissemination of knowledge about *Artemisia* and what challenges, resistance, and innovations accompanied this process. I also conducted interviews with patients, healers, and representatives of state health organizations, NGOs, and churches. In addition to these interviews, I undertook participant observations in patient self-help groups and ANAMED seminars. This core corpus of data was supplemented by investigations in other regions in Tanzania (Iringa, Uwemba, Tanga, Arusha Region, Dar es Salaam) and in Kenya (Athi River), interviews with experts, including representatives of the pharmaceutical industry for artemisinin production, the WHO, staff on the national malaria-control program, and commercial *Artemisia* farmers.

ANAMED's programmatic orientation is based upon Christian values (such as love thy neighbor and a sense of justice) and a scientifically oriented approach to traditional medicine; that is, one that is grounded in pharmacological and pharmacokinetic studies of efficacy (Hirt and M'Pia, 2000). By analyzing traditional knowledge of healing using scientific methods, this knowledge and the corresponding practices are legitimized such that they can be connected to indigenous, traditional (*dawa ya kienyeji/asili*), and biomedical fields (*dawa ya kisasa/hospitali*) (Meier zu Biesen, 2014). ANAMED understands healing as a form of naturopathic treatment (*matibabu asilia*)—premised on the idea that disease is of natural origin—that ordinary people can learn. This multidimensional approach serves as a legitimization strategy for these self-help groups, enabling them to perform both their moral claim to using the *Artemisia* plant and their trustworthiness to the broader public and their target groups.

The political idea of self-help and independency (*kujitegemea*) is not new to the Tanzanian context. With the Arusha Declaration of 1967, the then-incumbent president Julius Nyerere (1962–85) put forward his ideas about African socialism: one of his key messages was the policy of "trusting one's own strength" (*siasa ya kujitegemea*). The return to traditional medicine as a central concept of Tanzanian socialism was part of a strategy in pursuit of self-reliance. In a similar vein, ANAMED also endeavors to develop a self-sufficient healthcare provision. Dr. Faraji, a doctor and state-recognized health advisor, and Mrs. Hensch, a nurse from Germany, are the co-founders of ANAMED Tanzania. As becomes apparent in their statements, ANAMED's efforts to promote medicinal plants are associated with a demonopolization of medical knowledge:

ANAMED operates on various levels: There is the grassroots level, where we reach households that are able to treat diseases without any expenditures. The next level is where participants in seminars learn how to produce medicines derived from medicinal plants themselves. And then there is the level of medical facilities and institutions. There, patients can decide for themselves whether or not they take biomedical or our herbal medicines, or whether they combine the two. Two-thirds of expenditure on healthcare can be saved through the dissemination of our knowledge. In particular with malaria. People have many resources here but often lack the corresponding knowledge, which we try to revitalize.

(Mrs. Hensch, ANAMED)

Malaria disproportionately affects the rural poor, who are more vulnerable to infection. If you go into the communities, you find many sick people who die because they have no dispensary within reach and no access to ACTs. We teach them how to obtain first aid with the *Artemisia* plant. What we teach corresponds to the reality of life for people. In the publication of our recipes, we want to provide protection against the patenting of Galenic processes for obtaining medicines. Our results of research are to be understood as the good of the general public. Our approach is to educate patients; because there are many ways they can help themselves.

(Dr. Faraji, ANAMED)

The aspects of healthcare touched on here are of crucial import for ANAMED's operations. Despite extensive—and reformed—global health strategies, malaria remains one of the deadliest epidemics in Tanzania.* The reasons for this are complex: the rapid spread of resistance among malaria parasites to drugs, a lack of medicines and vaccines, and dysfunctional peripheral structures of healthcare systems following periods of structural adjustment have all been blamed for the persistence of malaria (Ginsburg and Deharo, 2011; Nsimba and Kayombo, 2008). After years of new approaches and holistic innovations in malaria control, influential donors and major health agencies have returned to advocating top–down programs (cf. Nájera et al., 2011). As a result, the reliance on medical technology (again) plays a key role in the design of malaria control programs (Cueto, 2013: 30f). On the global level, experts claim that the problem of malaria could be solved by implementing well-designed magic bullets. Besides insecticide-treated mosquito nets, vaccines, and genetically altered mosquitoes, access to ACTs is considered the most important factor in determining whether patients receive effective therapy (WHO, 2012, 2015).

On the local community level, however, the situation is more complicated. For efficient treatment, access to drug supply is only *one* factor among many. Studies have shown that public health systems often do not meet the requirements for sufficient treatment due to a lack of medical equipment (Gerrets, 2010; Kelly and Beisel, 2011). Patients are, furthermore, particularly vulnerable to malarial infection due to co-existing diseases, such as HIV/AIDS (De Ridder et al., 2008). Taking this local perspective further into account, certain available data have profound implications on the viability of malaria control programs. First, the demographics confirm that malaria is primarily a disease of the poor and malnourished (Kamat, 2013; Willcox et al., 2004b). Second, about 70% of the population in rural Tanzania has to cover a distance of up to 10 km to access basic healthcare (Hetzel et al., 2008; Kinung'hi et al., 2010). Third, although ACTs are now provided free of charge in government health facilities in Tanzania, in practice, the dispensing of ACTs remains erratic. On top of this, timely access to authorized ACT providers is below 50% (Bhatt et al., 2015; Khatib et al., 2013). One of the developments of greatest consequence is the increased circulation of counterfeit medicines. Encouraged by the high cost and poor availability of artemisinin, several surrogate medicines are in use throughout Tanzania—with fatal results (Maude et al., 2010). Moreover, resistance to artemisinin is a very real danger and has been interpreted as an early warning sign that new classes of antimalarials are needed (Dondorp, 2010; Weathers et al., 2014b). Conventional drug development, however, is slow and expensive; it takes up to 15 years for a medicine to make the journey from the late discovery phase to completed clinical trial and up to US$800

---

* More than 40% of all outpatient visits in Tanzania are attributable to malaria, resulting in an estimated 10–12 million clinical cases of malaria annually and up to 80,000 malaria-related deaths among all age groups (WHO, 2015). Children under five years of age and pregnant women are particularly vulnerable. It is estimated that malaria costs the Tanzanian government US$240 million every year in lost GDP (Kamat, 2013).

million to develop a new drug (Wells, 2011). Consequently, there is no evidence that coverage by high-quality ACTs can be guaranteed or sustained.

The strategic orientation of global malaria control programs scarcely takes this fact into account. Demand for improved access to medical care appears desirable at first glance; however, on closer inspection it seems that this would generate a series of disadvantageous consequences. Access to ACTs is currently realized through the sales and marketing division responsible for medicines produced in "the North" (Price, 2013). Countries affected by malaria—predominantly in "the South"—are thus forced to import these medicines and subvent them through funds provided by external financing bodies. Neither countries such as Tanzania nor private households are in a position to finance these medicines through their own means (Tibandebage et al., 2016). The need for access to these medicines leads countries afflicted by malaria in the current global political and economic balance of power into a position of chronic dependency on external funding bodies (Greene, 2011; Meier zu Biesen, 2013).

The concept of "structural violence" was coined within the framework of criticism of the global political and economic conditions relating to the persistence of epidemic diseases (Farmer, 1996). Prioritizing technology-oriented and biomedically based interventions in malaria control programs could be comprehended in this sense as a continuation of this form of structural violence. Instead of tackling the conditions facilitating economic dependency (property rights, monopolies), these underwent even further consolidation (see Petryna et al., 2006). A further negative consequence emerges from the internal logic of these programs: alternative approaches and solutions are marginalized. This observation becomes all the more politically explosive when we take into account what we know about patient behavior in sub-Saharan Africa: a number of studies demonstrate that medicinal plants play an important role in the treatment of systemic infections (for Tanzania, cf. Kayombo et al., 2012; Mboera et al., 2007).*

It is in this context that ANAMED, RITAM activists, and other researchers make their claim to consider all options for the development of new antimalarials. Wells (2011) and Willcox et al. (2011c) plead that we re-evaluate the "smash and grab" approach of randomly testing purified natural products and replace it instead with a patient data-led approach. In contrast, the parallel development of standardized phytomedicines can be achieved faster, more cheaply, and more sustainably for remote areas. The savings in both time and cost stem from the fact that substantial experience of human use increases the chances that a remedy will be safe, and that the necessary precautions will be known.

In addition to this, ANAMED considers treatment using the whole extract of the *Artemisia* plant as an opportunity to halt the development of drug resistance

---

* (Statistical) information supplied relating to the use of traditional medicine always remains somewhat vague, as only occasionally does it specify what type of traditional medicine is in use or how often patients actually make use of medicines on offer. A systematic overview of the literature on the application (incidence, frequency, use) of medicinal plants for treating malaria in developing countries is provided by Willcox and Bodeker (2004) and Willcox et al. (2004b).

(cf. Elfawal et al., 2012).* Furthermore, as a result of difficulties in accessing basic medical care, self-medication remains a widespread practice in malaria treatment—irrespective of socio-economic status (cf. Feierman, 1985; Kinung'hi et al., 2010). Frequently, sufferers apply antipyretic substances left over from a previous treatment. In many cases, these are past their use-by date or the amount is too small to cover a compete course of treatment. In light of these shortcomings, a number of studies have called for domestic self-treatment of malaria to be improved (Hetzel et al., 2008)—an approach pursued by ANAMED, through its seminar work in particular.

2.3 MEDICINE THROUGH SELF-PLANTATION: SEMINARS AND GREEN PHARMACIES

The most important channels for the dissemination of knowledge of *Artemisia* are the week-long seminars led by local ANAMED staff. Candidates for the seminars are recommended following discussions with representatives of the local health-care administration (*serekali ya kijiji*) in which the seminar is to take place. Those selected are considered to be able to profit in a particular way from the seminars: members of HIV/AIDS groups, NGOs, communal groups, churches, medical personnel, and research institutions that seek to support their countries in the treatment of malaria. Participants receive advice on sowing, cultivating, harvesting, processing, and applying the *Artemisia* plant.

In Tanzania, ANAMED works closely with the African Inland Church Tanzania (AICT), which, with over 300 local churches, is one of the most well-represented churches in the Mara Region. Under the aegis of the AICT, which has its headquarters in Musoma, a health service known as HUYAMU (*Huduma ya Afya Mara/Ukerewe*) was founded in 1999 to provide medical treatment alongside Christian and spiritual assistance in the form of mobile village clinics for the Mara Region and Ukerewe. These village clinics have frequently had an inspirational effect on members of the village community by encouraging them to acquire new opportunities for medical self-sufficiency through ANAMED.

The Christian religious orientation of this NGO has meanwhile helped to counteract prejudices against traditional healing practices. In Tanzania, ANAMED profits from this alliance with the AICT, which enables it to function as an authority because this platform makes it easier for ANAMED to communicate an alternative opinion—in concrete terms, the message that self-medication using medicinal plants is possible. The proactive stance taken by the AICT on medicinal plants is rooted,

---

* It has often been suggested (cf. Hirt and Lindsey, 2006a,b) that raw extracts of artemisinin are a natural ACT; however, this remains to be demonstrated (Wells 2011). It is difficult to design a study to show whether use of *Artemisia* tea promotes resistance to artemisinin. There are arguments on both sides, but no evidence (see in detail Willcox et al. 2004a; Willcox, 2009). The danger expressed by public health experts, that *Artemisia* decoctions may facilitate the generation of resistance due to the combination of underdosing with monotherapy, needs to be balanced against the observation that this decoction has been used in China for centuries without creating resistance.

according to my observations, among other things in passages from the Bible that express great respect toward herbal medicine (Meier zu Biesen, 2014). In addition, ANAMED takes into account that in terms of forces at play in the field of public health services, church organizations such as dioceses and missions assume an important role in the provision of healthcare. In the face of underrepresentation of government-provided healthcare services in Tanzania, medical care is often provided by NGOs, self-help groups, or religious movements (Dilger, 2012). Churches often have a better infrastructure than the formal health system, especially in remote areas (Willcox et al., 2011a).

In accordance with organization policy, each participant is required to pay an allowance of roughly US$2 to attend a seminar. This concept leads not infrequently to ANAMED staff encountering resistance from seminar participants. Reluctance to pay for training stems less from the cost itself than from prevailing policy in Tanzania on workshops. Tanzania has experienced an enormous increase in the number of civil–social organizations that provide their staff with the opportunity to take part in workshops in order to train in specific topical areas. NGO staff are supposed to be involved in improving the conception and communication of topics that aim to foster the country's sustainable development (*maendeleo*). Workshops have since established themselves in many organizations and now provide Tanzanian workers with an opportunity to attain both prestige and income. Participation in training courses is compensated by a day-wage (*posho*). As Green (2003)—on the basis of the critical concept of development developed by Ferguson (1990) and Escobar (1995)—has documented for the Tanzanian context, this manner of training is, as a rule, restricted to a particular elite, which is lured by increasingly higher day-wages (cf. Marsland, 2007: 763). Due to the fact that only certain NGOs with the requisite funds and their members can take part in training courses, a large proportion of the Tanzanian population is barred from these developments. Furthermore, the situation whereby the motivation to participate is induced to a great extent by financial incentives generates dependencies that make a critical insight into development processes in the country impossible. In contrast to most of the workshops run throughout Tanzania, ANAMED structures its seminars so that access is possible for all levels of society. The idea behind participants paying an allowance to attend is not just to cover the costs of the training course, but is also intended to shape participants' own self-concept and encourage them to pursue training and contribute to the development of their country of their own volition.

ANAMED seminars consist of theoretical (medicinal plants, hygiene, Galenism, ecology, solar energy) and practical elements (pharmaceutical production, plant cultivation, quality assurance of self-made pharmaceuticals). For *Artemisia*, the courses deal in great depth with therapeutic as well as agronomical aspects, as these are considered prerequisites for people growing crops. For those who do not want to or cannot cultivate *Artemisia*, there is now a cross-regional network of groups that allows members to procure the raw material from member producers.

Following the course, participants are able to register as ANAMED members. A certificate authorizes them to apply *Artemisia* therapeutically. This license is only valid for a limited period of time and is renewed as long as a member has complied with the rules pertaining to the application of the medicine. The matter of controlling and checking these procedures has been critically questioned repeatedly by representatives of the state health institutions in Mara, which are subject to the chair of the district medical officer (DMO) or the regional medical officer (RMO). The RMOs are certainly aware of the potential the distribution of *Artemisia* has for healthcare policy, however, the authorities wanted to be certain that the integration of the plant be accompanied by detailed monitoring and examination.

As I have shown elsewhere (Meier zu Biesen, 2013), the successful adoption of *Artemisia* implies a great many complex, pre-existing structures and circumstances. This includes, for example, the fundamental ability for the process whereby *Artemisia* is introduced to be incorporated from the perspective of the population (which uses the medicine). Moreover, it would be necessary to ensure that actors participating in the process of introduction pursue strategic and sustainable goals. In this manner, traditional medicinal plant uses have strongly influenced the ways in which new medical products have been adopted. This is strengthened all the more when the exclusion of certain elements—such as the association with traditional medicine and witchcraft—is achieved by way of classifications pertaining to standards. Moreover, there are (historically speaking) many health/political debates about the standardization of traditional medicines and efforts to integrate alternative medicine approaches into the healthcare provided to the Tanzanian population, originating both with the World Health Organization (WHO) and the Tanzanian government (cf. Mbwambo et al., 2007; Meier zu Biesen et al., 2012). Of greater centrality are the alliances that emerge within the context of the adoption process; for example, with regional district authorities or the native clergy, which take root only slowly as a result of lengthy processes of negotiation.

2.4 EFFICACY AND COST-EFFECTIVENESS

An important element of the seminar courses is the "rational" discussion of the potential and limitations of *Artemisia* therapy together with the differentiation between various clinical pictures of disease. ANAMED recommends that *Artemisia* not be deployed prophylactically, urging instead that it only be used to treat patients who have tested positive for malaria.

The patient observations and clinical studies upon which ANAMED bases its claims—those that also comprise the core of the reverse pharmacology approach—in which *Artemisia* was applied in tea form, achieved a 70%–100% recovery rate; the recrudescence rate was 39%.* Further sources state that teas

---

* The best results were achieved using the method in which 1 liter of boiling water was poured over 5 g (corresponding to 1 adult dose) of the dried plant material; the tea covered and left to brew for 15 min, strained, and then divided up into 4 rations of 250 mm each, to be taken over the course of a day. Patients were treated with this daily dose over a period of 7 days (see Willcox et al. 2004b). Other studies suggest that using powder derived from the whole *Artemisia* plant seems more effective than the infusion (Onimus et al., 2013). According to Willcox et al. (2011c), the optimal method of preparing and dosing the tea has yet to be defined.

prepared with care and on the basis of correct information are able to inhibit the parasite (parasite clearance) as early as the first day following treatment.*

Although ANAMED publicizes that *Artemisia* can cure a malaria infection, staff are well aware of the complexity of the concept of "cure" within the context of malaria, which must necessarily take into account the partial immunity of patients, the determination of parasite- and symptom-free status, as well as side effects, errors in treatment, and secondary effects (Willcox et al., 2011b).

During interviews with patients, I was particularly attentive to the terminology used by interviewees to describe their experiences of therapy. In most cases, patients attributed a healing (*kupona, kutibu kabisa*) effect to the *Artemisia* plant when treated for malaria (and not, for instance, in the context of HIV/AIDS) (see Meier zu Biesen, 2014). Malaria patients considered themselves to be healed when clinical evidence attested that there were no longer any plasmodia circulating in their blood. Patients without access to medical institutions or facilities equated being cured with the absence of symptoms such as fever, vomiting, headache, and aching limbs. The following statements made by both consumers of *Artemisia* and activists exemplify those of many patients who reported on their experiences with this herbal remedy. It becomes evident during the course of the statements how, as a result of *Artemisia* therapy, malaria symptoms disappear, symptoms of other diseases go into remission, relapse, or do not occur, and the patient's health improves after taking the tea:

> I was regularly ill with malaria (*ya kawaida nimepata homa ya malaria*). I continued to treat this with Fansidar®. But this medication ceased to work. I started to try out the tea made of *Artemisia* (*chai ya Artemisia*). The bitter taste was a great challenge (*uchokozi*). I brewed the tea every day and soon noticed that I was less tired. Usually during a bout of malaria I was plagued by pains (*uchungu*) and unable to work. The tea gave me strength (*nimepata nguvu*). After one week the malaria test was negative. I was healed (*nimepona kabisa*).
>
> (Patient)

> I used to take SP [sulfadoxine/pyrimethamine], but I suffered from side effects (*madhara*). I was weak for days on end. With *Artemisia* I regained my health quickly. That is what motivates the people here, this difference to the modern medicine, especially in cases when they are able to continue in their day-to-day work (*biashara ya kawaida*).
>
> (Patient)

> As we introduced *Artemisia* to the community, many asked themselves whether this was not traditional medicine (*dawa ya kienyeji*). At this point I explained to them that *Artemisia* is the same medicine as that in the [antimalarial] tablets (ACTs, *dawa ya mseto*). It took

---

* There are only a few well-controlled studies examining the extraction, recovery, and stability of the many compounds in *Artemisia* tea infusions (cf. Heide, 2006; Mueller et al., 2000, 2004; Ogwang et al., 2012; Räth et al. 2004). Willcox et al. (2011a), Van der Kooy and Verpoorte (2011), and Van der Kooy and Sullivan (2013) have collated all information available on the traditional *Artemisia* formulation. Weathers et al. (2014a) have provided an extensive review of clinical trials using *Artemisia* tea infusion therapy. The review by Van der Kooy and Sullivan (2013) contains proposals for future research.

a while before this education took hold. A new idea has to be brought from one neighbor to the next. Those who had taken *Artemisia* are full of hope and they are the ones who spread this message. The tea can be prepared by patients themselves *(wanajitengeneza wao wenyewe)*. The people here can grow *Artemisia* at home *(wanalima Artemisia nyumbani kwao)*. Those are the advantages of *Artemisia (faidha kuhusu Artemisia)* ... The disadvantage of *Artemisia (hasara katika Artemisia)* is the bitter *(chungu)* taste and some have difficulties completing the course of treatment *(wanashindwa kumuliza kipimo)*. They too feel better after a short while *(wanaskia ahueni)* and then simply stop drinking the tea *(wanaacha kunywa chai ya Artemisia)*. But we encourage patients to finish the entire dose. (ANAMED Tanzania)

As we first introduced *Artemisia* it was difficult. The community thought that we just wanted to do business *(kutafuta hela nyingi)* ... they had concerns *(mashaka)*. The people gradually decided to accept *Artemisia* for various reasons. The first was the low price *(kwanza garama niko chini)*. And no one complained about side effects *(haiwasumbui haiwapi hivu kusikia kuwashwa madhara)*. This medicine is a natural medicine *(dawa ya asili)* because it emerged from research *(zinatokea na utafiti)*, it has a high standard *(kiwango)*. (ANAMED Tanzania)

The concept of efficacy plays a central role in the legitimization of a therapy. Medical–anthropological studies show that research into the efficacy of a medicine is a complex process, both within the context of biomedical procedures and in terms of the behavior exhibited by patients—in particular with a view to the difference between subjective perception and intended effect. This complexity is particularly apparent in the different meaning attributed to the concept of efficacy (Young, 1983). Effective medicines are (in biomedical terms) substances tested in laboratories that induce physiological changes. Whether a medicine is actually efficacious depends upon a number of factors. For example, medicines that are labeled as efficacious but are expensive are unavailable to a large proportion of the population in developing countries, and are therefore inefficient. As a result, evidence of the clinical efficacy of an antimalarial medicine under controlled testing conditions and efficacy under real-life conditions are considered two different types of efficacy (Amin et al., 2004; Waldram, 2000). Kamat (2009) has ostensibly crystallized various factors for the Tanzanian context that should define the assessment of the efficiency of antimalarial medicines. The essential factors named in this approach are: national regulations (and their frequent changes) relating to the fight against malaria, the configuration of the local market for medicine (pharmaceuticals, traditional remedies), the reputation these medicines enjoy on a local level, price, regulations governing application, known side effects, and accessibility.*

---

* The challenge of efficacious malaria treatment is coupled closely with diagnostic testing procedures. Many patients are treated with antimalarial medicines although they have tested negatively for the pathogen (cf. Talisuna and Meya, 2007; Yeung et al., 2004). A study conducted in three major Tanzanian hospitals furnishes evidence that more than half of the antimalarial tablets prescribed are given to patients who have never had traces of plasmodia in their blood. This practice originates from an era in which antimalarial tablets were inexpensive and it was considered better to treat a febrile illness than to let a malaria infection take its course without treatment and run the risk of a serious illness.

In the following section, I will deal with the economic aspects of *Artemisia* therapy and, in so doing, with a central aspect of the question relating to the accessibility of antimalarial medicines. This involves first and foremost the socio-economic implications of the type of self-treatment propagated by ANAMED: local production of an antimalarial medicine.

2.5 ARTEMISININ: AN EFFECTIVE INGREDIENT AT COST PRICE?

Epidemiological, medical, and entomological aspects of malaria have been thoroughly researched. The same applies to the devastating socio-economic consequences recorded for countries in which malaria is endemic. Scientists often see the cause and, above all, the reasons for the persistence of the disease as closely correlated to poverty, which, in turn, exacerbates the disease's effects (Packard, 2009; Sachs and Malaney, 2002).

From an economic perspective, it is first the following aspect that makes *Artemisia* noteworthy as an antimalarial substance: the tea can be acquired by the local population cheaply or can be produced at no expense at home. A week-long course (35 g) of *Artemisia* powder costs US$0.50 and is thus considerably cheaper than the standard therapies—less than 1% of the price of imported antimalarial drugs (Hirt and Lindsey, 2009). A single *Artemisia* plant can produce enough leaves to treat up to 10 adult malaria patients. Vegetative propagation can yield enough leaves to treat thousands of patients, as a parent plant can produce up to 1000 stem cuttings. A single plant produces around 200 g of dried leaves. Therefore, 1000 plants produce around 200 kg of *Artemisia* leaves. Given an adult dose of 35 g, this means a potential yield that could treat 5714 patients (Hirt and Lindsey, 2006b).

Local efficiency discourses on this form of treatment have shown that the mere availability and inexpensive access afforded by the possibility of cultivating the plant at home are to be interpreted as important reasons for the acceptance of this substance. The attributes accorded to *Artemisia* tea are those of low cost (*siyo ghali sana*) or no cost (*bure*). *Artemisia* is also associated with a form of naturopathy (*matibabu asilia*), which means that, compared with pharmaceuticals, the tea treatment embodies the ideas of "naturalness" and "harmlessness." In addition, it is possible for metaphors connoted with strength (in terms of efficaciousness) to crystallize. Above all, the fact that no dangerous side effects (*madhara*) are attributed to the tea, and that patients not only experience fast relief from malaria symptoms following application, but also perceive the tea as fortifying (*unapata nguvu*), constitute assumptions related to efficacy.

In areas where Chinese medicine (*dawa ya Kichina*) retains a good reputation, the plant's association with China has influenced acceptance positively. The *Artemisia* plant also provides a prime example of the phenomenon whereby medical goods "from far away" (*inatoka mbali kwenu*) can assume a particular potency in people's expectations. In addition, the fact that antimalarial medicines recommended by the WHO consist of artemisinin promotes trust in the substance—at least among those aware of this connection.

ANAMED places great value on its members getting involved in the dissemination of knowledge about *Artemisia* and its application in their local communities. This idea is realized first and foremost through the formation of small, local groups. These groups operate at grassroots level and are entwined in their local communities. I would like to report on one such group of motivated producers in the following section.

2.6 WHEN PHARMACEUTICAL EXPECTATIONS FAIL AND PEOPLE ACT: GRASSROOTS RESPONSES TO HEALTH CRISES

All truth passes through three stages.
First, it is ridiculed.
Second, it is violently opposed.
Third, it is accepted as being self-evident.

(Schopenhauer, 1788–1860)

The local ANAMED group in Kigera-Etuma (*kikundi cha anamed Kigera-Etuma*) has set itself the goal of organizing *Artemisia*-based malaria treatment for the whole village. In addition to traditional healers (*waganga wa kienyeji*), the community in Kigera-Etuma has access to a Catholic health center, the House of Mercy (*mji wa huruma*), which provides villagers with basic medical care. The next nearest health-care facilities are in the city of Musoma, some 40 km away.

In Kigera-Etuma, malaria remains one of the most significant health issues, as the proximity to Lake Victoria provides ideal breeding conditions for the *Anopheles* mosquito. Most malaria researchers and policy makers agree that delay in seeking appropriate medical treatment for (especially childhood) falciparum malaria often results in severe complications, if not death. Falciparum malaria can progress rapidly, particularly in non-immune patients (Willcox et al., 2011c). Not surprisingly, the WHO has emphasized the importance of early diagnosis and effective treatment as one of the key factors in preventing high levels of malaria-related deaths. As is the case for other parts of the country, however, the village of Kigera-Etuma suffers from poor infrastructure, making prompt distribution of ACTs difficult.

Especially in situations of medical emergency in rural Tanzania when health facilities are out of reach or have run out of ACTs, ANAMED partners are often the only source of antimalarial treatment.* *Artemisia* plays a role, not only in the treatment, but also in the temporary control of malaria, largely through the prevention of coma until the infected person reaches a hospital or clinic stocked with ACTs (see Weathers et al., 2014b). ANAMED's experience suggests that immediate application of the medicinal plant can contribute to stronger immunity, increase excretion levels of the pyrogens (fever-inducing substances), and thereby have a fever-reducing effect. This, in turn, relieves the circulatory system and thus provides enough time to get the patient to a hospital. Moreover, it has been argued that when the supply of ACTs is insufficient, it makes sense to save ACTs for young children—among whom the mortality rate is particularly high—and use *Artemisia* extracts to treat adult patients, who are at lower risk (in high-transmission areas where residents develop partial immunity) (see Willcox et al., 2011a).

---

* Other factors underlying treatment delays in Tanzania, such as user fees, are discussed by Kamat (2013).

The village center of Kigera-Etuma sprawls along the main road with its many small shops (*maduka*) and market (*soko*). The path leading to the meeting point of the local ANAMED group passes through thick bush. It was only with the help of one of the group's members that I was able to reach the small brick house, in whose backyard—a peaceful space shaded by trees—the group conducted its meetings. The members meet three times a week in order to tend to the *Artemisia* field they hold in common. The group's proximity to Lake Victoria is one particular characteristic. Fertile soil on the shores of the lake and unrestricted access to fresh water have enabled the group to cultivate a lush plantation of *Artemisia*. Planted between palms and high-growing banana plants, the *Artemisia* plants are irrigated via furrows with water from the river and receive a continuous supply of high-quality water. As we strolled through the plantation, staff showed me their plants, which measured up to 2 m in height and demonstrated a vibrant green color. At the time of my visit, the plants were just about to bloom and had thus reached the harvesting stage, at which the highest possible yields of the sesquiterpene artemisinin may be preserved.

In March 2007, the group had already harvested 550 plants, from which it had managed to extract over 35 kg of *Artemisia* raw material. By using vegetative propagation, it was possible to maintain the plantation's considerable size the following year without expending capital for purchasing new seeds. When it first commenced its activities, the group sold *Artemisia* tea on a local scale. Over time, it has cultivated channels to sell *Artemisia* raw material that reach as far as the Kenyan city of Migori. The most effective advertising proved to be production of a high-quality *Artemisia* powder and professional production techniques: clean, well-sealed packaging with labels detailing the name, tea dosage, and use-by date as well as the contact details of the group.

The group is now networked with the medical facilities and organizations of the AICT and has secured itself a further market for its tea. As well as two dispensaries (in Bunda and Musoma), the Mkula Hospital in Mkula and the Kolandoto Clinic in Shinyanga are associated with the AICT and offer a large range of plant-based medicines produced by local ANAMED groups.

Although the idea of using medicinal plants enjoys a long tradition in Tanzania, in particular for chronic illnesses, the transmission of *new*—or traditionally secret—knowledge about medicinal plants such as *Artemisia* (cultivation, processing, dosage, and application) is innovative and now, finally, available. Through this transmission of knowledge—also to healthcare organizations and NGOs—new configurations as well as new social and cultural developments have taken shape to the extent that *Artemisia* is now being used within the context of medical self-help and, as such, is able to contribute to the development of independency in the treatment of systemic infections. New knowledge of *Artemisia* and its application harbors potential—in particular in the context of Tanzania's economic realities—above all because the therapeutic use of the medicinal plant permits a form of therapy open to patients of all income levels. Through scientifically grounded seminars, this concept has undergone professional upgrading largely at grassroots level—as intended by ANAMED. This form of professionalization plays, in turn, a major role in the levels of trust and determination to acquire knowledge that the local population demonstrates, respectively, toward ANAMED activists and the contents of these courses. At the same time, various actors—such as government and church representatives—have assumed a critical stance toward ANAMED and have, for example, followed the implementation of the medicinal plant with skepticism.

I preceded this section with a quote from Arthur Schopenhauer in which he speaks of various stages that new ideas have to pass through before they are accepted generally. The stages traversed by the "new" idea of *Artemisia*—a naturopathic-oriented malaria treatment based upon a locally producible plant—since its introduction appear to illustrate Schopenhauer's aphoristically pointed theory of stages, in particular because the adoption of *Artemisia* was fraught with opposition; ideological reservations about the new remedy were present in various phases of the medicine's adoption. At the end of the day, it was its strategic classification as a "natural medicine" that made it possible to overcome opposition: opponents of the therapy were thus not obliged to transfer their conceptions of occult and harmful practices associated with the (polemical) category of "traditional medicine" onto *Artemisia*.

Also implicit in ANAMED's efforts are power relations, above all because the instrumentalization of Western science can also be used to paternalistic effect. From a historical perspective, the aims of Christian missionaries were entwined with the adoption of medical products because religious ideas were particularly well suited to transportation via medical provision. On the one hand, medical provision was regarded as the continuation of the work of Jesus Christ—and ANAMED formulates its self-image in this precise manner. On the other, medical provision had the power to penetrate communities that resisted Christianity and served as a suitable instrument for countering "heathen belief," above all because the superiority of European medicine was interpreted as proof of the superiority of Western "rational" thought over African "superstition" (cf. Langwick, 2011). ANAMED also refers to the notion of rationality, which in this context refers to the ability to gauge a chain of cause and effect and thus refers to the relationship between an aim (treating disease) and the means used to achieve this aim (scientific investigation of drug recipes). To this effect, ANAMED achieves its continuing (intellectual) superiority over local medical practices not only by means of its explicitly religious–missionary character, but above all through a powerful construct of "scientific truth"—based upon supposed scientific rationality—coupled thereto (on this perspective, see Latour, 1999). ANAMED lays claim to a "scientific neutrality" coupled to the distribution of the plant by producing social reality—in this case, knowledge about health, risks, and disease (on the politics of biopower, see Foucault, 1998).

NGOs such as ANAMED operating in the health sector direct their attentions differently toward individuals and collectives and try to exert influence on their actions, via strategies of self-authorization, for example, which target funding and support for technologies of the self that are tied to the goals of government, they are involved in the production of specific subjects. With reference to the Foucauldian concept of governmentality, which analyzes power relations on the levels of both macrosociety and the everyday, Schlüter (2007) analyzes the work of NGOs in the area of health in Tanzania. According to Schlüter, this concept is particularly suitable for researching NGOs because it comprises specific characteristics of governance. In this sense, "leadership" stands for both the task of leading others (with either more or less strict mechanisms of force) and the manner of behaving in a field of multiple possibilities. An act of exercising power comprehended in this way—as a "creation of probabilities"—suggests many various forms of action and fields of practice aimed at diverse ways of controlling and leading others. Such a perspective on techniques of leading others and the self, Schlüter argues, permits us to view NGO work in Tanzania as a "form of governance." It is possible to recognize a strategy field of power relations defined accordingly in ANAMED: The organization does not just operate on the basis of (explicit or implicit) bans on courses of action, for example, the use of traditional medicine according to Christian values and biomedical concepts of efficacy, but rather on the basis of its power to motivate individuals and communities to behave in a particular way. ANAMED lays claim to leadership in the sense of initiating and inducting others. According to my observations, however, the contents of ANAMED's own aims do not necessarily constitute hegemonic instruments of power forced upon the population. In practice, this model has rather emerged as a conceptional reference point—above all because it is grounded in voluntariness and consensus—that also assists other project partners to formulate and, above all, *legitimate* their strategies. Conflict between local practices and national legislation is, in a manner of speaking, ANAMED's field of efficacy. Only by illustrating the dynamics and relationships of efficacy within this field has it proven possible to trace the contours of this NGO's activities.

2.7 CONCLUSION

Social scientists have revealed that, during the past decade, global health has developed, above all, into an ethical issue. Ethical questions concern, for example, the effects of global standardizing of health programs on local practices and on the question of how these real-life situations in turn influence global discourses and technologies and construct assumptions made about efficacy (cf. Fassin, 2007; Biehl, 2008; Lock and Nichter, 2002; Whyte et al., 2006). Intricately tied to this is the question of the ability of states, global health organizations, and pharmaceutical companies to exert power and influence that affects the efficacy of health interventions, in particular taking into consideration that the biomedically based system of healthcare—as a field of expertise—looks back on a long historical record of subordinating naturopathic practices (Hayden, 2007; Langwick, 2011). With the increasing development of global ways of viewing health and traditional medicine, "the local" has begun to enjoy a new level of visibility and is now interpreted as an expression of cultural authority and cultural rights. At the same time, that which is "traditional" has been identified (then as now) as a source of negative—in the sense of harmful—impacts on health. *Artemisia* is currently at the forefront of the heated debate between biomedical, technocentric control strategies and the holistic approach of traditional medicinal systems (e.g., in the WHO Position Statement on the recommendation *not* to use non-pharmaceutical forms of *Artemisia*) (WHO, 2012).

What challenges and contradictions surface when a non-biomedical theory of disease and therapeutic practice encounters the (bureaucratic) logic of responsibility and corresponding demands for evidence? What form do the political interests and power relations that constitute a medical field as "traditional," "natural," or "local" assume? What factors affect the extent to which governments secure measures and resources that preserve the health of their populations? How do global structures, such as the Trade-Related Aspects of Intellectual Property Rights Agreement on the prospecting of plants and patent rights, affect access to life-saving medicines?

Global health scholarship frequently invests its energies in developing models of optimal interventions and in identifying programs that supposedly "work" and that might therefore be scaled up across a range of often widely divergent contexts. The politics of "humanitarianism and emergency" (Bornstein and Redfield, 2011; Fassin and Pandolfi, 2010) are exclusively related to technological interventions. In the context of malaria, approaches based on technical fixes that are (again) becoming increasingly vertical, intervention-specific, and donor-funded—such as the large-scale deployment of ACTs—are seen as a way forward in disease control and elimination; calls for more integrated approaches in community-based studies and the related concerns they raise have been ignored (Kamat, 2013).

Despite the gradual globalization of (Asian) traditional medicine, which has been explicitly placed on the global agenda as a direct answer to medicines (i.e., artemisinin-based antimalarials), traditional medicine has never been systematically applied or strategically focused as a tool for medical emergency settings. While many laudable efforts to engage civil society and activists have ensued, a strong biomedical orientation remains pervasive, casting community engagement as politically necessary but "scientifically" irrelevant. As hopes for a magic bullet reign and the power of "data" undergoes increasingly intense fetishization in global health debates (Adams, 2013), the visions of technocrats tend to outweigh other forms of meaningful evidence. As noted by Biehl (2016), in this "emergency modality of intervention," such approaches create new modes of intervention, which encourage technical responses without much concern for ongoing living conditions and with limited impact on vulnerability to epidemic diseases such as malaria.

While these are important arguments on the level of institutional biosecuritization, Biehl argues that this body of work largely ignores local perspectives. Furthermore, the trope of "security" functions through a largely westernized notion of governance and biopower. The crucial and important question here is as follows: What other forms of securitization exist elsewhere—such as the strategic use of traditional medicine—and what clashes occur when they come into contact with one another through global health interventions?

This chapter has tackled the cognitive dimension between evidence-based facts versus local representations, the danger of (over-emphasizing) evidence-based interventions in global health, and what can be measured versus *what matters* and what is locally represented. It has dealt with issues relevant to current debates in global health relating to demands for efficacy and access to therapeutic care in severe epidemics. In so doing, I was concerned with the question of what opportunities Tanzania's local population actually has to withdraw from the power-related processes of healthcare provision mentioned herein or to help shape these actively: What creative, new arrangements are able to emerge when the medicinal plant *Artemisia* is used independently in its non-pharmaceutical form which, according to several actors, is the precondition for fully exploiting its potential? I have also discussed what impact neoliberal demands have on the efficacy of global health institutions and which pathways individuals and communities choose in order to acquire knowledge and skills, to make their own decisions about what they consider to be evident, and to what extent they operate as agents in the sense that they actively participate in their own healthcare provision as informed, self-confident, and creative agents. In particular, I raised the question of whether an improvement in healthcare for patients can be achieved primarily through access to ACTs as the most important means of overcoming therapeutic marginality—a question implied in the critique of the pharmaceuticalization of healthcare provision (cf. Petryna et al., 2006)—or if other factors, such as patient care in the context of humanitarian intervention and through the strategic use of herbal medicines via self-plantation, are also significant factors that expand patients' power to act.

Using the example of reverse pharmacology approaches, I have demonstrated how knowledge derived from Chinese medicine underwent revitalization and how the Tanzanian population deploys this newly acquired knowledge and absorbs it into local methods of medical treatment.

Current research on malaria concentrates first and foremost on inexpensive methods of synthesizing the agent artemisinin. In particular, with respect to competition through the pharmaceutical industry, the high production costs of ACTs together with the alarming increase in drug resistance has led to increasing interest in developing an effective antimalarial phytomedicine. As we are now in a position to build upon both the foundation provided by profound insights into local and global (artemisinin) production conditions and a now considerable amount of research on the medicinal plant *Artemisia*—which gained popularity through the Nobel Prize for Medicine in 2015—we can investigate in greater depth the local possibilities that might open up when *Artemisia* is developed as a professionalized phytomedicine. Further research that situates itself at the nexus between traditional consumption of medicinal plants and their pharmacological investigation in interdisciplinary collaborations has the potential to transcend the Tanzanian context and provide insights relevant to health policy in other countries.

REFERENCES

Adams V. 2013. Evidence-based global public health: Subjects, profits, erasures. In J. Biehl, and A. Petryna (eds) *When People Come First: Critical Studies in Global Health.* Princeton University Press, Princeton, 54–90.

Aftab T., Ferreira J.F.S., Masroor M., Khan A., and Naeem M. 2014. *Artemisia annua – Pharmacology and Biotechnology.* Springer, Heidelberg.

Amin A., Hughes D.A., Marsh V., et al. 2004. The difference between effectiveness and efficacy of antimalarial drugs in Kenya. *Tropical Medicine and International Health* 9(9): 967–974.

Bhatt S., Weiss D.J., Cameron E., et al. 2015. The effect of malaria control on *Plasmodium falciparum* in Africa between 2000 and 2015. *Nature* 526(7572): 207–2011.

Biehl J. 2008. Drugs for all: The future of global AIDS treatment. *Medical Anthropology* 27 (2): 99–105.

Biehl J. 2016. Theorizing global health. *Medicine Anthropology Theory* 3(2): 127–142.

Bornstein E. and Redfield P. 2011. *Forces of Compassion. Humanitarianism between Ethics and Politics.* School for Advanced Research Advanced Seminar Series. School for Advanced Research, Sante Fe.

Craig S. 2015. "Slow" medicine in fast times. In *Savage Minds.* Notes and Queries in Anthropology. http://savageminds.org/2015/06/23/slow-medicine-in-fast-times, retrieved on 12 November 2016).

Cueto, M. 2013. Malaria and the global health in the twenty-first century. In J. Biehl and A. Petryna (eds) *When People Come First. Critical Studies in Global Health.* Princeton University Press, Princeton, 30–53.

De Ridder S., Van der Kooy F., and Verpoorte R. 2008. *Artemisia annua* as a self-reliant treatment for malaria in developing countries. *Journal of Ethnopharmacology* 120(3): 302–314.

Dilger H. 2012. Targeting the empowered individual: Transnational policy making, the global economy of aid, and the limitations of biopower in Tanzania. In H. Dilger, A. Kane, and S.A. Langwick (eds) *Medicine, Mobility, and Power in Global Africa.* Indiana University Press, Indianapolis, 60–91.

Dondorp A.M., Yeung S., White L. Nguon C., Day N.P., Socheat D., and Von Seidlein L. 2010. Artemisinin resistance: Current status and scenarios for containment. *Nature Reviews Microbiology* 8(4): 272–280.

Elfawal M.A., Towler M.J., Reich N.G., Golenbock D., Weathers P.J., and Rich S.M. 2012. Dried whole plant *Artemisia annua* as an antimalarial therapy. *PLoS One* 7(12): 1–7.

Escobar A. 1995. *Encountering Development: The Making and Unmaking of the Third World.* Princeton University Press, Princeton.

Etkin N.L., do Céu de Madureira M., and Burford G. 2004. Guidelines for ethnobotanical studies on traditional antimalarials. In M. Willcox, G. Bodeker, and P. Rasoanaivo (eds) *Traditional Herbal Medicines for Modern Times. Traditional Medicinal Plants and Malaria.* CRC Press, Boca Raton, 217–230.

Farmer P. 1996. On suffering and structural violence: A view from below. In A. Kleinman, V. Das, and M. Lock (eds) *Social Suffering.* University of California Press, Berkeley, 261–283.

Farmer P. 1997. Social scientists and the new tuberculosis. *Social Science and Medicine* 44(3): 347–358.

Fassin D. 2007. *When Bodies Remember: Experiences and Politics of AIDS in South Africa.* University of California Press, Berkeley.

Fassin D. and Pandolfi M. 2010. *Contemporary States of Emergency. The Politics of Military and Humanitarian Interventions.* MIT Press, Cambridge, MA.

Feierman S. 1985. Struggles for control: The social roots of health and healing in modern Africa. *African Studies Review* 28(2/3): 73–147.

Ferguson J. 1990. *The Anti-Politics Machine: 'Development', Depoliticization and Bureaucratic Power in Lesotho.* Cambridge University Press, Cambridge.

Foucault M. 1998. *The History of Sexuality. Vol 1: The Will to Knowledge.* Penguin, London.

Gaudillière J.-P. 2014. An Indian path to biocapital? Traditional knowledge inventories, intellectual property and the reformulation of Ayurvedic drugs. *East Asian Science, Technology and Society* 8(4): 391–415.

Gerrets R. 2010. *Globalizing International Health: The Cultural Politics of 'Partnership' in Tanzanian Malaria Control.* New York University, New York.

Ginsburg H. and Deharo E. 2011. A call for using natural compounds in the development of new antimalarial treatments: An introduction. *Malaria Journal* 10(1): 1–7.

Graz B., Kitua A.Y., and Malebo H.M. 2011. To what extent can traditional medicine contribute a complementary or alternative solution to malaria control programmes? *Malaria Journal* 10(6): 1–7.

Green M. 2003. Globalizing development in Tanzania. Policy franchising through participatory project management. *Critique of Anthropology* 23(2): 123–143.

Greene J.A. 2011. Making medicines essential: The emergent centrality of pharmaceuticals in global health. *BioSocieties* 6(1): 10–33.

Hayden C. 2007. A generic solution? Pharmaceuticals and the politics of the similar in Mexico. *Current Anthropology* 48(4): 475–95.

Heide L. 2006. Artemisinin in traditional tea preparations of *Artemisia annua. Transactions of the Royal Society of Tropical Medicine and Hygiene* 100(8): 802.

Hetzel M.W., Dillip A., Lengeler C., et al. 2008. Malaria treatment in the retail sector: Knowledge and practices of drug sellers in rural Tanzania. *BMC Public Health* 8(1): 1–11.

Hirt H.M. and Lindsey K. 2006a. *Artemisia Annua Tea – A Revolution in the History of Tropical Medicine.* Anamed – Aktion Natürliche Medizin, Winnenden.

Hirt H.M. and Lindsey K. 2006b. *Rich Artemisia instead of Poor (So-called Pure) Artemisinin!* Anamed – Aktion Natürliche Medizin, Winnenden.

Hirt H.M. and Lindsey K. 2009. *A-3: From Research to Experience.* Anamed – Aktion Natürliche Medizin, Winnenden.

Hirt H.M. and M'Pia B. 2000. *Natürliche Medizin in den Tropen.* Anamed – Aktion Natürliche Medizin, Winnenden.

Hsu E. 2006. The history of Qinghao in the Chinese materia medica. *Transactions of the Royal Society of Tropical Medicine and Hygiene* 100(6): 505–508.

Kamat V.R. 2009. Cultural interpretations of the efficacy and side effects of antimalarials in Tanzania. *Anthropology and Medicine* 16(3): 293–305.

Kamat V.R. 2013. *Silent Violence. Global Health, Malaria, and Childhood Survival in Tanzania.* The University of Arizona Press, Tucson.

Kayombo E.J., Uiso F.C., and Mahunnah R.L.A. 2012. Experience on healthcare utilization in seven administrative regions of Tanzania. *Journal of Ethnobiology and Ethnomedicine* 8(1): 1–8.

Kelly A.H., and Beisel U. 2011. Neglected malarias: The frontlines and back alleys of global health. *BioSocieties* 6(1): 71–87.

Khatib R.A., Selemani M., Mrisho G.A., et al. 2013. Access to artemisinin-based antimalarial treatment and its related factors in rural Tanzania. *Malaria Journal* 12(155): 1–8.

Kinung'hi S.M., Mashauri F., Mwanga J.R., et al. 2010. Knowledge, attitudes and practices about malaria among communities: Comparing epidemic and non-epidemic prone communities of Muleba District, North-western Tanzania. *BMC Public Health* 19(10): 1–11.

Langwick S. 2011. *Bodies, Politics, and African Healing. The Matter of Maladies in Tanzania.* Indiana University Press, Bloomington.

Latour B. 1999. *Pandora's Hope. Essays on the Reality of Science Studies.* Harvard University Press, Cambridge.

Li Q. and Weina P. 2010. Artesunate: The best drug in the treatment of severe and complicated malaria. *Pharmaceuticals* 3(7): 2322–2332.

Lock M. and Nichter M. 2002. Introduction: From documenting medical pluralism to critical interpretations of globalized health knowledge, policies and practices. In M. Lock and M. Nichter (eds) *New Horizons in Medical Anthropology. Essays in Honour of Charles Leslie*. Routledge, New York, 1–34.

Marsland R. 2007. The modern traditional healer: Locating "hybridity" in modern traditional medicine, Southern Tanzania. *Journal of Southern African Studies* 33(4): 751–765.

Maude R., Woodrow C.J., and White J. 2010. Artemisinin antimalarials: Preserving the "magic bullet." *Drug Development Research* 71(1): 12–19.

Mboera L.E.G., Makundi E.A., and Kitua A.Y. 2007. Uncertainty in malaria control in Tanzania: Crossroads and challenges for future interventions. *American Journal of Tropical Medicine and Hygiene* 77(6_suppl): 112–118.

Mbwambo Z.H., Mahunnah R.L., and Kayombo E.J. 2007. Traditional health practitioner and the scientist: Bridging the gap in contemporary health research in Tanzania. *Tanzania Health Research Bulletin* 9(2): 115–120.

Meier zu Biesen C. 2010. The rise to prominence of *Artemisia annua L.* – the transformation of a Chinese plant to a global pharmaceutical. *African Sociological Review* 14(2): 24–46.

Meier zu Biesen, C. 2013. *Globale Epidemien – Lokale Antworten. Eine Ethnographie der Heilpflanze Artemisia annua in Tansania*. Campus Verlag, Frankfurt.

Meier zu Biesen C. 2014. Notions of efficacy around a Chinese medicinal plant: *Artemisia annua* – an innovative AIDS therapy in Tanzania. In R. van Dijk, H. Dilger, M. Burchardt, and T. Rasing (eds) *Saving Souls, Prolonging Lives: Religion and the Challenges of AIDS Treatment in Africa*. Ashgate Publishers, London, 271–295.

Meier zu Biesen C. 2017. From coastal to global: The transnational flow of Ayurveda and its relevance for Indo-African linkages. *Global Public Health* (RGPH), forthcoming.

Meier zu Biesen C., Dilger H., and Nienstedt T. 2012. *Bridging Gaps in Health Care And Healing: Traditional Medicine and the Traditional Health Care Sector in Zanzibar*. Freie Universität Berlin, Berlin.

Mueller M., Karhagomba I.B., Hirt H.M., and Wemakor E. 2000. The potential of *Artemisia annua* as a locally produced remedy for malaria in the tropics: Agricultural, chemical and clinical aspects. *Journal of Ethnopharmacology* 73(3): 487–493.

Mueller M., Runyambo N., Wagner I., Borrmann S., Dietz K., and Heide L. 2004. Randomized controlled trial of a traditional preparation of Artemisia Annua L. (Annual Wormwood) in the treatment of malaria. *Transactions of the Royal Society of Tropical Medicine and Hygiene* 98(5): 318–321.

Nájera J., González-Silva M., and Alonso P.L. 2011. Some lessons for the future from the global malaria eradication programme (1955–1969). *PLoS Medicine* 8(1): 1–7.

Nsimba S.E.D. and Kayombo E.J. 2008. Sociocultural barriers and malaria health care in Tanzania. *Evaluation and the Health Professions* 31(3): 318–322.

Ogwang P. E., Ogwal J.O., Kasasa S., Olila D., Ejobi F., Kabasa D., and Obua C. 2012. *Artemisia annua* L. Infusion consumed once a week reduces risk of multiple episodes of malaria: A randomised trial in a Ugandan community. *Tropical Journal of Pharmaceutical Research* 11(3): 445–453.

Onimus M., Carteron S., and Lutgen P. 2013. The surprising efficiency of *Artemisia annua* powder capsules. *Medicinal and Aromatic Plants* (2/3): 1–4.

Packard R. 2009. "Roll back malaria, roll in development"? Reassessing the economic burden of malaria. *Population and Development Review* 35(1): 53–87.

Petryna A., Lakoff A., and Kleinman A. 2006. *Global Pharmaceuticals: Ethics, Markets, Practices*. Duke University Press, Durham.

Pordié L. and Gaudillière J.-P. 2014. The reformulation regime in drug discovery: Revisiting polyherbals and property rights in the Ayurvedic industry. *East Asian Science, Technology and Society* 8(1): 57–79.

Price R.N. 2013. Potential of artemisinin-based combination therapies to block malaria transmission. *Journal of Infectious Diseases* 207(11): 1627–1629.

Räth K., Taxis K., Walz G., Gleiter C.H., Li S.-M., and Heide L. 2004. Pharmacokinetic study of artemisinin after oral intake of a traditional preparation of *Artemisia annua* L. (annual wormwood). *American Journal of Tropical Medicine and Hygiene* 70(2): 128–132.

Sachs J. and Malaney P. 2002. The economic and social burden of malaria. *Nature* 415(6872): 680–685.

Schlüter N. 2007. Krankheit, armut und marginalität. Ethnographie einer lokalen AIDS-NGO in Tansania. Master Thesis. Institut für Europäische Ethnologie der Humboldt-Universität zu Berlin.

Schopenhauer A. 1788–1860. *Aphorismen*. Insel-Verlag, Leipzig.

Talisuna A.O., and Meya D.N. 2007. Diagnosis and treatment of malaria. *British Medical Journal* 334(24): 377–386.

Tibandebage P., Wangwe S., Mackintosh M., and Mujinja P.G.M. 2016. Pharmaceutical manufacturing decline in Tanzania: How possible is a turnaround to growth? In M. Mackintosh, G. Banda, P. Tibandebage, and W. Wamae (eds) *Making Medicines in Africa. The Political Economy of Industrializing for Local Health*. International Political Economy Series. Springer, Heidelberg, 45–64.

Van der Kooy F. and Sullivan S.E. 2013. The complexity of medicinal plants: The traditional *Artemisia annua* formulation, current status and future perspectives. *Journal of Ethnopharmacology* 150(1): 1–13.

Van der Kooy F. and Verpoorte R. 2011. The content of artemisinin in the *Artemisia annua* tea infusion. *Planta Medica* 77(15): 1754–1756.

Waldram J.B. 2000. The efficacy of traditional medicine: Current theoretical and methodological issues. *Medical Anthropology Quarterly* 14(4): 603–625.

Weathers P., Reed K., Hassanali A., Lutgen P., and Engeu P.O. 2014a. Whole plant approaches to therapeutic use of Artemisia annua L. (Asteraceae). In T. Aftab, J.F.S. Ferreira, M. Masroor, A. Khan, and M. Naeem (eds) *Artemisia annua – Pharmacology and Biotechnology*. Springer, Heidelberg, 51–75.

Weathers P., Towler M., Hassanali A., Lutgen P., and Engeu P.O. 2014b. Dried-leaf *Artemisia annua*: A practical malaria therapeutic for developing countries? *World Journal of Pharmacology* 3(4): 39–55.

Wells T. 2011. Natural products as starting points for future anti-malarial therapies: Going back to our roots? *Malaria Journal* 10(3): 1–12.

White N.J. 2008. Qinghaosu (Artemisinin): The price of success. *Science* 320(5874): 330–334.

WHO. 2012. WHO position statement on effectiveness on non-pharmaceutical forms of *Artemisia annua* l. against malaria. http://www.who.int/malaria/diagnosis_treatment/position_statement_herbal_remedy_artemisia_annua_l.pdf?ua=1, retrieved on 13 July, 2016.

WHO. 2015. Word malaria report 2015. http://www.who.int/malaria/publications/world-malaria-report-2015/report/en/, retrieved on 22 November, 2016.

Whyte S.R., Whyte M.A., Meinert L., and Kyaddondo B. 2006. Treating AIDS: Dilemmas of unequal access in Uganda. In A. Petryna, A. Lakoff, and A. Kleinman (eds) *Global Pharmaceuticals: Ethics, Markets, Practice*. Duke University Press, Durham, 240–262.

Willcox M. 2009. *Artemisia* species: From traditional medicines to modern antimalarials – and back again. *Journal of Alternative and Complementary Medicine* 15(2): 101–109.

Willcox M., Benoit-Vical F., Fowler D., et al. 2011b. Do ethnobotanical and laboratory data predict clinical safety and efficacy of anti-malarial plants? *Malaria Journal* 10(1): 1–9.

Willcox M. and Bodeker G. 2004. Frequency of use of traditional herbal medicines for the treatment and prevention of malaria: An overview of the literature. In M. Willcox, G. Bodeker, and P. Rasoanaivo (eds) *Traditional Herbal Medicines for Modern Times. Traditional Medicinal Plants and Malaria.* CRC Press, Boca Raton, 161–185.

Willcox M., Bodeker G., and Rasoanaivo P. 2004b. *Traditional Herbal Medicines for Modern Times. Traditional Medicinal Plants and Malaria.* CRC Press, Boca Raton.

Willcox M., Burton S., Oyweka R., Namyalo R., Challand S., and Lindsey K. 2011a. Evaluation and pharmacovigilance of projects promoting cultivation and local use of *Artemisia annua* for malaria. *Malaria Journal* 10(84): 1–24.

Willcox M., Graz B., Falquet J., Diakite C., Giani S., and Diallo D. 2011c. A "reverse pharmacology" approach for developing an anti-malarial phytomedicine. *Malaria Journal* 10(8): 1–10.

Willcox M., Rasoanaivo P., Sharma V.P., and Bodeker G. 2004a. Comment on: Randomized controlled trial of a traditional preparation of *Artemisia annua* L. (annual wormwood) in the treatment of malaria. *Transactions of the Royal Society of Tropical Medicine and Hygiene* 98(12): 755–756.

Yeung S., Pongtavornpinyo W., Hastings I.M., Mills A.J., and White N.J. 2004. Antimalarial drug resistance, artemisinin-based combination therapy, and the contribution of modeling to elucidating policy choices. *American Journal of Tropical Medicine and Hygiene* 71(2_suppl): 179–186.

Young A. 1983. The relevance of traditional medical cultures to modern primary health care. *Social Science and Medicine* 17(16): 1205–1211.

CHAPTER 3

Use of *Artemisia annua* L. in the Treatment of Diseases—An Update

M. Naeem, Tariq Aftab, Asfia Shabbir, and M. Masroor A. Khan

CONTENTS

3.1 INTRODUCTION

Malaria remained a major scourge of mankind until the Chinese introduced artemisinin to the world as a remedy to this disease. Pinato and Stebbing (2015) reported that various species of *Artemisia* had been well known to the Chinese for thousands of years for its pharmacological properties. In his manuscript *A Handbook of Prescriptions for Emergencies*, the Chinese physician and alchemist Ge Hong (284–346 AD) described in detail the use of *Artemisia annua* (sweet wormwood or, in Chinese, *Qinghao*) as a remedy against malarial fevers. *A. annua* L. has been used for over two millennia in traditional Chinese medicine for a variety of complaints—from skin disorders to high blood pressure and malaria. The genus is native to Asia (Pinato and Stebbing, 2015). The antimalarial drug, artemisinin, was discovered by Tu Youyou, a Chinese scientist, who was awarded half of the 2015 Nobel Prize in Medicine for her discovery. Following the discovery of artemisinin as an antimalarial drug, treatments containing an artemisinin derivative (artemisinin combination therapies, ACTs) have now been standardized for the treatment of *Plasmodium falciparum* malaria worldwide.

3.2 PHYTO-CONSTITUENTS OF THE PLANT

A. annua is rich in sesquiterpenes, flavonoids, coumarins, essential oil, palmitic acid, stigmasterol, β-sitosterol, aurantiamide acetate, annuadiepoxide, and β-glucosidase III, among other constituents.

The sesquiterpenes include arteannuin, artemisinin A (qinghaosu I), artemisinin B (qinghaosu II), deoxyartemisinin (qinghaosu III), qinghaosu IV, qinghaosu V, qinghaosu VI, arteannuin C, artemisinic acid (qinghao acid), methyl artemisinate, artemisinol, annulide, friedelin, and friedelan-3β-ol.

The flavonoids include chrysosplenol-D, artemetin, casticin, cirsilineol, axillarin, cirsiliol, tamarixetin, rhamnetin, cirsimaritin, rhamnocitrin, chrysoeriol, kaempferol, quercetin, luteolin, and patuletin.

Artemisia's coumarins include scopoletin, coumarin, 6-methoxy-7-hydroxycoumarin, and scoparone, while its oil contains camphor, β-caryophyllene, iso-Artemisia ketone, β-pinene, bornyl acetate, carveol, benzyl isovalerate, β-farnesene, copaene, γ-muurolene, fenchone, linalool, isoborneol, α-terpineol, borneol, camphene, myrcene, limonene, and γ-terpineol (http://www.chineseherbshealing.com/wormwood-herb/).

3.3 PHARMACOLOGICAL ACTIONS OF *A. ANNUA*

Researchers have gathered comprehensive information about the medicinal properties of *A. annua* that have allowed it to be used against various diseases (Alesaeidi and Miraj, 2016). *A. annua* is commonly used for its antimalarial, immunosuppressive, anti-inflammatory, anticancer, antihypertensive, antioxidative, antimicrobial, antiparasitic, insecticidal, and antiviral properties (Figure 3.1).

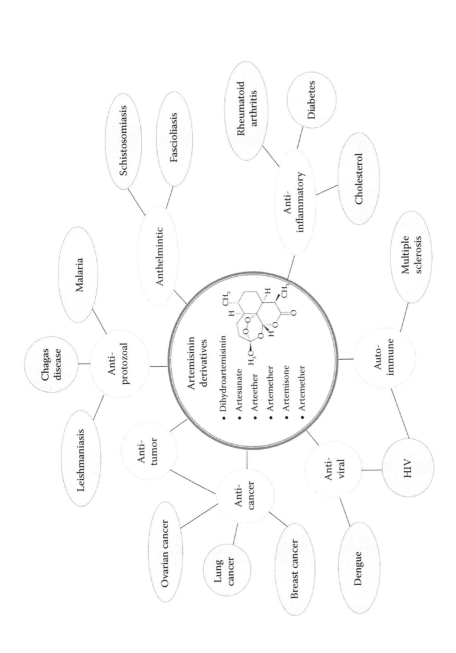

Figure 3.1 Diagrammatic representation outlining biological activities of artemisinin and their potential applications in various human diseases.

3.3.1 Antimalarial Activity

Artemisinin, a compound extracted from *Artemisia*, is of great importance for its different therapeutic and antimalarial activities (Kiani et al., 2016). Artemisinin and its derivatives exert their effect by interfering with the plasmodial hemoglobin catabolic pathway and inhibition of heme polymerization. *In vitro* experimental study shows the inhibition of digestive vacuole proteolytic activity in the malaria parasite by artemisinin (Ferreira, 2004). *Ex vivo* experiments have also shown accumulation of hemoglobin in the parasites treated with artemisinin, suggesting inhibition of hemoglobin degradation. Artemisinin was found to be a potent inhibitor of heme polymerization activity mediated by *Plasmodium yoelii* lysates as well as *P. falciparum* histidine-rich protein II (Penissi et al., 2006).

3.3.2 Anti-Inflammatory Activity

The aqueous methanolic extract of *A. annua* was shown to possess anti-inflammatory activity when studied using carrageenan- and egg albumin-induced acute rat paw edema, and cotton pellet- and grass pith-induced chronic inflammation models. The extract, at a dose of 200 g/kg, has been found to possess significant anti-inflammatory activity on the tested experimental models. It exhibited its maximum anti-inflammatory effects against carrageenan- and egg albumin-induced rat paw edema of 55.4% and 53.2%, respectively, after 5 h.

In chronic models, at a dose of 200 g/kg, the extract resulted in a 60% reduction in granuloma weight. This effect was comparable to that of diclofenac sodium, a non-steroidal anti-inflammatory agent. Phytochemical results also suggest that the triterpenoids, flavonoid, polyphenols, and coumarin present in the plant extract inhibit the development of maximum edema response in acute and chronic models (Patočka and Plutarb, 2003).

3.3.3 Immunosuppressive Activity

A. annua has been widely used in traditional Chinese medicine to treat autoimmune diseases such as systemic lupus erythematosus and rheumatoid arthritis. Ethanolic extracts of *A. annua* significantly suppressed concanavalin A (Con A)- and lipopolysaccharide (LPS)-stimulated splenocyte proliferation *in vitro* in a concentration-dependent manner. The ethanolic extract of the plant could suppress the cellular and humoral responses. *A. annua* has immunosuppressive activity in the treatment of some autoimmune diseases (Verdian-Rizi, 2009).

3.3.4 Antihypertensive Activity

Artemisia extracts can be envisaged as anti-hypertensives and cardiovascular protectants. Extract of *A. annua* reduces blood pressure, prevents elevation of glycosylated hemoglobin levels, and possesses a hypoliposis effect, in addition to protecting against body weight loss in diabetic animals. Aqueous extracts (100 and 200 mg/kg) significantly inhibited phenylephrine-induced contraction and potentiated the endothelium-dependent relaxation of rat aortic rings in Krebs solution (Allen et al., 1997).

3.3.5 Antiparasitic Activity

A study of the effects of artemisinin on *Neospora caninum*, a protozoal parasite infecting a wide range of mammals and causing abortion in cattle, was performed (Verdian-Rizi, 2009). Cultured host cells (Vero cells or mouse peritoneal macrophages) were infected with *N. caninum* tachyzoites and supplemented with concentrations of 20, 10, 1, 0.1, and 0.01 µg/ml artemisinin. At 20 or 10 µg/ml for 11 days, artemisinin eliminated all microscopic foci of *N. caninum* completely. At 1 µg/ml for 14 days, the same result was observed. Over shorter periods, 0.1 µg/ml artemisinin reduced the intracellular multiplication of *N. caninum* tachyzoites ($p < 0.05$).

Pretreatment of host cells had no effect on this multiplication. There was no apparent toxicity to host cells in long-term studies. The effect of artemether was tested against the larval stages of *Schistosoma mansoni* from the time of skin penetration to the early adult liver stage in mice and hamsters. The animals did not develop schistosomiasis if treated with artemether during the first month after infection. The parasite was especially susceptible during the third and fourth weeks after infection, with treatment resulting in reductions of 75.3%–82.0% in parasite burden compared with the non-treated controls. Animals subjected to various schedules of repeated treatment experienced cure rates of 97.2% to 100%. The chemotherapy of leishmaniosis is handicapped by drug resistance, especially against sodium antimony gluconate.

Artemisinin showed anti-leishmanial activity against both promastigotes and amastigotes, with IC50 values of 160 and 22 µM, respectively, and with a high safety index (>22-fold) (Ferreira, 2004; Wilcox et al., 2004; Verdian-Rizi, 2009).

3.3.6 Antimicrobial Activity

Extract of *A. annua* had the beneficial effect of controlling microbial infections. Antimicrobial effects of methanol and ethanol extracts of *A. annua* leaves against *Staphylococcus aureus* PTCC 1431, *Salmonella enterica* PTCC 1231, *Klebsiella pneumoniae* PTCC 1053, *Shigella dysenteriae* PTCC 1188, and *Escherichia coli* PTCC 1399 were reported (Tajehmiri et al., 2014).

Different extracts and metabolites isolated from *A. annua* have been tested *in vitro* against several Gram-positive and Gram-negative bacteria and fungi, and have demonstrated activity against some microorganisms; notably, *Bacillus subtilis*, *Salmonella enteritidis*, and *Candida albicans*. The essential oil obtained from the plant also shows significant antimicrobial activity against all the microorganisms tested, except *Pseudomonas aeruginosa*. Further study has revealed that it exerts its maximum activity against the fungal microorganisms *Saccharomyces cerevisiae* (MIC = 2 mg/ml) and *Candida albicans* (MIC = 2 mg/ml). The oil was also found to have moderate inhibitory activity against *Staphylococcus aureus* and *Escherichia coli*, with MIC values of 32 mg/ml and 64 mg/ml, respectively (Meier, 2011).

3.3.7 Insecticidal Activity

An experiment was carried out by Zamani et al. (2011) on the effects of different concentrations (15%, 11%, 8%, 5.5%, and 4%) of essential oil of *A. annua* on 17-day-old larvae of the Indian meal moth *Plodia interpunctella* (Hübner) under laboratory conditions. The LC_{25}, LC_{50}, and LC_{75} were estimated to be 6.0%, 8.4%, and 11.3%, respectively, after 24 hours. Sub-lethal doses of essential oil reduced adult emergence, longevity of male and female insects, fecundity, and female fertility. The energy reserves (i.e., protein, carbohydrate, and lipid contents) of treated larvae were significantly reduced in comparison with the controls. Insect mortality was independent of sex.

3.3.8 Antioxidant Activity

According to various research studies, experimental models of antioxidant potency demonstrated that the leaves and inflorescences had the highest percentage of protein, crude fat, and *in vitro* digestible fractions but the lowest levels of detergent fibers. These tissues also had the highest composition of the major elements as well as manganese and copper (Verdian-Rizi, 2009). Their relatively high amino acid and vitamin profiles equally reflect a desirable nutritional balance in addition to their high antioxidant capacities (Nofal et al., 2009). It is established that *A. annua* is a good reservoir of nutrients and antioxidants that might favor its use by humans as a potential herbal tonic.

3.4 THERAPEUTIC SPECTRUM OF *A. ANNUA* L. DERIVATIVES: ACTIVITY AGAINST INFECTIOUS DISEASES

Since the beginning of the 1950s, a constant effort has been made to identify novel drugs from natural sources with the aim of developing efficient treatments against different human diseases (Seeff et al., 2001). Nature has played a valuable role in achieving this objective. Inspired by this objective, the western medicinal approach has been to identify active molecules occurring in commonly used medications derived from traditional medicine. This has resulted in the rediscovery of highly effective drugs such as camptothecin (O'Leary and Muggia, 1998) paclitaxel, and artemisinin among a long list of natural products (Phillipson, 1994; Morris et al., 2006). Amidst the unexpected success of artemisinin in combating malaria, several other novel pharmacological activities attributable to well-known properties of the sesquiterpene lactone structure, with its oxepane ring and endoperoxide bridge, have been revealed. Besides *Plasmodium*, artemisinin and its derivatives (artemisinins) exert profound activities on other protozoans (*Leishmania, Trypanosoma, Amoebas,*

Neospora caninum, and *Eimeria tenella*), trematodes (*Schistosoma*, liver flukes), and viruses (human cytomegalovirus, hepatitis B and C viruses). Studies have demonstrated beneficial effects of *Artemisia* in the treatment of diseases of South African people, such as HIV, arthritis, bladder/urinary tract infections, borreliosis, diabetes, epilepsy, eye infections, and epistaxis (bleeding from the nostrils). The following section describes the activity of *Artemisia* against major human diseases.

3.4.1 Malaria

A report on the use of *A. annua* L. infusion for malaria prevention by a Ugandan community states that around 60% of the 350 million annual cases of clinical malaria and 80% of deaths due to malaria occur in sub-Saharan Africa (world malaria report, 2005). Malaria and related fevers are major causes of poverty, work absenteeism, and poor performance in schools (Vitor-Silva et al., 2009). Other plants of this family that do not contain artemisinin, such as *A. absinthium, A. herba-alba, A. apiacea, A. ludoviciana, A. abrotanum*, and particularly *A. afra*, have excellent antimalarial properties. The latter is widely used for this application in South Africa and Tanzania. A dose of 24 mg/kg of *Artemisia* powder was much more effective in reducing parasitemia in mice infected with *Plasmodium chabaudi* than an equivalent dose of pure artemisinin (Elfawal et al., 2012). The Worcester Institute reported that the consumption of dried whole plant *A. annua* delayed the appearance of malaria drug resistance and overcame resistance to artemisinin (Elfawal et al., 2014; Weathers et al., 2014).

3.4.2 Leishmaniasis

Leishmaniasis is a serious health problem worldwide, especially in India and South Asia (Murray et al., 2005). The WHO estimates that there are 12 million people with undetected *Leishmania* infection and resistance to conventional treatment with pentavalent antimony and other drugs, such as amphotericine B and miltefosine. This protozoan infects its host via hematophagous mosquitoes, causing cutaneous and mucosal infections and, in some cases, such as infections with the genus *Donovani* (*D. infantum* and *archibaldi*), with visceral consequences that can be lethal for the host (Herwaldt, 1999; Palumbo et al., 2009). One study found artemisinin to be active against the promastigote and amastigote forms with IC_{50} values of 166 and 20 µM, respectively, through its ability to induce apoptosis in the parasite (Sen et al., 2007). *In vitro* infection with *L. major* promastigotes was reduced after treatment with artemisinin, with an IC_{50} of 0.75 µM. Moreover, artemisinin and artemether have demonstrated leishmanicidal activity against mastigotes in murine macrophages, with IC_{50} values of 30 µM and 3 µM, respectively (Yang and Liew, 1993). C-9-artemisinin analogues are more active against *L. donovani* if derivatization is carried out at the β-position rather than the α-position. It should be pointed out that the endoperoxide bridge of artemisinin-related compounds is also crucial to their leishmanicidal activity (Avery et al., 2003; Menon et al., 2006).

3.4.3 Artemisinin Drugs in Multiple Sclerosis (Experimental Autoimmune Encephalomyelitis)

Multiple sclerosis (MS) is a class of autoimmune disease that affects the central nervous system, and which is mediated by both T helper (Th) 17- and Th1-type cells. The main symptom of patients with MS is progressive paralysis. Experimental allergic encephalomyelitis (EAE) is a well-established murine model to study the pathogenesis of MS and also used for drug screening. In 2001, Zuo's group conducted an experiment to prove the immunosuppressive functions of novel artemisinin derivatives including SM735, SM905, and SM934 *in vitro* and *in vivo*. Later, Zhao et al. (2012) and Li et al. (2013) studied the immunosuppressive properties of SM933, another artemisinin derivative discovered by Zuo's group. It was established that SM933 exerted unique anti-inflammatory properties through regulatory mechanisms involving the NF-kB and Rig-G/JAB1 signaling pathways, leading to amelioration of EAE (Wang et al., 2007).

3.4.4 Anthelmintic and Antitrematode Effects

As mentioned in the Bible (Revelation 8:10), *A. annua* possesses anthelmintic properties known since antiquity. One of the partners of IFBV-BELHERB in Senegal has noticed that *A. annua* is effective in the treatment of schistosomiasis (bilharziosis), which is the second-most important disease in tropical Africa. The derivatives of artemisinin have also shown efficacy against some trematodes, including *Fasciola hepatica*. Trematode infections negatively affect human and livestock health, and threaten global food safety. Triclabendazole and praziquantel are the only drugs approved as anthelmintics for human trematodiasis. Crude extract of *A. annua* was tested against adult *Schistosoma mansoni* under *in vitro* conditions (Mesa et al., 2015).

In an *in vitro* study, artemether (at $10–30\,\mu g/mL$) caused severe tegument lesions in adult *Fasciola gigantica* 24 hours after the treatment. The effect of artemether on the tegument was comparable to that of triclabendazole, and both affected immune modulation, osmoregulation, and nutrient absorption of *F. gigantica* (Shalaby et al., 2009). In sheep infected with *F. hepatica*, a single intravenous dose of 40 mg/kg artesunate reduced egg count and worm burden by 69% and 77%, respectively, while an intramuscular dose of artemether at 40 mg/kg reduced fecal egg count (eggs per gram, EPG) and worm burden by 97.6% and 91.9%, respectively (Keiser et al., 2010).

Another study carried by Cala et al. (2014) evaluated the anthelmintic activity of *A. annua* crude extracts *in vitro* and compared the most effective extract with artemisinin in sheep naturally infected with *Haemonchus contortus*. *A. annua* leaves extracted with water, aqueous 0.1% sodium bicarbonate, dichloromethane, and ethanol were evaluated *in vitro* by the egg hatch test (EHT) and by the larval development test (LDT) using *H. contortus* for the bicarbonate extract only. *A. annua* extract dose ($1.27\,\mu g/mL$) inhibited 99% of egg hatching (LC_{99}) of gastrointestinal nematodes of sheep.

3.4.5 Chagas disease (*Trypanosoma cruzi*)

It has been demonstrated that low doses of artemisinin are able to inhibit the development of *T. cruzi* and *T. brucei rhodesiense* by inhibiting ATPase activity. Nevertheless, further studies are needed to better assess the influence of artemisinin on membrane pumps regulating cellular calcium in these parasites (Mishina et al., 2007). In 2013, the University of Cumana in Venezuela evaluated the impact of *A. annua* on epimastigotes of *T. cruzi*. After a seven-day treatment, the density of parasites significantly decreased. Two types of infusions were used: one with leaves from plants grown in Venezuela, the other with leaves from Luxembourg. A dose-dependent effect on proliferation was noticed. The infusion from Luxembourg had a stronger effect, although seeds for the two locations were of the same origin.

3.4.6 Immunity and HIV

Artesunate exerted *in vitro* antiviral activity against HIV-1 at nanomolar concentrations, partially inhibiting the cell cycle of two strains of this virus (Ba-L and NL4-3) (Efferth et al., 2002). Although the anti-HIV activity of artemisinin and its derivatives has not been studied in humans, clinical evidence has been gathered on the effect of artesunate on HIV in co-infections with *P. falciparum* (Birku et al., 2002). Several trial runs by IFBV-BELHERB and its partners in the Democratic Republic of the Congo, Uganda, and India have reported preliminary results indicating that *A. annua* raises CD4+ levels. Low CD4+ values are indicative of a depressed immune system, as often observed with HIV. Several informal claims have been made in Africa that *A. annua* tea infusions are also able to inhibit HIV. In addition, the University of Leiden ran *in vitro* trials with *A. annua* and *A. afra* on validated cellular systems for anti-HIV activity and found it to be effective (Lubbe et al., 2012).

3.4.7 Cancer

Artemisinins have been shown to decrease cell proliferation, reduce angiogenesis, and trigger apoptosis in cancer cells (Efferth et al., 2001, 2003; Krishna et al., 2008; Sertel et al., 2013). Several studies have revealed higher stability and better efficacy of artemisinin-derived synthetic dimers in the treatment of cancer (Paik et al., 2006; Stockwin et al., 2009; Gong et al., 2013). In other studies, artemisinins were shown to have value in cancer drug development by inducing growth arrest at various stages of the cell division cycle (Willoughby et al., 2009). In human colon cancer cell lines (HT29), artemisinins were found to activate BAX to induce the release of cytochrome c, leading to apoptosis in cancer cells (Riganti et al., 2009). Artemisinin reduced the levels of HIF-1α and VEGF in mouse embryonic stem cells, suggesting its utility as an antiangiogenic agent (Wartenberg et al., 2003). In human melanoma cells (A375M), artemisinin reduced MMP2 levels more than threefold and blocked cell migration (Buommino et al., 2009). On the other hand, DHA decreased MMP2 and/or MMP9 levels, and suppressed metastasis of human ovarian cancer cells (HO8910PM) and human pancreatic cancer cells (BxPC-3 and PANC-1) via NF-κB inhibition (Wang et al., 2011; Wu et al., 2012). Further, several research studies have shown that artemisinin–transferrin conjugates displayed higher anticancer efficacy than artemisinins alone (Nakase et al., 2008, 2009; Lai et al., 2009). They reported that targeted magnetic nanoliposomes containing artemisinin and holo-transferrin were produced, and *in vitro* studies were carried out on their anticancer effect on breast cancer cells (MCF-7). *In vivo* studies were also carried out on mice suffering from breast cancer (Gharib et al., 2015). Artemisinin–transferrin exhibits increased specificity toward cancer cells and minimizes potential side effects on normal mammalian cells.

In summary, artemisinins have been shown to inhibit tumor growth, induce growth cycle arrest, promote apoptosis, negate angiogenesis and tissue invasion of tumors, and inhibit cancer metastasis. A case report described a 70% reduction in the size of laryngeal squamous cell carcinomas in patients treated with artesunate for two months (Singh and Verma, 2002). Furthermore, artesunate was found to control tumor growth and improve survival of patients with metastatic uveal melanoma (Berger et al., 2005). In a randomized controlled trial, artesunate-treated patients showed improved short-term and one-year survival rates, and prolonged time to cancer progression when compared with the control group, with negligible side effects observed (Zhang et al., 2008).

3.4.8 Dengue and Other Viruses

A. annua possess a wide variety of antiviral effects, particularly against the *Herpes simplex* virus, *Corona* virus and dengue virus. In 2014, a female patient in Vanuatu with dengue fever, and her relatives, were cured after consumption of an *A. annua* infusion (P. Lutgen, personal communication). In this regard, some data on an *in vitro* study had already been published (Zandi et al., 2011). These findings, however, still need to be confirmed by clinical trials in accordance with the WHO protocol.

3.4.9 Diabetes

A study was conducted to evaluate the efficacy of *A. annua* (Kaysom) extract supplementation in reducing the metabolic abnormalities accompanying alloxan-induced diabetes in male albino rats. In comparison with control rats, diabetic rats showed a marked decline ($p < 0.01$) in serum insulin levels, body weight (4.98%), total proteins, albumin, globulin, and HDL accompanied by marked elevation ($p < 0.01$) of fasting blood glucose levels, HOMAIR, ASAT, ALAT, GGT, urea, creatinine, uric acid, serum TC, TG, LDL, VLDL, and ratios of TC/HDL and LDL/HDL (risk factors). Supplementation of diabetic rats with the plant extract significantly normalized most of the studied parameters (Helal et al., 2014).

Another species of *Artemisia, A. herba-alba*, is widely used in Iraqi folk medicine for the treatment of diabetes mellitus, probably because *A. herba-alba* reduces blood sugar. Oral administration of 0.39 g/kg body weight of the aqueous extract of the leaves or bark produced a significant reduction in blood glucose level (Al-Khazraji et al., 1993).

3.4.10 Cholesterol

Artemisia species effect decreases in lipid levels similar to those obtained with statins. Thus, for example, aqueous extracts of *A. sieberi* significantly decrease levels of cholesterol, HDL, LDL, and triglycerides in diabetic rats (Mansi and Lahham, 2008). Similar pharmacological effects of *A. annua* have also been observed by researchers at the Université des Montagnes in Cameroon (R Chougouo, personal communication).

3.4.11 Amelioration of Collagen-Induced Arthritis

Researchers have found that the water-soluble artemisinin derivative SM905 (obtained from *A. annua*) improves collagen-induced arthritis by suppressing inflammatory and Th17 responses (De Jesus, 2003). Their experimental approaches used type II bovine collagen (CII) to induce arthritis in DBA/1 mice. SM905 was given orally either before (one day before booster immunization) or after disease onset (throughout the 14 days after booster immunization).

Disease incidence and severity were monitored using mRNA expression of pro-inflammatory mediators, as determined by real-time PCR; purified T cell proliferation, assessed by [3H]-thymidine incorporated assay; and Th17/Th1/Th2-type cytokine production, measured by ELISA (Wang et al., 2008). Oral treatment with SM905 delayed disease onset, reduced arthritis incidence and severity, and reduced the expression of pro-inflammatory cytokines, chemokines, and chemokine receptors in draining lymph nodes (Wang et al., 2008).

3.5 CONCLUSION

Having been used in Chinese medicine for two millennia, some of the gems of traditional medicine's treasure chest have been rediscovered in recent years. Artemisinin is certainly one of the most promising natural products investigated in the past couple of decades. With regard to malaria, artemisinin has the potential to considerably improve the desperate situation that the world is facing. Fortunately, the value of this compound is not limited to the treatment of malaria, and a wealth of studies have demonstrated the activity of artemisinin and its derivatives against cancer, viruses, parasites, schistosomiasis, and various autoimmune diseases. Therefore, the present article provides an update on the exploitation of derivatives of *Artemisia* in various other deadly human diseases besides malaria. In this era, when many scientists are searching for compounds with increased specificity to their molecular and cellular targets, awareness of artemisinin has surged due to the compound's multifunctionality. This class of compounds seems to have several targets that are important in different diseases. Since natural products have evolved in plants as chemical weapons to protect against infections by bacteria, viruses, and other microorganisms, multifunctional molecules seem to be more versatile and, therefore, more successful in protecting plants from environmental harm than monospecific molecules.

However, the exact signaling molecules or receptive substances that directly interact with artemisinins remain undetermined. It is likely that multiple key signaling molecules are involved, considering that artemisinins can exert broad-spectrum inhibitory effects on several major signaling pathways. Identification of these artemisinin-binding targets is imperative to further understand which chemical functional groups in artemisinins are responsible for the modes of action in different disease conditions. Further, we have extensive *in vitro* and *in vivo* preclinical data in the present literature in support of therapeutic applications for artemisinins, especially artesunate, in a variety of human disease conditions. However, despite the established safety record of artemisinins, clinical studies of artesunate in non-malaria-related diseases are still very limited, and this is an area that demands rigorous and intensive research.

REFERENCES

Alesaeidi S. and Miraj S. 2016. A systematic review of anti-malarial properties, immunosuppressive properties, anti-inflammatory properties, and anti-cancer properties of *Artemisia annua. Electronic Physician* 25(8): 3150–3155.

Al-Khazraji S., al-Shamaony L., and Twaij H. 1993. Hypoglycaemic effect of *Artemisia herba alba*. I. Effect of different parts and influence of the solvent on hypoglycaemic activity. *Journal of Ethnopharmacology* 40(3): 163–166.

Allen P.C., Lydon J., Danforth H.D. 1997. Effects of components of *Artemisia annua* on coccidia infections in chickens. *Poultry Science* 76(8): 1156–1163.

Avery M.A., Muraleedharan K.M., Desai P.V., Bandyopadhyaya A.K., Furtado M.M., and Tekwani B.L. 2003. Structure-activity relationships of the antimalarial agent artemisinin. 8. design, synthesis, and CoMFA studies toward the development of artemisinin-based drugs against leishmaniasis and malaria. *Journal of Medicinal Chemistry* 46(20): 4244–4258.

Berger T.G., Dieckmann D., Efferth T., Schultz E.S., Funk J.O., Baur A., and Schuler G. 2005. Artesunate in the treatment of metastatic uveal melanoma—first experiences. *Oncology Reports* 14(6): 1599–1604.

Birku Y., Mekonnen E., Björkman A., and Wolday D. 2002. Delayed clearance of *Plasmodium falciparum* in patients with human immunodeficiency virus co-infection treated with artemisinin. *Ethiopian Medical Journal* 40(Suppl 1): 17–26.

Buommino E., Baroni A., Canozo N., Petrazzuolo M., Nicoletti R., Vozza A., and Tufano M.A. 2009. Artemisinin reduces human melanoma cell migration by down-regulating $\alpha V\beta 3$ integrin and reducing metalloproteinase 2 production. *Investigational New Drugs* 27(5): 412–418.

Cala A.C., Ferreira J.F.S., Carolina A., Chagas S., Gonzalez J.M., Rodrigues R.A.F., Foglio M.A., et al. 2014. Anthelmintic activity of *Artemisia annua* L. extracts *in vitro* and the effect of an aqueous extract and artemisinin in sheep naturally infected with gastrointestinal nematodes. *Parasitology Research* 113(6): 2345–2353.

De Jesus L. 2003. Effects of artificial polyploidy in transformed roots of *Artemisia annua* L. Doctoral dissertation, Worcester Polytechnic Institute, Worcester.

Efferth T., Dunstan H., Sauerbrey A., Miyachi H., and Chitambar C. 2001. The anti-malarial artesunate is also active against cancer. *International Journal of Oncology* 18(4): 767–773.

Efferth T., Marschall M., Wang X., Huong S.M., Hauber I., Olbrich A., et al. 2002. Antiviral activity of artesunate towards wild-type, recombinant, and ganciclovir-resistant human cytomegalo viruses. *Journal of Molecular Medicine (Berlin)* 80(4): 233–242.

Efferth T., Sauerbrey A., Olbrich A., Gebhart E., Rauch P., Weber H.O., et al. 2003. Molecular modes of action of artesunate in tumor cell lines. *Molecular Pharmacology* 64(2): 382–394.

Elfawal M.A., Towler M.J., Reich N.G., Golenbock D., Weathers P.J., and Rich S.M. 2012. Dried whole plant *Artemisia annua* as an antimalarial therapy. *PloS One*. 7(12): e52746.

Elfawal M.A., Towler M.J., Reich N.G., Weathers P.J., and Rich S.M. 2014. Dried whole-plant *Artemisia annua* slows evolution of malaria drug resistance and overcomes resistance to artemisinin. *Proceedings of the National Academy of Sciences of the United States of America* 112(3): 821–826.

Ferreira J.F. 2004. *Artemisia annua*: The hope against malaria and cancer. *Proceedings of the Jan 15-17/2004 meeting*. Mountain State University, Beckley, WV.

Gharib A., Faezizadeh Z., Mesbah-Namin S.A.R., and Saravani R. 2015. Experimental treatment of breast cancer-bearing BALB/c mice by artemisinin and transferrin-loaded magnetic nanoliposomes. *Pharmacognosy Magazine* 11(Suppl 1): 117–122.

Gong Y., Gallis B.M., Goodlett D.R., Yang Y., Lu H., Lacoste E., et al. 2013. Effects of transferrin conjugates of artemisinin and artemisinin dimer on breast cancer cell lines. *Anticancer Research* 33(1): 123–132.

Helal E.G., Abou-Aouf N., Khattab A.M., and Zoair M.A. 2014. Anti-diabetic effect of *Artemisia annua* (Kaysom) in alloxan-induced diabetic rats. *The Egyptian Journal of Hospital Medicine* 57: 422–431.

Herwaldt B.L. 1999. Leishmaniasis. *Lancet* 354(9185): 1191–1199.

Keiser J., Veneziano V., Rinaldi L., Mezzino L., Duthaler U., and Cringoli G. 2010. Anthelmintic activity of artesunate against *Fasciola hepatica* in naturally infected sheep. *Research in Veterinary Science* 88(1): 107–110.

Kiani B.H., Suberu J., and Mirza B. 2016. Cellular engineering of *Artemisia annua* and *Artemisia dubia* with the rol ABC genes for enhanced production of potent anti-malarial drug artemisinin. *Malaria Journal* 15(1):252. doi:10.1186/s12936-016-1312-8.

Krishna S., Bustamante L., Haynes R.K., and Staines H.M. 2008. Artemisinins: Their growing importance in medicine. *Trends in Pharmacological Sciences* 29(10): 520–527.

Lai H., Nakase I., Lacoste E.P.S.N., and Sasaki T. 2009. Artemisinin–transferrin conjugate retards growth of breast tumors in the rat. *Anticancer Research* 29(10): 3807–3810.

Li X., Li T.T., Zhang X.H., Hou L.F., Yang X.Q., Zhu F.H., Tang W., and Zuo J.P. 2013. Artemisinin analogue SM934 ameliorates murine experimental autoimmune encephalomyelitis through enhancing the expansion and functions of regulatory T cell. *PLoS One* 8(8): e74108.

Lubbe A., Seibert I., Klimkait T., and van der Kooy F. 2012. Ethnopharmacology in overdrive: The remarkable anti-HIV activity of *Artemisia annua*. *Journal of Ethnopharmacology* 141(3): 854–859.

Mansi K. and Lahham J. 2008. Effects of *Artemisia sieberi* Besser (*A. herba-alba*) on heart rate and some hematological values in normal and alloxan-induced diabetic rats. *Journal of Basic and Applied Sciences* 4(2): 57–62.

Meier zu Biesen C. 2011. The rise to prominence of *Artemisia annua* L.—The transformation of a Chinese plant to a global pharmaceutical. *African Sociological Review/Revue Africaine de Sociologie* 14(2): 24–46.

Menon R.B., Kannoth M.M., Tekwani B.L., Gut J., Rosenthal P.J., and Avery M.A. 2006. A new library of C-16 modified artemisinin analogs and evaluation of their anti-parasitic activities. *Combinatorial Chemistry and High Throughput Screening* 9(10): 729–741.

Mesa L.E., Lutgen P., Velez I.D., Segura A.M., and Robledo S.M. 2015. *Artemisia annua* L., potential source of molecules with pharmacological activity in human diseases. *American Journal of Phytomedicine and Clinical Therapeutics* 3(5): 436–450.

Mishina Y.V., Krishna S., Haynes R.K., and Meade J.C. 2007. Artemisinins inhibit *Trypanosoma cruzi* and *Trypanosoma brucei rhodesiense in vitro* growth. *Antimicrobial Agents and Chemotherapy* 51(5): 1852–1854.

Morris T., Stables M., and Gilroy D.W. 2006. New perspectives on aspirin and the endogenous control of acute inflammatory resolution. *Scientific World Journal* 6: 1048–1065.

Murray H.W., Berman J.D., Davies C.R., and Saravia N.G. 2005. Advances in leishmaniasis. *Lancet* 366(9496): 1561–1577.

Nakase I., Gallis B., Takatani-Nakase T., Oh S., Lacoste E., Singh N.P., et al. 2009. Transferrin receptor-dependent cytotoxicity of artemisinin–transferrin conjugates on prostate cancer cells and induction of apoptosis. *Cancer Letters* 274(2): 290–298.

Nakase I., Lai H., Singh N.P., and Sasaki T. 2008. Anticancer properties of artemisinin derivatives and their targeted delivery by transferrin conjugation. *International Journal of Pharmaceutics* 354(1): 28–33.

Nofal S.M., Mahmoud S.S., Ramadan A., Soliman G.A., and Fawzy R. 2009. Anti-diabetic effect of *Artemisia judaica* extracts. *Research Journal of Medicine and Medical Sciences* 4(1): 42–48.

O'Leary J. and Muggia F.M. 1998. Camptothecins: A review of their development and schedules of administration. *European Journal of Cancer* 34(10): 1500–1508.

Paik I.-H., Xie S., Shapiro T.A., Labonte T., Narducci Sarjeant A.A., Baege A.C., and Posner G.H. 2006. Second generation, orally active, antimalarial, artemisinin-derived trioxane dimers with high stability, efficacy, and anticancer activity. *Journal of Medicinal Chemistry* 49(9): 2731–2734.

Palumbo E. 2009. Current treatment for cutaneous leishmaniasis: A review. *American Journal of Therapeutics* 16(2): 178–182.

Patočka J, Plucarb B. 2003. Pharmacology and toxicology of absinthe. *Journal of Applied Biomedicine*. 1: 199–205.

Penissi A.B., Giordanob O.S., Guzmánc J.A., Rudolphd M.I., and Piezzia R.S. 2006. Chemical and pharmacological properties of dehydroleucodine, a lactone isolated from *Artemisia douglasiana* Besser. *Molecular Medicinal Chemistry* 10: 1–11.

Phillipson J.D. 1994. Natural products as drugs. *Transactions of the Royal Society of Tropical Medicine and Hygiene* 88(Suppl 1): S17–S19.

Pinato D.J. and Stebbing J. 2015. *Artemisia*: A divine dart against cancer? *Lancet Oncology* 16(7): 759–760.

Riganti C., Doublier S., Viarisio D., Miraglia E., Pescarmona G., Ghigo D., and Bosia A. 2009. Artemisinin induces doxorubicin resistance in human colon cancer cells via calcium-dependent activation of HIF-1α and P-glycoprotein overexpression. *British Journal of Pharmacology* 156(7): 1054–1066.

Seeff L.B., Lindsay K.L., Bacon B.R., Kresina T.F., and Hoofnagle J.H. 2001. Complementary and alternative medicine in chronic liver disease. *Hepatology* 34(3): 595–603.

Sen R., Bandyopadhyay S., Dutta A., Mandal G., Ganguly S., Saha P., and Chatterjee M. 2007. Artemisinin triggers induction of cell-cycle arrest and apoptosis in *Leishmania donovani* promastigotes. *Journal of Medical Microbiology* 56 (9): 1213–1218.

Sertel S., Plinkert P., and Efferth T. 2013. Activity of artemisinin-type compounds against cancer cells. In H. Wagner and G. Ulrich-Merzenich (eds) *Evidence and Rational Based Research on Chinese Drugs*. Springer, Vienna, 333–362.

Shalaby H.A., El Namaky A.H., and Kamel R.O.A. 2009. *In vitro* effect of artemether and triclabendazole on adult *Fasciola gigantica*. *Veterinary Parasitology* 160(1): 76–82.

Singh N.P. and Panwar V.K. 2006. Case report of a pituitary macroadenoma treated with artemether. *Integrative Cancer Therapies* 5(4): 391–394.

Singh N.P. and Verma K.B. 2002. Case report of a laryngeal squamous cell carcinoma treated with artesunate. *Archive of Oncology* 10(4): 279–280.

Stockwin L.H., Han B., Yu S.X., Hollingshead M.G., ElSohly M.A., Gul W., et al. 2009. Artemisinin dimer anticancer activity correlates with heme-catalyzed reactive oxygen species generation and endoplasmic reticulum stress induction. *International Journal of Cancer* 125(6): 1266–1275.

Tajehmiri A., Issapour F., Moslem M.N., Lakeh M.T., and Kolavani M.H. 2014. *In vitro* antimicrobial activity of *Artemisia annua* leaf extracts against pathogenic bacteria. *Advanced Studies in Biology* 6(3): 93–107.

Verdian-Rizi M.R. 2009. Chemical composition and antimicrobial activity of the essential oil of *Artemisia annua* from Iran. *Pharmacognosy Research* 1(25): 21–24.

Vitor-Silva S., Roberto R.L., Tamam P., Marcus L. 2009. Malaria is associated with poor school performance in an endemic area of the Brazilian Amazon. *Malaria Journal* 8(1): 230.

Wang J.X., Tang W., Zhou R., Wan J., Shi L.P., Zhang Y., et al. 2008. The new water-soluble artemisinin derivative SM905 ameliorates collagen-induced arthritis by suppression of inflammatory and Th17 responses. *British Journal of Pharmacology* 153(6): 1303–1310.

Wang Z., Qiu J., Guo T.B., Liu A., Wang Y., Li Y., and Zhang, J.Z. 2007. Anti-inflammatory properties and regulatory mechanism of a novel derivative of artemisinin in experimental autoimmune encephalomyelitis. *Journal of Immunology* 179(9): 5958–5965.

Wang S.J., Sun B., Cheng Z.X., Zhou H.X., Gao Y., Kong R., et al. 2011. Dihydroartemisinin inhibits angiogenesis in pancreatic cancer by targeting the NF-κB pathway. *Cancer Chemotherapy and Pharmacology* 68(6): 1421–1430.

Wartenberg M., Wolf S., Budde P., Grunheck F., Acker H., Hescheler J., et al. 2003. The antimalaria agent artemisinin exerts antiangiogenic effects in mouse embryonic stem cell-derived embryoid bodies. *Laboratory Investigation* 83(11): 1647–1655.

Weathers P.J., Towler M., Hassanali A., and Lutgen P. 2014. Dried-leaf *Artemisia annua*: A practical malaria therapeutic for developing countries? *World Journal of Pharmacology* 3(4): 39–55.

Willcox M., Bodeker G., Bourdy G., Dhingra V., Falquet J., Ferreira J.F., et al. 2004. *Artemisia annua* as a traditional herbal antimalarial. *Traditional Medicinal Plants and Malaria* 43–59.

Willoughby J.A., Sundar S.N., Cheung M., Tin A.S., Modiano J., and Firestone G.L. 2009. Artemisinin blocks prostate cancer growth and cell cycle progression by disrupting Sp1 interactions with the cyclin-dependent kinase-4 (CDK4) promoter and inhibiting CDK4 gene expression. *Journal of Biological Chemistry* 284(4): 2203–2213.

Wu B., Hu K., Li S., Zhu J., Gu L., Shen H., et al. 2012. Dihydroartiminisin inhibits the growth and metastasis of epithelial ovarian cancer. *Oncology Reports* 27(1): 101–108.

Yang D.M. and Liew F.Y. 1993. Effects of qinghaosu (artemisinin) and its derivatives on experimental cutaneous leishmaniasis. *Parasitology* 106(1): 7–11.

Zamani S., Sendi J.J., and Ghadamyari M. 2011. Effect of *Artemisia annua* L. (Asterales: Asteraceae) essential oil on mortality, development, reproduction and energy reserves of *Plodia interpunctella* (Hübner) (Lepidoptera: Pyralidae). *Journal of Biofertilzer Biopesticide* 2(1): 105.

Zandi K., Teoh B., Sam S., Wong P., Mustafa M., and Abubakar S. 2011. Antiviral activity of four types of bioflavonoid against dengue virus type-2. *Virology Journal* 8(1): 560–564.

Zhang Z., Yu S., Miao L., Huang X., Zhang X., Zhu Y., et al. 2008. Artesunate combined with vinorelbine plus cisplatin in treatment of advanced non-small cell lung cancer: A randomized controlled trial. *Journal of Chinese Integrative Medicine* 6(2): 134.

Zhao Y.G., Wang Y., Guo Z., Dan H.C., Baldwin A.S., Hao W., and Wan Y.Y. 2012. Dihydroartemisinin ameliorates inflammatory disease by its reciprocal effects on Th and regulatory T cell function via modulating the mammalian target of rapamycin pathway. *The Journal of Immunology* 189(9): 4417–4425.

CHAPTER 4

Current Perspectives and Future Prospects in the Use of *Artemisia annua* for Pharmacological and Agricultural Purposes

Ebiamadon Andi Brisibe

CONTENTS

4.1 INTRODUCTION

Within the past five decades or so, an innocuous plant of Asian origin called annual wormwood, or *Artemisia annua* L., has attracted a very high degree of research attention because of the inspiring discovery that it produces a complex phytochemical repertoire of more than 600 biologically active secondary metabolites, including terpenoids, essential oil constituents, and numerous phenolic acids, flavonoids, minerals, vitamins, and amino acids in its aerial portions (Brown 2010, Brisibe et al. 2009, Bhakuni et al. 2001), which are of immense medicinal importance. Truly an intriguing plant, *A. annua* has fundamentally changed the way parasitic diseases, especially malaria, are treated around the globe due to its immense importance in controlling numerous internal and external microbial parasites that afflict both humans and livestock (Ferreira et al. 2010, Brisibe et al. 2014). The plant is generally recognized as safe for human consumption and has been used in traditional Chinese medicine for the treatment of fevers and chills associated with malaria dating as far back as 168 BC (Weathers et al. 2014a). *A. annua* is currently the major (and only) source of the highly potent and effective artemisinin drug, which is the basis for a class of key ingredients widely used for the production of the first-line antimalarial drugs that are needed for treatment of the vector-borne lethal falciparum malaria in the most affected countries of Asia and Africa that are resistant to other traditional therapies. As a result of this high demand for the drug, efforts have been intensified around the globe in recent years to elevate its production through several *in planta* mechanisms including but not necessarily limited to overexpression or downregulation of genes associated with artemisinin biosynthesis in the plant. Sadly, however, not much has been achieved through these pathways.

A cadinane-type sesquiterpene lactone with a crucial endoperoxide bridge, artemisinin is produced and sequestered in glandular trichomes which are found on leaves, floral buds, and flowers (Ferreira and Janick 1995, Tellez et al. 1999) of *A. annua*. In spite of its enormous pharmacological importance, in both human and animal healthcare (Efferth et al. 2011), its availability, especially as a key active ingredient in the production of the world's most effective malarial therapies, artemisinin-based combination therapies (ACTs), which contain an artemisinin-derived semisynthetic drug in combination with a secondary antimalarial drug partner(s) (to prevent artemisinin drug resistance from emerging), such as lumefantrine, piperaquine, mefloquine, trimethoprim, and so on, is limited not only by low yield of the compound in *A. annua*, but even more so by the uncertainty of farmers and producers, as currently plant-derived artemisinin stands to be relegated to the background by its bioengineered counterpart.

4.2 PRODUCTION AND THERAPEUTIC USES OF ARTEMISININ

As a very important pharmacological molecule capable of saving the lives of millions of malaria patients in different parts of the world, it is not surprising that several attempts have been made to synthesize artemisinin through organic means (Zhu and Cook 2012, Schmid and Hofheinz 1983). However, the process is very complex with many reaction steps and typically results in low yields. The chemical analogues produced are thus not economically competitive with that synthesized naturally in *A. annua* (Ferreira et al. 2010), which means its extraction from the dried leafy biomass of the plant using an appropriate hydrocarbon (usually either an initial hexane or petroleum ether) solvent extraction step before proceeding to purifying the product by partitioning the first extract in solvents of different polarities before using chromatography to separate the different fractions, invariably appears to be the most viable option for producing cheap and large quantities of the drug.

Presently, increased cultivation of the crop in small-holder fields in Asia and Africa, which are usually less than 1 ha, and the improvement of extraction methods appear to be the most effective strategies for producing this important sesquiterpenoid compound, which is highly potent and effective against all *Plasmodium* species, including multidrug-resistant strains.

In spite of their pharmacological importance, a major shortcoming in the production of sesquiterpenoid compounds via whole plants, especially in *A. annua*, is the relatively lengthy growing cycle required to obtain appreciable yields (g/100 g dry weight). So far, the best commercial varieties of the crop yield only *ca.* 1.5% artemisinin or slightly more, and agricultural yields seldom exceed 70 kg/ha (Kumar et al. 2004). Generally, the period from time of planting to artemisinin extraction from the plant is approximately 5–8 months. Not surprisingly, the yields derived from dried leafy biomass after such a relatively lengthy period are considered low for commercial production, where a full ton of dried leafy biomass can only produce between 6–18 g of purified artemisinin (Brisibe et al. 2012), which is quite small for a compound of such immense importance in saving numerous lives. This low yield thus appears to be one of the most intractable problems related to the production and use of artemisinin-derived drugs against malaria, especially in Africa where the cultivation of the crop is new and not yet ubiquitous.

Apart from malaria, artemisinin and its semisynthetic derivatives, including dihydroartemisinin, artesunate, and artemether, for example, have also received global attention lately because of their potential therapeutic effects in the treatment of several other ailments, including hard-to-treat diseases such as those caused by a wide variety of viruses like HIV (Lubbe et al. 2012), hepatitis B, C, and many others (Efferth et al. 2008). Artemisinin-derived drugs are also effective against several neglected tropical parasitic diseases including toxoplasmosis, trypanosomiasis, cryptosporidiosis, amoebiasis, giardiasis, and leishmaniasis (Brisibe and Chukwurah 2014), as well as chronic trematodal food-borne parasitic diseases such as schistosomiasis, clonorchiasis, and fascioliasis that afflict more than 1 billion people every year (Ferreira et al. 2011). Artemisinin and/or its semisynthetic derivatives have also been reported to have several other therapeutic uses. For example, recent studies have indeed shown promise for their use as antifungal, antibacterial (Roth and Acton 1989), anti-inflammatory (Magalhães et al. 2012), and antitumor agents. A recent study that evaluated the anti-inflammatory, antioxidant, and antimicrobial effects of artemisinin extracts generated from *A. annua* using four different solvents demonstrated that artemisinin has potent effects against periodontopathic bacteria. Artemisinin and its semisynthetic derivatives have also attracted much scientific interest lately due to their use for the treatment of human and animal cancer cell lines (Breuer and Efferth, 2014, Tilaoui et al. 2014, Efferth et al. 2011, Firestone and Sundar, 2009), since they have been found to be highly cytotoxic against cancerous growth. For example, artesunate, tested in 55 cell lines at the National Cancer Institute of the United States of America was reported to be most active against leukemia and colon cancer cell lines with mean GI 50 values of 1.11 ± 0.56 μM and 2.13 ± 0.74 μM, respectively. Compared with established antitumor drugs, artesunate performed quite as effective as the conventional treatments (Efferth et al. 2001). Artesunate inhibits angiogenesis and vascular endothelial growth factor (VEGF) production in chronic myeloid leukemia (CML) K 562 cells *in vitro*. It decreased the VEGF level in CML by 2 μM/L. An *in vivo* study that evaluated the angiogenic effects of artesunate in chicken chorial-lantoic membrane neovascularization model demonstrated that artesunate decreased the angiogenic activity in a dose-dependent manner signifying that artesunate could be a potential antileukemic agent (Huan-huan et al., 2004; Zhou et al., 2007). In addition, rats treated with a single dose of 50 mg/kg body weight of 7,12-dimethylbenz(a) anthracene (DMBA), to induce the development of breast cancer, indicated that such animal models (provided with rat chow containing 0.02% artemisinin and monitored for breast cancer), after 40 weeks showed oral artemisinin treatment to have significantly (p < 0.02) delayed and in some cases prevented outright the development of breast cancer compared to the control animals. Moreover, breast tumors in the artemisinin-treated rats were fewer (p < 0.02) and smaller in size (p < 0.05) when compared with the controls, which suggests that artemisinin could be used as a potent cancer chemopreventive agent (Lai and Singh 2006). Although numerous studies have demonstrated that artemisinin and its semisynthetic derivatives effectively slow the proliferation of cancer cells and nhibit growth promoting pathways between cells, further research would still need to be performed to specifically determine the mechanisms used by these sesquiternoid compounds to produce the anticancer effects.

Now aside from the therapeutic uses of artemisinin outlined previously, recent research has continued to buttress the need for further investigations on the use of *A. annua* derived drugs. For example, ethanol extracted from the plant showed an immunosuppressive effect on autoimmune diseases such as lupus erythematosus and rheumatoid arthritis (Zhang and Sun 2009), while SM905, a water-soluble artemisinin derivative, ameliorates collagen-induced arthritis by suppression of inflammatory and Th17 responses. Oral treatment with SM905 not only delayed disease onset, reduced arthritis incidence and severity, but also suppressed the enhanced expression of pro-inflammatory cytokines, chemokines, and chemokine receptors in draining lymph nodes. In established arthritis, SM905 profoundly inhibited disease progression, reduced IL-17A and RORgt mRNA expression, and suppressed pro-inflammatory mediator expression in arthritic joints (Wang et al. 2008). Similarly, another artemisinin semisynthetic derivative, artesunate, has been reported to block the production of IL-1b, IL-6, and IL-8 from TNF-a-stimulated human rheumatoid arthritis fibroblast-like synoviocytes (Xu et al. 2007). In addition, artesunate inhibits lipopolysaccharide-induced production of TNFa, IL-6, and nitric oxide (NO), and expression of toll-like receptor 4 (TLR4) and TLR9 from macrophages (Li et al. 2006, 2008). The exact molecular mechanism that mediates these anti-inflammatory effects by artesunate has not been unequivocally determined. However, there is some evidence pointing to the inhibition of nuclear factor NF-kB transcriptional activity by artesunate and other artemisinin derivatives (Li et al. 2006, 2008; Wang et al. 2007). More recently, artesunate has been found to ameliorate experimental allergic airway inflammation, probably via negative regulation of the PI3K/Akt pathway and the downstream NF-kB activity. These findings provide a novel therapeutic value for artesunate in the treatment of allergic asthma (Cheng et al. 2011).

In another recent study, artemisinin has equally been found to possess remarkably strong activity against *Helicobacter pylori*, the pathogen responsible for gastroduodenal infections that has been critically implicated in the pathogenesis of active and chronic gastritis, peptic ulcer, and gastric carcinoma (Malfertheiner et al. 2005, Marshall et al. 2001) in humans. The organism changes the gastric epithelium directly through bacterial toxicity and indirectly via inflammation-mediated damage (Andersen 2007). For many centuries, several plant materials have been successfully used in different cultures around the world to cure stomach ailments. It is not surprising, therefore, to rationalize that some bioactive principles derived from plants could act via the anti-*H. pylori* mechanism as well. Interestingly, a large-scale random screening of natural product-based molecules derived from *A. annua* has demonstrated a remarkably high and hitherto unknown antibiotic effect of artemisinin and several of its semisynthetic analogues including, β-artecyclopropylmether, β-arteether, α-artemether, β-artemether, and β-artefurfurylether against *H. pylori* in experimental trials (Goswami et al. 2012).

4.3 BIOTECHNOLOGICAL STRATEGIES FOR ENHANCING ARTEMISININ PRODUCTION

Considering that *A. annua* is presently the only viable source of artemisinin, there is understandably a great degree of interest in enhancing its production. And although effective, the agronomic platform as the main production strategy seems unlikely to solve the problem of global artemisinin availability due to the boom-and-bust cycle that its production has become associated with lately. As there are growing concerns that the current artemisinin supply chain will be unable to meet future requirements, it is obvious that there is need for an additional source of artemisinin, that provides a consistent, reliable, and inexpensive supply. Consequently, a multifaceted approach using several strategies including utilization of the advanced techniques emerging from modern plant biotechnology and classical molecular biology, such as increasing *in planta* production systems through a genetic approach as demonstrated previously (Debruner et al. 1996, Delabays 1994), must be adopted. The recent identification of the loci associated with artemisinin production in *A. annua* (Graham et al. 2010) is also a step in the right direction, which will greatly enhance molecular breeding of highly improved lines for cultivation. Some other biotechnological strategies that would enhance the production of artemisinin include industrial fermentation, where artemisinic acid can be produced in *Nicotiana benthamiana* (van Harpen et al. 2010) or production of artemisinin precursors in heterologous systems such as microorganisms (Paddon et al. 2013, Ro et al. 2006, Teoh et al. 2006), and the semisynthesis of artemisinin from two of its precursors, artemisinic acid and dihydroartemisinic acid, which are usually discarded in the extraction process (Brisibe et al. 2008a). Some of these strategies are highlighted in the next section.

4.3.1 *In Vitro* Production of Artemisinin in Callus and Cell Suspension Cultures of *A. annua*

There are a few reports which have demonstrated that artemisinin can be produced successfully in callus and cell suspension cultures of *A. annua* though the artemisinin yields have been quite abysmal (Jaziri et al. 1995). However, some other investigations have proven that artemisinin production can be enhanced by the presence of roots (Fulzele et al. 1991, Ferreira and Janick 1996). For example, it has been reported that 0.287% dry weight of artemisinin can be produced in a hormone-free medium when root production was maximized, though no artemisinin or its immediate precursors were detected in the roots (Ferreira and Janick 1996). These results are an indication that differentiated shoot cultures could serve as high-value products for pharmaceutical use, since they contain artemisinin levels comparable with those observed during agricultural production. However, the low biomass produced makes it definite that tissue cultures might not be a suitable strategy for commercial exploitation of artemisinin. Aside from these, other procedures based on the transformation of *A. annua* plants using *Agrobacterium rhizogenes* have led to the formation and maintenance of hairy roots that can produce higher levels of artemisinin than the normal roots in *in vitro* cultures. De Jesus-Gonzalez and Weathers (2003) have reported the production of several stable tetraploid hairy root clones of *A. annua.* Some of these were discovered to have given significantly more artemisinin content than the diploid parent (Clone YUT 16). Wang and colleagues (2001) equally found that adjusting the light spectrum in transformed hairy root cultures of *A. annua* has a significant influence on biomass yield and artemisinin content.

These authors demonstrated that a high biomass yield of 5.73 g dry weight/l and an artemisinin content of 31 mg/g plant material could be obtained under red light at 660 nm, which are 17% and 67%, respectively, higher than those obtained under white light. In a related study, Liu and others (2002) found that light irradiation influenced the growth and production of artemisinin in transformed hairy root cultures of *A. annua* too. When hairy roots were cultured under illumination of 3000 lux for 16 h using several cool-white fluorescent lamps, a dry weight of 13.8 mg/l and an artemisinin concentration of 244.5 mg/l, respectively, were obtained. In addition, artemisinin content can also be successfully increased by about 57% through simply regulating the ratio of nitrate to ammonium and total initial nitrogen concentrations in transformed hairy root cultures (Wang and Tan 2002). These results are useful and encouraging so far, but a lot more effort needs to be made using the whole plant as well as some specific tissues.

4.3.2 Metabolic Engineering of the Artemisinin Biosynthesis Pathway

In recent years, the use of genetic engineering techniques to alter the metabolic pathway of artemisinin biosynthesis in transgenic *A. annua* has been attempted (Arsenault et al. 2008, Liu et al. 2011). This has been achieved mainly through the introduction of key genes encoding for enzymes regulating the biosynthetic pathway leading to the formation of artemisinin *in planta*. In this connection the role of certain genes, especially those involving key enzymes in the biosynthesis of artemis-inin such as farnesyl diphosphate synthase (FDPS) and amorpha-4,11-diene synthase (ADS) readily comes to mind. It could be speculated that genes controlling these key enzymes can be manipulated such that the enzymes become overexpressed in *A. annua*. Alternatively, other enzymes which are involved in pathways competing for precursors of artemisinin, for example, squalene synthase (SQS) can be inhibited through genetic engineering such that the genetically modified plants produce more artemisinin.

Efforts are equally geared toward the development of transgenic plants by intro-ducing the gene for artemisinin production (from *A. annua*) into a much faster-growing plant species, for example, chicory or tobacco (*Nicotiana tabacum*) with a proportionately higher leaf biomass, possibly, to enhance higher artemisinin yield at a very low cost. Such efforts already appear to be largely rewarding as demonstrated recently where the introduction of a gene into *N. tabacum* resulted in the expression of an active enzyme and the accumulation of the first dedicated precursor of arte-misinin (amorpha-4, 11-diene) ranging from 0.2 to 1.7 ng/g fresh weight of leaf tissue (Wallaart et al. 2001). In an alternative approach, Malhotra and colleagues (2016) at the International Centre for Genetic Engineering and Biotechnology (ICGEB) in India and in the USA, identified key genes in *A. annua* that could be transferred to and stably expressed in edible plants to optimize artemisinin yields. Essentially, they achieved high levels of artemisinin biosynthesis in *N. tabacum* plants by engineering two metabolic pathways targeted to three different cellular compartments (chloro-plast, nucleus, and mitochondria). The doubly transgenic plants showed a three-fold enhancement of isopentenyl pyrophosphate (IPP), and transgenes AACPR, DBR2, and CYP71AV1 targeted to chloroplasts resulted in higher expression and an effi-cient photo-oxidation of DHAA to artemisinin. Other studies have also trans-formed a cDNA encoding cotton FDPS, under the control of CaMV 35S promoter, into *A. annua* via *A. tumefaciens* or *A. rhizogenes*. By overexpressing FDPS, a key enzyme in the biosynthesis of artemisinin, the content of the sesquiterpenoid antimalarial drug was increased by about 0.8%–1% dry weight in the transgenic plants (Chen et al. 2000). Lately, *N. benthamiana* has also been deployed at commer-cial scale for rapid production of several pharmaceutical precursors of artemisinin (van Harpen et al. 2010), further opening up the vista of opportunities that can be utilized for the production of this essential antimalarial drug.

4.3.3 Up-Scaling of Semisynthesis of Artemisinin in Microbial Systems

Surprisingly, this feature does not seem to be unique to plants alone. Recent advances using recombinant microbes circumvented the poor performance of plant terpene cyclases by expressing a codon-optimized fold (Martin et al. 2003). In a remarkable series of metabolic engineering experiments, these authors used engi-neered mevalonate pathway genes from yeast, which was about 30 to 90 times more efficient than the normal pathway in *E. coli*. This combined approach high-lights an increased production of amorpha-4,11-diene by approximately 1000 fold

(Martin et al. 2003), which taken further into the pathway would possibly lead to the production of artemisinic acid. In a more facile approach, a cytochrome P450 monooxygenase gene (CYP71AV1) isolated directly from glandular trichomes of *A. annua* (Teoh et al. 2006) and inserted in yeast cells performed a three-step oxidation of amorpha- 4,11-diene that allowed its conversion into artemisinic acid in yields that appear suitable for large-scale fermentation (Ro et al. 2006). These authors successfully added or tweaked a dozen genes in yeast in commercial fermentation tanks to produce artemisinic acid. Coming on the footsteps of this development, it is of special pharmacological interest that efforts are currently underway to optimize the CYP71AV1 gene expression system in several prokaryotic strains in order to sustain high-level production of amorpha-4,11-diene that can be easily converted to artemisinic acid, which can be subsequently oxidized to yield artemisinin (Hale et al. 2007). The hallmark in all of these studies was the desire to modify the genomes of bacteria and yeast which can be fermented in huge bioreactors to yield a plentiful and inexpensive supply of artemisinic acid. This metabolically synthesized artemisinic acid can be obtained easily through a simple purification process, which can be converted to artemisinin through a few inexpensive chemical steps in the laboratory. The artemisinin thus produced can be further converted through simple downstream chemistry into derivatives such as dihydroartemisinin, artesunate, or artemether for possible integration with other antimalarial drugs for the production of low-cost, life-saving ACTs with a great impact on malaria mortality or the treatment of several neglected parasitic diseases in the tropics.

Production of artemisinin in large fermentation vessels through microbial engineering and simple chemistry may pave the way for an industrial process capable of supplementing the global supply of the drug from a second source, independent of the uncertainties associated with artemisinin production in the plant (Paddon et al. 2013), which have had a highly negative impact both on the producers and on health outcomes. This approach came as a promise to increase supplies of high-quality artemisinin and, overall, lower the cost of ACTs in the near future (Ro et al. 2006). Sadly, however, several years down the road this has not yet materialized because the science-related logistics are still beset with a lot of problems, as the process has only recently moved into commercial production and distribution. Consequently, it is expected that botanical production from *A. annua* will remain a crucial source of artemisinin for the foreseeable future, although promise of the arrival of semisynthetic artemisinin to international commerce has put considerable pressure on prices of the plant-derived compound. Meanwhile, factory produced *ex planta*-derived artemisinin, when the process becomes commercially successful, could serve as a supplemental source of the drug, and not necessarily as the single silver bullet that would solve all the problems associated with the production and supply of artemisinin. This is especially so as artemisinin semisynthesized in microbial systems in fermentation tanks might be no cheaper than the *in planta*-derived version. In this regard, artemisinin derived from *ex planta* sources could be used to simply smoothen deficits that are presently experienced in agricultural production. This will be inevitable, as the loss of a child every 40 seconds to malaria (Bowles et al. 2008, Sachs and Malaney 2002) in parts of the world should prompt everyone to focus on enhancing the present supply of plant-derived artemisinin by cultivating genetically improved varieties and increasing the land area dedicated to the crop.

4.3.4 Enhanced Semisynthesis of Artemisinin through Conversion of Sesquiterpenic Precursors of Artemisinin

Apart from the extraction of artemisinin from dried leafy biomass of *A. annua* and biotechnological means for enhancing its production in microbial systems, a new and efficient method being touted involves the establishment of commercial-scale extraction of artemisinic acid and dihydroartemisinic acid, two major sesquiterpenoid precursors of artemisinin, which have been identified from most commercial cultivars of the plant. A Chinese cultivar which was cultivated in 2006 in a West Virginia field in the United States of America and analyzed for artemisinin, dihydroartemisinic acid, and artemisinic acid by HPLC throughout the growing season showed a peak in artemisinin production between August 28 and September 1 (Ferreira 2008). These plants had 0.93% artemisinin, 1.6% dihydroartemisinic acid, and 0.28% artemisinic acid, respectively. The author suggested that artemisinin production could at least be doubled by using both dihydroartemisinic acid and artemisinic acid that is usually eliminated in the by-product of artemisinin production (Jorge Ferreira, pers. comm.). Approximate quantifications for these sesquiterpenoid precursors indicate that there were about 24% of dihydroartemisinic acid and 5% of artemisinic acid, respectively, from the high-artemisinin-containing cultivar, Artemis (Ferreira, unpublished). Unfortunately, both dihydroartemisinic acid and artemisinic acid, which are usually extracted with refluxing in the extraction solvent, are presently discarded in the artemisinin purification steps, where artemisinin is pooled into nonpolar fractions. It is, therefore, of immense economic importance that methods for extraction and conversion of both dihydroartemisinic acid and artemisinic acid into artemisinin are optimized. This can potentially increase the final artemisinin profile derivable from a given quantity of dried leafy biomass by approximately 30% (Brisibe et al. 2008b), especially against the backdrop of a recent finding where a photochemistry-based method, developed by researchers at the Max Plank Institute transformed dihydroartemisinin into artemisinin without enzymes, rather, just with the use of light and oxygen (Lévesque and Seeberger 2012). It is obvious, therefore, that this approach is feasible and can be used to increase artemisinin production from the crop.

4.4 COMMON PHARMACOLOGICAL AND AGRICULTURAL USES OF *A. ANNUA* AND SOME OF ITS SECONDARY METABOLITES

Having highlighted some of the different systems that can be used for the enhanced production of artemisinin and the therapeutic uses of this sesquiterpenoid compound, it is equally important to state that the artemisinin used presently in the formulation of ACTs for treatment of malaria is often not cost-effective due to the very expensive process used to extract and purify it from *A. annua*. Presently, more than 95% of artemisinin is produced in China and Vietnam while the reminder comes from a few other sources in India and Africa. However, as reliable as these drugs (especially the ACTs for malaria treatment) could be, they are not readily available to all patients in malaria-endemic regions due either to the low yield of artemisinin from the native plant as outlined previously or the high cost of the treatments, making them inaccessible to most patients afflicted with malaria. Consequently, in many regions of sub-Saharan Africa with a high incidence rate of malaria, local populations, against the advice of WHO, continue to drink extracts of *A. annua* leaves as a tea (or infusion) or consume the dried leaves of the plant directly in a porridge mixture not only in the treatment of malaria fever but also for the treatment of anorexia, insomnia, anemia, a lack of appetite, flatulence, stomach ache, jaundice, indigestion, diabetes, sickle-cell anemia, testosterone-induced benign prostatic hyperplasia (Brisibe, unpublished), obesity (Baek et al. 2015), and HIV (Lubbe et al. 2012). Another important factor responsible for this development is that a large part of the population in many developing countries, especially in rural areas of Asia, Africa, and Latin America, do not have confidence in orthodox or Western-based drugs, as all their healthcare needs have always been met by herbal remedies (Brisibe et al. 2009b). Understandably, therefore, both hot water extracts and the dried leaves of the plant consumed in the porridge mixture would contain not only artemisinin but other bioactive compounds, including polymethoxylated flavonoids such as artemetin, casticin, chrysosplenetin, chrysosplenol-D, cirsilineol, and eupatorine, and more than a dozen other sesquiterpenes that abound in the leaves, which have been indicated as important compounds with antimalarial (Elfawal et al. 2014, Willcox 2009) and potentially anticancer activities. Synergistic benefits may also be derived from the presence of other antimalarial compounds such as dehydrosilibin and dimethylallyl campferide. Aside from this, it has been reported that the traditional *A. annua* tea therapy contained artemisinin as well as some antioxidant compounds mostly flavonoids (Chukwurah et al. 2014, Willcox et al. 2007, Rath et al. 2004). In addition to their bioavailability, these compounds, such as phenols, saponins, flavonoids, alkaloids, and tannins, act to reduce parasitemia independent of artemisinin (Liu et al. 1992), perhaps by inducing an oxidative stress. The presence of other compounds in *A. annua* leaves has thus raised suspicion as to the possibility of their synergistic role with artemisinin in malaria and cancer treatment (Ferreira et al. 2010). These *in planta* constituents potentiate and enhance the overall activity of artemisinin (Elford et al. 1987), the reason given for the long-term use of the plant as a tea in China even before the discovery of artemisinin (Ferreira et al. 2010). Consequently, given the complex nature of *A. annua* and the many bioactive components and nutrients present in its tissues (Bhakuni et al. 2001, Brisibe et al. 2009), it would be simplistic to consider the consumption of either the traditional tea or whole plant material essentially as a monotherapy; a fear expressed by many people, which is understandable. However, this worry appears to be misplaced. Some recent studies have demonstrated that there may be less chance of resistance occurring from the combined use of numerous plant constituents, which enhances the overall activity of artemisinin and can prevent *Plasmodium* or any other microbial parasite from developing resistance to the compound (Elfawal et al. 2014). In fact, some of these recent studies indicate that treatment with the whole *A. annua* plant provides a multicomponent defense system that is a more efficient means of slowing resistance development than a comparable dose of the purified monotherapeutic artemisinin, which is not only expensive (Elfawal et al. 2015) but also inefficient (Kangethe 2016). On account of the fact that *A. annua* has such broad potential therapeutic power against *Plasmodium* parasites, it could effectively be used as a cost-effective means for malaria treatment in many developing countries (Kangethe 2016).

Strictly speaking, *A. annua* is an annual weed with an aggressive and vigorous growth habit. A sweet smelling and highly aromatic plant, it is considered to have originated and occurs naturally as part of the steppe vegetation in Northern China (Ferreira et al. 2005). However, it now grows effectively in many other climatic conditions. In Asia, for example, it is well distributed and extends as a native into Southern Siberia, Vietnam, and Northern India. Outside of Asia, the plant has adapted ubiquitously to different growth conditions as seen in many parts of Europe, the United States, Australia, and Argentina (Ferreira et al. 2005). In Africa, it has been introduced into commercial-scale cultivation in Tanzania, Kenya, Uganda, and Madagascar within the past couple of decades and quite recently in Nigeria (Brisibe et al. 2012), where evaluation of optimal agronomic practices and mass selection for late flowering and high-artemisinin-yielding lines were evaluated using seeds from Brazil, China, Vietnam, India, Germany, and the United States with interesting results. During these trials, the hybrid populations from Brazil originated plants that had a growth span of about 192 days before flowering and were up to 2.84 m in height (Figure 4.1) with an average leaf biomass yield of 324 g/plant and artemisinin concentrations as high as 1.0975% (on a g/100 g dry weight basis) under humid lowland tropical conditions (Brisibe et al. 2012).

Now considering that some plant secondary metabolites appear to have a more synergistic effect when provided as an extract or powder rather than in a purified form, an edible form of *A. annua* leafy biomass via a compacted capsule in combination with an ACT partner has also been offered as a reliable, safe, and

Figure 4.1 Full-grown *A. annua* plants in the field.

inexpensive mode to deliver the drug. In fact, it would be very tempting to consider the whole plant treatment as an alternative delivery mechanism for artemisinin. This is supported by the recent findings of Weathers and colleagues (2011) and Elfawal et al. (2014, 2015) which have provided strong evidence to suggest that other parasite-killing substances present in the whole plant material may be acting through their potentiation of artemisinin, which renders whole plant consumption as an innovative plant-based artemisinin combination therapy (pACT). In one of their novel studies, Elfawal et al. (2014) actually demonstrated the efficacy of the whole plant, not as a tea or an infusion, but as a malaria therapy, and found it to be more effective than a comparable dose of pure artemisinin in a rodent malaria model. Furthermore, they also observed that development of stable resistance to the whole plant treatment by the Plasmodium parasite was achieved three times more slowly than stable resistance to pure artemisinin. Weathers and others (2011) also showed that mice fed with dried whole plant material of *A. annua* had about 40 times more artemisinin in their bloodstream than those fed with a corresponding amount of the pure drug. This amount exceeded eight-fold the minimum concentration of serum artemisinin (10 µg/l) required against *P. falciparum*, which suggests that the active ingredients contained in the whole plant material were delivered faster and in greater quantity than those from pure drug treatments. In fact, in a human trial in Kenya (Table 4.1, cited from Weathers et al. 2014b), dried *A. annua* leaf tablets (delivered as pACT), fed to 48 malaria patients, yielded results similar to trials with pure artemisinin, but much less artemisinin was required when the drug was delivered as dried leaves (ICIPE, 2005).

Though plant-based supply of artemisinin is not in consonance with the preference of modern pharmaceutical industry for single-ingredient drugs, this method would dramatically reduce the cost of healthcare not only in developing countries, such as those in sub-Saharan Africa where malaria is endemic, but perhaps also in more developed societies, where a holistic approach to disease treatment with herbal products has recently become fashionable. There are several examples that illustrate the synergistic benefits of drug delivery using complex botanical materials in preference to that of an isolated form (Raskin et al. 2002). The author, therefore,

Table 4.1 Kenyan Human Trial Data for Orally Delivered Dried Leaf *A. annua*

Artemisinin Dose (mg)		No. of Patients	Leaf DW (g)		% Recrudescence
Day 1	Days 2–6		Day 1	Days 2–6	
7.4 × 2	3.7 × 2	12	2	1	25
11.1 × 2	7.4 × 2	12	3	2	9.1
14.8 × 2	11.1 × 2	12	4	3	16.7
18.5 × 2	14.8 × 2	12	5	4	9.1
Compared to orally delivered pure artemisinin					
Day 1	Days 2–7				
2500	227 NA 24				

Note: Plant-based artemisinin combination therapy (dried leaf A. annua tablets, ea 500 mg, 3.7 mg artemisinin/tablet); NA: not available.

is completely in agreement with the proposal of Weathers and colleagues (2011) that the loading of capsules with compacted *A. annua* leaf powder of a known dosage to which the ACT secondary drug partner can be added or administered separately or even the compression of the dried leaf powder as tablets (Weathers et al. 2014b; Figure 4.2), could be other cost-effective, inexpensive, and reliable methods of artemisinin delivery in resource-poor setting, especially in Africa where the scourge has its highest toll of mortality. The processing facilities for such inexpensive artemisinin delivery methods could be centered within an area where local farmers currently grow the plant such that the entire process could be self-sustaining. This proposition will not only strengthen local healthcare, as confirmed by the efficacy of the plant-derived artemisinin and the ethanol extract from locally cultivated plants in *in vitro* evaluation studies, but also the local economy of the area, especially in Africa, where the crop is grown.

Taken together, these observations have such strong support that apart from the use of WHO-recommended ACTs, some researchers have vigorously campaigned in favor of either re-establishing the use of traditional *A. annua* tea (Magahalaes et al. 2016, Van der Kooy and Verpoorte 2011) or using dried whole leaves as compressed tablets (Figure 4.2) or pACTs (Elfawal et al. 2012, 2014, 2015; Weather et al. 2011; Brisibe and Chukwurah 2014); with the caveat that the plant material used has high or clinical levels of artemisinin in remote areas where malaria is endemic. Considering that the onset of cerebral malaria and malaria-induced coma is fast and the nearest medical facility or hospital could be as far as a journey of 2–3 days away, the use of the plant material (in whole or as *A. annua* tea therapy) should be investigated scientifically and, hopefully, permitted to sustain a malaria patient to reach a health center stocked with antimalarial drugs (Ferreira et al. 2010).

These treatment methods will not only save precious lives but have several advantages. First, they are quite inexpensive. Second, they are in forms that most resource-poor societies can rely on. It is conceivable that the cultivation of *A. annua* must be encouraged in many developing countries, not solely for the purpose of artemisinin

Figure 4.2 Three granulation sizes of *A. annua* leaves; two were pressed into tablets. Granulation size L to R: 2.0 mm (could not form into tablets), 0.6 mm into tablets, approximately 100–150 μm into tablets. Tablet weights are approximately 0.17 and 0.30 g, respectively; tablet dimensions were 10 mm diameter, 3 mm maximum depth (Weathers et al. 2014b).

extraction, but also for the significance of the plant in its multi-purpose therapeutic potential and holistic treatment of malariaand a variety of other neglected tropical diseases and ailments as outlined in this Chapter, must be encouraged.

In an exciting recent study, dried leaves of *A. annua* were included in the diet preparations of streptozocin-induced diabetic rats, resulting not only in a gradual but significant reduction in their blood glucose levels, but also a significant effect in controlling the loss of body weight (Brisibe et al. submitted). Though it may be quite difficult to attribute this observation to the function(s) of any of the chemical constituents of the plant, it is interesting that an earlier pharmacological study demonstrated that Tarralin™, an ethanolic extract of *Artemisia dracunculus* L., decreases hyperglycemia in rodents with chemically induced diabetes, as well as those genetically prone to diabetic conditions showing insulin resistance (Ribnicky et al. 2006). In that study, blood glucose concentrations, blood insulin levels, and phosphoenolpyruvate carboxykinase (PEPCK) expression in healthy animals were not altered. Consequently, the authors attributed the effects observed with Tarralin™ to several distinct modes of action useful for the improvement of complications associated with diabetes; one of which was the presence of more than one active component in the plant extract. This is, perhaps, equally the case in the present study. In fact, the aerial parts of *A. annua* are known to contain a diverse and extensive portfolio of biologically active chemicals including terpenoids, flavonoids (such as luteolin, apigenin, and peduletin), coumarins (such as scopoletin and tomentin), steroids, phenolics, purines, terpenes (such as costunolide), lipids, and aliphatic compounds (Brisibe et al. 2009, Bhakuni et al. 2001). Not surprisingly, any of these compounds or a combination of some acting in synergy may have been responsible for the antihyperglycemic activities observed in this study, since most of these compounds are known as scavengers of oxidative stress and have been well documented to possess antidiabetic activities (Houstis et al. 2006, Okada et al. 1995, Ahmed et al. 2000, Neogi et al. 2003, Jung et al. 2006) with a positive effect in alleviating diabetes as well as reducing its secondary complications.

It is commonly known that hyperglycemia increases the generation of free radicals by glucose auto-oxidation, and this increment of free radicals is what usually leads to organ damage. Consequently, besides the influence that could be ascribed directly to phytochemical constituents in the plant material, it is equally possible that the reduction in hyperglycemia seen in this study with dietary supplements of *A. annua* leaves could be attributable to a reversal in the destruction of the β-cells produced in the islets of Langerhans in the pancreas, which could be associated with the ability of the large amount of antioxidants produced by *A. annua* (Brisibe et al. 2009) to mop up all free radicals and impede the resulting destructive oxidative process. Indeed, free radical scavengers would protect DNA from damage and may improve cell survival. This notion is consistent with the results demonstrated earlier (Oboh 2005, Akah and Okafor 1992, Atangwho et al. 2007) following the administration of aqueous leaf extracts of *Telfairia occidentalis* and *Vernonia amygdalina*, two local medicinal plants that are known to have both antihyperglycemic and hypolipidemic properties in experimental animals and so could be used in the management of diabetes mellitus.

Aside from the reasons listed previously, the effects seen with *A. annua* in the study cited earlier could equally be attributed to any of the many other explanations which have been proposed as being responsible for the action of antidiabetic plants (Bnouham et al. 2002). First, antidiabetic plant materials may stimulate insulin secretion from pancreatic β-cells and induce regeneration, revitalization and/or hyperplasia of these cells. Second, antidiabetic plant extracts can equally act by imitating an "insulin-like action." It is equally possible that antidiabetic plants may be acting through a supply of several key minerals, such as copper, zinc, magnesium, manganese, and calcium, which are seen in high concentrations in *A. annua* leaves, to the islet of Langerhans β-cells, or they can also decrease the level of glucagon, induce a decrease in the intestinal absorption of glucose, and/or reduce the peripheral use of glucose. Moreover, they could equally act on liver enzymes causing stimulation of glucogenesis and/or inhibition of glycogenolysis (Bnouham et al. 2002).

Regardless of the particular metabolic pathway responsible for the effects seen in the study, the data recorded there clearly demonstrate the beneficial effects of pulverized dried leaves of *A. annua* on hyperglycemia in albino rats. Coupled with clinical trials, which are obviously necessary, it may be possible to offer this plant material as a useful dietary supplement for the management of blood glucose levels in diabetic patients. However, it is a common practice to examine levels of biochemical and hematological indices in animals as these are equally considered very important means of assessing their health status at any particular point in time. With this in mind, our study also examined the influence of the plant material on biochemical and hematological parameters in chemically induced diabetic and anemic rats. Surprisingly, there were no abnormalities in the kidney and liver function tests among the different groups of streptozotocin-induced rats treated with the plant material though significant decreases were observed in the hematological parameters studied in all the groups initially treated with phenylhydrazine when compared with the untreated control (Brisibe et al. submitted). This is expected, as phenylhydrazine is an oxidant that destroys red blood cells through its effect on enzymes involved in energy metabolism. It primarily damages mature red blood cells by creating oxidative stress, which induces an erythropoenia or denatures the hemoglobin in erythrocytes by binding to its proteins, thus creating a state of anemia. The data generated in this earlier study provide distinctive evidence that dietary preparations of *A. annua* have the ability to also significantly improve some hematological indices in anemic rats. This observation is reflected by the consistently increased levels in the packed cell volume (PCV), hemoglobin (Hb) concentration, and red blood cell (RBC) count, which might not also be unconnected with the dietary phytochemical composition of the plant material. Nutritional characterization of *A. annua* herbage demonstrated that the leaves are rich in proteins, digestible carbohydrates, fat, amino acids, and a full complement of minerals such as zinc, copper, calcium, magnesium, and so on. Some of these and other constituents, including iron, flavonoids, and the antioxidant vitamins A, C, and E, which are commonly deficient in many plant-based diets, are abundant in *A. annua* leaves (Brisibe et al. 2009). Interestingly, these are well-established haematopoietic factors that have direct influence on the production of blood in the bone marrow. Iron, for instance, plays a critical role in the transport of oxygen throughout the body and in cellular processes of growth and division. Its deficiency results in a decrease in the hemoglobin concentration in the red blood cells, which leads to anemia. Iron-deficiency anemia, a major health problem, is usually a nutritional disorder afflicting large population groups in different parts of the world. It is especially widespread among vulnerable infants, adolescent girls, and pregnant women, particularly in populations that feed largely on plant food sources with poor iron content or monotonous cereal-based diets. Remarkably, as demonstrated in this study *A. annua* leaves can be used as an alternative treatment therapy for anemia (Brisibe et al. submitted). Moreover, some of the amino acids derived from *A. annua* including alanine, arginine, aspartate, cystine, glycine, glutamine, histidine, leucine, isoleucine, lysine, methionine, serine, tryptophan, threonine, phenylalanine, valine, and tyrosine that were found in fairly high concentrations could also be used for the synthesis of the β-globin chains of haemoglobin and this could equally have contributed to an increase in the levels of haemoglobin observed in the study. This observation has enormous potential on the pathogenesis of sickle-cell diseases in humans where *A. annua* leaf extracts were observed to provide sickle erythrocytes with important protection against oxidative stress.

Over and above all of the diseases and ailments that can be treated with *A. annua*, and especially as the incidence of HIV/AIDS becomes more prevalent in different parts of the world with varying consequences, a lot of new drugs (both natural and through organic synthesis) are being evaluated in the fight against the scourge. So far, *A. annua* has been identified as one of the few medicinal plants to show great promise in this regard (Lubbe et al. 2012), perhaps because its tea infusion stimulates white blood cell activities and has normalizing effects on immune functions of AIDS patients with highly compromised immune systems. It will be a major pharmacological novelty once the anti-HIV effects of *A. annua* have been unequivocally confirmed in humans.

In recent years, there has been an increase in the number of scientific investigations that have validated the potential of *A. annua* and its extracts, both as dietary feed supplement and to treat a variety of animal diseases, including cancer (Almeida et al. 2012; Brisibe et al. 2008b, 2009; Cherian et al. 2013; Drăgan et al. 2010; Ferreira 2009; Turnet and Ferreira 2005; Breuer and Efferth 2014). One such ailment affecting poultry, that results in severe economic loss to farmers, is coccidiosis (caused by the *Eimeria* species, belonging to the same protozoan class as *Plasmodium*), which has also been treated, along with other parasitic diseases affecting poultry, with the dried leafy biomass of *A. annua* (Almeida et al. 2012, Brisibe et al. 2008a). The results of these and other studies generally indicate that the dietary addition of artemisinin and/or dried leaves of *A. annua* protects against weight losses and reduces oocyst output and lesion scores after oral infection with coccidia. One major advantage of artemisinin is its high margin of safety. Even at high oral single doses, it seems to have only few adverse side effects in broiler chickens. Bearing in mind the potential effect of *A. annua* against bacteria and parasites, dried plant material or plant extract could possibly be used in the control of several other diseases in the production of small livestock. Equally of notable significance as protozoa, trematode infections in livestock have also been demonstrated to be of huge veterinary importance (Ferreira et al. 2011). Presently, an estimated 350 million cattle and 250 million sheep are at risk of acquiring *Fasciola* species with an estimated annual loss of more than US\$2–3 billion (Keiser et al. 2008). *Fasciola* is highly adaptable and can affect different types of livestock including buffalos, goats, and several other species (including humans) that can serve as intermediary hosts, making its control very difficult (Lin et al. 2007), especially in light of growing concerns of resistance development to synthetic drug molecules, such as triclabendazole, in cattle and sheep (Moll et al. 2000). This has highlighted the need to search for alternative trematocidal treatments, and invariably, artemisinin readily comes to mind on the basis of the inspiring discoveries outlined previously.

Apart from these commonly known traditional uses of *A. annua* leaf herbage in the treatment of several human and livestock diseases, many research groups and stakeholders are presently seeking for alternative uses and therapies for *A. annua* that are efficient, affordable, accessible, and widely available. Some of the highlighted uses so far include to boost immune systems, to produce scopoletin (by extraction), as an insect repellent, as a perfume from essential oils, and as flavorings for alcoholic beverages. The plant also has other important roles in agriculture, including serving as an effective plant growth inhibitor with great potential as an organic herbicide (Abate et al. 2011), as a pesticide in stored grains (Brisibe et al. 2011), and as ectoparasite repellant for fishery (Ekanem and Brisibe 2010), respectively. Essential oil blends derived from *A. annua* have also shown promising results with respect to reduction of *Clostridium perfringens* colonization and proliferation (Ivarsen et al. 2010, Timbermont et al. 2010) within the gastrointestinal tract of broiler chickens. All of these are protective capacities conferred on the plant since it produces a complex phytochemical repertoire of more than 600 biologically active secondary metabolites including the sesquiterpenoid artemisinic compounds, essential oil constituents, and numerous phenolic acids, flavonoids, minerals, vitamins, and amino acids in its aerial portions (Bhakuni et al. 2001, Brisibe et al. 2009). This, in itself, is not surprising, as many natural products in plants are multifunctional molecules that protect their primary hosts from bacterial, viral, and other microbial infections, or even from herbivores such as insects and worms. Against this backdrop, therefore, there should be no single usage intended for *A. annua*, but a range of treatment possibilities provided by the plant ingredients with potential benefits to human and animal health.

In fact, A. *annua* and artemisinin uses for the livestock industry are currently in expansion, based on current reports of the anti-protozoal, antibacterial and antioxidant activities of the plant, its extracts, and its essential oil. Some animal parasites effectively controlled with *A. annua*, its essential oil, and artemisinin include *Eimeria* that causes coccidiosis (Brisibe et al., 2008a; Almeida et al., 2012), the trematodal blood fluke *Schistosoma* spp. (Lescano *et al.*, 2004), and bacteria (Juteau et al., 2002). In a collaborative study between Nigerian, Brazilian, and US scientists, different tissues of *A. annua* were analyzed for its potential use in animal feed and scored high values for crude protein, antioxidant capacity, and as source of amino acids, with negligible amounts of anti-nutritive components such as phytate, tannin, and total oxalate (Brisibe *et al.*, 2009).

In addition to what has been previously outlined, *A. annua* is also a highly nutritive plant as it has been identified as a storehouse of several important nutrients and antinutrients. The leaves are rich in many nutritional components, and a full complement of minerals such as zinc, copper, calcium, magnesium, iron, and so on. Some of these, especially iron, flavonoids, and the antioxidant vitamins that are commonly deficient in many plant-based diets, are abundant in *A. annua* leaves (Brisibe *et al.*, 2009). Interestingly, these are well-established haematopoietic factors that have direct influence on the production of blood in the bone marrow. Iron, for instance, plays a critical role in the transport of oxygen throughout the body and in cellular processes of growth and division. Its deficiency results in a decrease in the hemoglobin concentration in the red blood cells, which leads to anemia. Iron-deficiency anemia, a major health problem, is usually a nutritional disorder afflicting large population groups in different parts of the world. It is especially widespread among vulnerable infants, adolescent girls, and pregnant women, particularly in populations that feed largely on plant food sources with poor iron content. Remarkably, as demonstrated elsewhere *A. annua* leaves can be used as an alternative treatment therapy for anemia. Moreover, some of the amino acids derived from *A. annua* including alanine, arginine, aspartate, cystine, glycine, glutamine, histidine, leucine, isoleucine, lysine, methionine, serine, tryptophan, threonine, phenylalanine, valine, and tyrosine that were found in fairly high concentrations (Brisibe et al. 2009) could also be used for the synthesis of the β-globin chains of hemoglobin and this could also have contributed to an increase in the levels of hemoglobin observed in the study. This observation has an enormous potential on the pathogenesis of sickle-cell disease in humans where *A. annua* leaf extracts were observed to provide sickle erythrocytes with important protection against oxidative stress (data submitted for publication elsewhere).

4.5 CONCLUDING REMARKS

There is a glaring indication all over the world today that orthodox medicine alone cannot provide all the answers and remedies to the numerous health-related problems that may befall mankind. Consequently, there is a rapidly growing demand today for natural products and plant-derived medicines throughout the world, particularly because of their potential to keep pathogenic strains sensitive. The concept of growing crops for health rather than for food or fiber is slowly changing plant biotechnology and medicine. Rediscovery of the connection between plants and human and animal health is responsible for launching a new generation of botanical therapeutics such as plant-derived pharmaceuticals, multicomponent botanical drugs, dietary supplements, and plant-based recombinant proteins (Raskin et al. 2002). Many of these products will soon complement conventional pharmaceuticals in the treatment, prevention, and diagnosis of diseases, while at the same time adding value to agriculture.

There is no doubt that *A. annua* is one such remarkable medicinal plant that will be at the forefront of the coming pharmaceutical and agricultural revolution. Not surprisingly, the spotlight on international malaria therapy is presently focused on the availability of artemisinin and the supply of ACTs from this seemingly simple yet versatile plant that is suddenly found at the forefront of global efforts aimed at the eradication of malaria. It is essential that the production of artemisinin from *A. annua* and its use as the key active ingredient in the manufacture of ACTs be seen as the central focus. However, the heavy demand placed on the plant due to its huge pharmacological benefits and broad spectrum of applications, especially in the treatment of malaria and several other diseases afflicting humans and livestock, would necessitate that its cultivation in different parts of the world be accelerated with much vigor.

REFERENCES

Abate S., Damtew Z., and Mengesha B. 2011. *Artemisia annua* as an alternative potential weed control option. *African Journal of Food, Agriculture, Nutrition, and Development* 11: 1–6.

Ahmed M., Akhtar M.S., Malik T. and Gilani A.H. 2000. Hypoglycaemic action of the flavonoid fraction of *Cuminum nigrum* seeds. *Phytotherapy Research* 14: 103–106.

Akah P.A. and Okafor C.I. 1992. Hypoglycaemic effect of *Vernonia amygdalina* Del in experimental rabbits. Plant Med. Res. 1: 6–10.

Almeida G.F., Horsted K., Thamsborg S.M., Kyvsgaard N.C., Ferreira J.F.S., and Hermansen J.E. 2012. Use of *Artemisia annua* as a natural coccidiostat in free-range broilers and its effects on infection dynamics and performance. *Veterinary Parasitology* 186: 178–187. doi:http://dx.doi.org/10.1016/j.vetpar.2011.11.058.

Andersen L.P. 2007. Colonization and infection by *Helicobacter pylori* in humans. *Helicobacter* 12 (Suppl 2): 12–15.

Arsenault P.R., Wobbe K.K., and Weathers P.J. 2008. Recent advances in artemisinin production through heterologous expression. *Current Medicinal Chemistry* 15: 2886–2896.

Atangwho I.J., Ebong P.E., Eyong E.U., Eteng M.U., and Uboh F.E. 2007. *Vernonia amygdalina* Del: A potential prophylactic antidiabetic agent in lipid complication. *Global Journal of Pure and Applied Sciences* 13: 103–106.

Baek H.K., Shim H., Lim H., Shim M., Kim C.-K., Park S-K., Lee Y.S., Song K-D., Kim S-J., and Yi SS. 2015. Anti-adipose effect of *Artemisia annua* on diet-induced-obesity mice model. *Journal of Veterinary Science* 16: 389–396.

Bhakuni R.S., Jain D.C., Sharma R.P., and Kumar S. 2001. Secondary metabolites of *Artemisia annua* and their biological activity. *Current Science* 80: 35–48.

Bnouham M., Mekhfi H., and Legssyer A. 2002. Medicinal plants used in the treatment of diabetes in Morocco. *International Journal of Diabetes and Metabolism* 10: 33–50.

Bowles D., Smallwood M., and Graham I. 2008. Fast track breeding of *A. annua*. Ph.D. Symposium of the Zurich-Basel Plant Science Center, 6th June, Switzerland.

Breuer E. and Efferth T. 2014. Treatment of iron-loaded veterinary sarcoma by *Artemisia annua*. *Natural Products and Bioprospecting* 4(2): 113–118. doi:10.1007/s13659-014-0013-7.

Brisibe E.A., Adugbo S.E., Ekanem U., Brisibe F., and Figueira G.M. 2011. Controlling bruchid pests of stored cowpea seeds with dried leaves of *Artemisia annua* and two other common botanicals. *African Journal of Biotechnology* 10: 9586–9592.

Brisibe E.A., Brisibe F., Agba D., and Abang A.E. 2017. Clonally propagated *Artemisia annua* and its remarkable attenuation of experimental oxidative stress-induced hyperglycaemia and anaemia burden in mammalian models (submitted).

Brisibe E.A. and Chukwurah P.N. 2014. Production of artemisinin *in planta* and in microbial systems need not be mutually exclusive. In: Aftab T., Ferreira J.F.S., Khan M.M.A. and Naeem M. (Eds.). *Artemisia annua – Pharmacology and Biotechnology.* Springer-Verlag, Berlin. DOI 10.1007/978-1-4614-5001-6_2.

Brisibe E.A., Udensi O., Chukwurah P.N., de Magalhäes P.M., Figueira G.M., and Ferreira J.F.S. 2012. Adaptation and agronomic performance of *Artemisia annua* L. under lowland humid tropical conditions. *Industrial Crops and Products* 39: 190–197. doi:http://dx.doi.org/10.1016/j.indcrop.2012.02.018.

Brisibe E.A., Umoren U.E., Brisibe F., Magalhaes P.M., Ferreira J.F.S., Luthria D., Wu X., and Prior P. 2009. Nutritional characterization and antioxidant capacity of different tissues of *Artemisia annua* L. *Food Chemistry* 115: 1240–1426.

Brisibe E.A., Umoren U.E., Owai P.U., Brisibe F. 2008a. Dietary inclusion of dried *Artemisia afnnua* leaves for management of coccidiosis and growth enhancement in chicken. *African Journal of Biotechnology* 7: 4083–4092.

Brisibe E.A., Uyoh E.A., Brisibe F., Magalhäes P.M., Ferreira J.F.S. 2008b. Building a golden triangle for the production and use of artemisinin derivatives against *Falciparum* malaria in Africa. *African Journal of Biotechnology* 7: 4884–4896.

Brown G.D. 2010. The biosynthesis of artemisinin (Qinghaosu) and the phytochemistry of *Artemisia annua* L. (Qinghao). *Molecules* 15: 7603–7698.

Chen D.H., Ye H.C., Li G.F. 2000. Expression of a chimeric farnesyl diphosphate synthase gene in *Artemisia annua* L. transgenic plants via Agrobacterium tumefaciens-mediated transformation. *Plant Science* 155: 179–185.

Cheng C., Ho W.E., Goh F.Y., Guan S., Kong L.R., Lai W., Leung B.P., and Wong W.S.F. 2011. Antimalarial drug artesunate attenuates experimental allergic asthma via inhibition of the phosphoinositide 3-kinase/Akt pathway. *PLoS ONE* 6(6): e20932. doi:10.1371/journal.pone.0020932.

Cherian G., Orr I.A., Burke I.C., and Pan W. 2013. Feeding *Artemisia annua* alters digesta pH and muscle lipid oxidation products in broiler chickens. *Poultry Science* 92: 1085–1090.

Chukwurah P.N., Brisibe E.A., Osuagwu A.N., and Okoko T. 2014. Protective capacity of *Artemisia annua* as a potent antioxidant remedy against free radical damage. *Asian Pacific Journal of Tropical Biomedicine* 4: S92–S98.

Debrunner N., Dvorak V., Magalhaes P.M., and Delabays N. 1996. Selection of genotypes of *Artemisia annua* L. for the agricultural production of artemisinin. In: Pank F (Ed.) *International Symposium on Breeding Research on Medicinal Plants*, Quedlinburg, pp. 222–225.

De Jesus-Gonzalez L. and Weathers P.J. 2003. Tetraploid *Artemisia annua* hairy roots produce more artemisinin than diploids. *Plant Cell Reports* 21: 809–813.

Delabays N. 1994. La domestication et l'amelioration genetique de l'*Artemisia annua* dans le cadre du developpement d'un nouveau medicament contre la malaria. *International Conference on Aromatic and Medicinal Plants.* 1–11.

Drăgan L., Titilincu A., Dan I., Dunca I., Drăgan M., and Mircean V. 2010. Effects of *Artemisia annua* and *Pimpinella anisum* on *Eimeria tenella* (Phylum Apicomplexa) low infection in chickens. *Science Parasitology* 11: 77–82.

Efferth T., Dunstan H., Sauerberry A., Miyachi H., and Chitambar C.R. 2001. Antimalarial artesunate is also active against cancer. *International Journal of Oncology* 18: 767–773.

Efferth T., Herrmann F., Tahrani A. and Wink M. 2011. Cytotoxic activity of secondary metabolites derived from *Artemisia annua* L. towards cancer cells in comparison to its designated active constituent artemesinin. *Phytomedicine* 18: 959–969.

Efferth T., Romero M.R., Wolf D.G., Stamminger T., Marin J.J.G., and Marschall M. 2008. The antiviral activities of artemisinin and artesunate. *Clinical Infectious Diseases: An Official Publication of the Infectious Diseases Society of America* 47: 804–811.

Ekanem A. and Brisibe E.A. 2010. Effects of ethanol extract of *Artemisia annua* L. against monogenean parasites of *Heteobranchus longifilis*. *Parasitology Research* 106: 1135–1139. doi:10.1007/s00436- 010-1787-0.

Elfawal M.A., Towler M.J., Reich N.G., Golenbock D., Weathers P.J., and Rich S.M. 2012. Dried whole plant *Artemisia annua* as an antimalarial therapy. *PLoS ONE* 7: e52746. doi:10.1371/journal.pone.0052746.

Elfawal M.A., Towler M.J., Reich N.G., Weathers P.J., and Rich S.M. 2014. Dried whole-plant Artemisia annua slows evolution of malaria drug resistance and overcomes resistance to artemisinin. *Proceedings of the National Academy of Science*. www.pnas.org/cgi/doi/10.1073/pnas.1413127112

Elfawal M.A., Towler M.J., Reich N.G., Weathers P.J., and Rich S.M. 2015. Dried whole-plant Artemisia annua slows evolution of malaria drug resistance and overcomes resistance to artemisinin. *Proceedings of the National Academy of Sciences of the United States of America* 112: 821–826. doi:10.1073/pnas.1413127112.

Elford B.C., Roberts M.F., Phillipson J.D., and Wilson R.J. 1987. Potentiation of the antimalarial activity of qinghaosu by methoxylated flavones. *Transactions of the Royal Society of Tropical Medicine and Hygiene* 81: 434–436. doi:10.1016/0035-9203(87)90161-1.

Ferreira J.F.S. 2008. Seasonal and post-harvest accumulation of artemisinin, artemisinic acid, and dihydroartemisinic acid in three accessions of *Artemisia annua* cultivated in West Virginia, USA. *Planta Medica* 74: 310–311(Abstract only).

Ferreira J.F.S. 2009. Artemisia species in small ruminant production: their potential antioxidant and anthelmintic effects. In: Morales, M. (Ed.), *Appalachian Workshop and Research Update: Improving Small Ruminant Grazing Practices*. Mountain State University/USDA, Beaver, WV, pp. 53–70.

Ferreira J.F.S. and Janick J. 1996. Roots as an enhancing factor for the production of artemisinin in shoot cultures of *Artemisia annua*. *Plant Cell Tissue and Organ Culture* 44: 211–217.

Ferreira J.F.S., Laughlin J.C., Delabays N., and Magalhães P.M. 2005. Cultivation and genetics of *Artemisia annua* L. for increased production of the antimalarial artemisinin. *Plant Genetic Resources: Characterization and Utilization* 3: 206–229.

Ferreira J.F.S., Luthria D.L., Sasaki T., and Heyerick A. 2010. Flavonoids from *Artemisia annua* L. as antioxidants and their potential synergism with artemisinin against malaria and cancer. *Molecules* 15: 3135–3170.

Ferreira J.F.S., Peaden P., and Keiser J. 2011. In vitro trematocidal effects of crude alcoholic extracts of *Artemisia annua, A. absinthium, Asimina triloba*, and *Fumaria officinalis*. *Parasitology Research* 109: 1585–1592. doi:10.1007/s00436-011-2418-0.

Firestone G.L. and Sundar S.N. 2009. Anticancer activities of artemisinin and its bioactive derivatives. *Expert Reviews in Molecular Medicine* 30(11): e32. doi:10.1017/S1462399409001239.

Fulzele D.P., Spipahimalani A.T., and Heble M.R. 1991. Tissue cultures of *Artemisia annua*: organogenesis and artemisini n production. *Phytotherapy Research* 5: 149–153.

Goswami S., Bhakuni R.S., Chinniah A., Pal A., Kar S.K., and Dasa P.K. 2012. Anti-*Helicobacter pylori* potential of artemisinin and its derivatives. *Antimicrobial Agents and Chemotherapy* 56: 4594–4607. doi:10.1128/AAC.00407-12.

Graham I.A., Besser K., Blumer S., Branigan C.A., Czechowski T., et al. 2010. The genetic map of *Artemisia annua* L. identifies loci affecting yield of the antimalarial drug artemisinin. *Science* 327: 328–331. doi:10.1126/science.1182612.

Hale H., Keasling J.D., Renninger N., and Diagana T.T. 2007. Microbially derived artemisinin: A biotechnology solution to the global problem of access to affordable antimalarial drugs. *American Journal of Tropical Medicine and Hygiene* 77: 198–202.

Hart S.P., Ferreira J.F.S., and Wang Z. 2007. Efficacy of wormwoods (*Artemisia* spp.) as an anthelmintic in goats. *Journal of Animal Science* 86: 92.

Houstis N., Rosen E.D., and Lander E.S. 2006. Reactive oxygen species have a causal role in multiple forms of insulin resistance. *Nature* 440: 944–948.

Huan-huan C., Li-Li Y., and Shang-Bin L. 2004. Artesunate reduces chicken chorioallantoic membrane neovascularization and exhibits antiangiogenic and apoptoic activity on human microvascular dermal endothelial cells. *Cancer Letters* 21(2): 163–171.

ICIPE, 2005. Whole-leaf Artemisia annua-based antimalarial drug: report on proof-of-concepts studies. (Unpublished report. Retrieved on July 20, 2013).

Ivarsen E., Frette X.C., Christensen K.B., Engberg R.M., Kjaer A., Jensen M., Grevsen K., and Christensen L.P. 2010. Evaluation of the efficiency of essential oil components from *A. annua* as an antimicrobial against *Clostridium perfringens* in poultry. Abstracts Book of 6th Conference on Medicinal and Aromatic Plants of Southeast European Countries (CMAPSEEC); Antalya, Turkey. *Pharmacognosy Magazine* 6 (Suppl.): 126.

Jaziri M., Shimonura K., Yoshimatsu K., Faucounnier M.L., Marlier M., and Homes J. 1995. Establishment of normal and transformed root cultures of *Artemisia arinua* L. for atemisinin production. *Journal of Plant Physiology* 145: 175–177.

Jung M., Park M., Lee H.C., Kang Y.H., Kang E.S., and Kim S.K. (2006) Antidiabetic agents from medicinal plants. *Current Medicinal Chemistry* 13: 1203–1218.

Juteau F., Masotti V., Bessiere J.M., Dherbomez M., Viano J. (2002). Antibacterial and antioxidant activities of Artemisia annua essential oil. *Fitoterapia* 73: 532–535.

Kangethe L.N. 2016. Resistance mitigating effect of *Artemisia annua* phytochemical extracts in cultures of *Plasmodium falciparum* and in *Plasmodium berghei* and *Plasmodium yoelii*. Doctorate degree thesis submitted to Jomo Kenyatta University of Agriculture and Technology, Kenya.

Keiser J., Rinaldi L., Veneziano V., Mezzino L., Tanner M., et al. 2008. Efficacy and safety of artemether against a natural Fasciola hepatica infection in sheep. *Parasitology Research* 103: 517–522.

Kumar S., Gupta S.K., Singh P., Bajpai P., Gupta M.M., Singh D., Gupta A.K., Ram G., Shasany A.K., and Sharma S. 2004. High yields of artemisinin by multi-harvest of *Artemisia annua* crops. *Industrial Crops and Products* 19: 77–90.

Lai H. and Singh N.P. 2006. Oral artemisinin prevents and delays the development of 7,12-dimethylbenz(a) anthracene (DMBA)-induced breast cancer in the rat. *Cancer Letters* 233(1): 43–48.

Lescano S.Z., Chieffi P.P., Canhassi R.R., Boulos M., Neto V.A. 2004. Anti-schistosomal activity of artemether in experimental Schistosomiasis mansori. *Rev. Saude Publ.* 38: 71–75.

Lévesque F. and Seeberger P.H. 2012. Continuous-flow synthesis of the antimalaria drug artemisinin. *Angewandte Chemie International Edition* 51: 1706–1709. doi:10.1002/anie.201107446.

Li B., Zhang R., Li J., Zhang L., Ding G., Luo P., He S., Dong Y., Jiang W., Lu Y., Cao H., Zheng J., and Zhou H. 2008. Antimalarial artesunate protects sepsis model mice against heat-killed *Escherichia coli* challenge by decreasing TLR4, TLR9 mRNA expressions and transcription factor NF-kB activation. *International Immunopharmacology* 8: 379–389.

Li W.D., Dong Y.J., Tu Y.Y., and Lin Z.B. 2006. Dihydroarteannuin ameliorates lupus symptom of BXSB mice by inhibiting production of TNF-a and blocking the signaling pathway NF-kB translocation. *International Immunopharmacology* 6: 1243–1250.

Liu B., Wang H., Du Z., Li G., and Ye H. 2011. Metabolic engineering of artemisinin biosynthesis in *Artemisia annua* L. *Plant Cell Reports* 30: 689–694.

Liu K.C.S.C., Yang S.L., Roberts M.F., Elford B.C., and Phillipson J.D. 1992. Antimalarial activity of *Artemisia annua* flavonoids from whole plants and cell-cultures. *Plant Cell Reports* 11: 637–640.

Liu C.Z., Guo C., Wang Y.C., and Ouyang F. 2002. Effect of light irradiation on hairy root growth and artemisinin biosynthesis of *Artemisia annua* L. *Process Biochemistry* 38: 581–585.

Lin R.Q., Dong S.J., Nie K., Wang C.R., Song H.Q., Li A.X., Huang W.Y., and Zhu X.Q. 2007. Sequence analysis of the first internal transcribed spacer of rDNA supports the existence of the intermediate Fasciola between F. hepatica and F. gigantica in mainland China. *Parasitology Research* 101(3): 813–817. doi:10.1007/s00436-007-0512-0.

Lubbe A., Seibert I., Klimkait T., and van der Kooya F. 2012. Ethnopharmacology in overdrive: The remarkable anti-HIV activity of *Artemisia annua*. *Journal of Ethnopharmacology* 14: 854–849. doi:10.1016/j.jep.2012.03.024.

Magalhães P.M., Dupont I., Hendrickx A., Joly A., Raas T., Dessy S., Sergent T., and Schneider Y.J. 2012. Anti-inflammatory effect and modulation of Cytochrome P450 activities by Artemisia annua tea infusions in human intestinal Caco-2 cells. *Food Chemistry* 134: 864–871.

Magalhães P.M., Figueira G.M., de Souza J.M., Ventura A.M.R., Ohnishi M.D.O., da Silva D.R., Lobo L.A.G., Eleres F.B., Libonati R., Willcox M., and Pimentel E.C. 2016. *Artemisia annua*: A new version of a traditional tea under randomized, controlled clinical trial for the treatment of malaria. *Advances in Bioscience and Biotechnology* 7: 545–563.

Malfertheiner P., Sipponen P., Naumann M., Moayyedi P., Mégraud F., Xiao S-D., Sugano K., and Nyrén O. 2005. *Helicobacter pylori* eradication has the potential to prevent gastric cancer: A state-of-the-art critique. *The American Journal of Gastroenterology* 100: 2100–2115.

Malhotra K., Subramaniyan M., Rawat K., Kalamuddin M., Qureshi M.I., Malhotra P., Mohmmed A., Cornish K., Daniell H., and Kumar S. 2016. Compartmentalized metabolic engineering for artemisinin biosynthesis and effective malaria treatment by oral delivery of plant cells. *Molecular Plant* 9: 1464–1477. doi:10.1016/j.molp.2016.09.013.

Marshall B.J. 2001. One hundred years of discovery and rediscovery of *Helicobacter pylori* and its association with peptic ulcer disease. In Mobley HLT, Mendz GL, Hazell SL (ed), *Helicobacter Pylori: Physiology and Genetics*. ASM Press, Washington, DC, pp. 19–24.

Martin V.J., Pitera D.J., Withers S.T., Newman J.D., and Keasling J.D. 2003. Engineering a mevalonate pathway in *Escherichia coli* for production of terpenoids. *Nature Biotechnology* 21: 796–802.

Moll L., Gaasenbeek C.P.H., Vellema P., and Borgsteede F.H.M. 2000. Resistance of *Fasciola hepatica* against triclabendazole in cattle and sheep in The Netherlands. *Veterinary Parasitology* 91(1–2): 153–158.

Neogi P., Lakner F.J., Medicherla S., Cheng J., Dey D., Gowri M., Nag B., Sharma S.D., Pickford L.B., and Gross C. 2003. Synthesis and structure–activity relationship studies of cinnamic acid-based novel thiazolidinedione antihyperglycemic agents. *Bioorganic & Medicinal Chemistry* 11: 4059–4067.

Oboh G. 2005. Hepatoprotective property of ethanolic and aqueous extracts of fluted pumpkin (*Telfairia occidentalis*) leaves against garlic-induced oxidative stress. *Journal of Medicinal Food* 8: 560–563.

Okada Y., Miyauchi N., Suzuki K., Kobayashi T., Tsutsui C., Mayuzumi K., Nishibe S., and Okuyama T. (1995) Search for naturally occurring substances to prevent the complications of diabetes. II. Inhibitory effect of coumarin and flavonoid derivatives on bovine lens aldose reductase and rabbit platelet aggregation. *Chemical & pharmaceutical Bulletin* 43: 1385–1387.

Paddon C.J., Westfall P.J., Pitera D.J., Benjamin1 K., Fisher, K. et al. 2013. High-level semi-synthetic production of the potent antimalarial artemisinin. *Nature* 496: 528–532.

Raskin I., Ribnicky D.M., Komarnytsky S., Ilic N., Poulev A., Borisjuk N., Brinker A., Moreno D.A., Ripoll C., Yakoby N., O'Neal J.M., Cornwell T., Pastor I., and Fridlender B. 2002. Plants and human health in the twenty-first century. *Trends in Biotechnology* 20: 522–531.

Rath K., Taxis K., Walz G., Gleiter C.H., Li S-M., and Heide L. 2004. Pharmacokinetic study of Artemisinin after oral intake of a traditional preparation of *Artemisia annua* L. (Annual wormwood). *American Journal for Tropical Medicine and Hygiene* 70: 128–132.

Ribnicky D.M., Poulev A., Watford M., Cefalu W.T., and Raskin I. 2006. Antihyperglycaemic activity of Tarralin™, an ethanolic extract of *Artemisia dracunculus* L. *Phytomedicine* 13: 550–557.

Ro D.K., Paradise E.M., Ouellet M., Fisher K.J., Newman K.L., Ndungu J.M., Ho K.A., Eachus R.A., Ham T.S., Kirby J., Chang M.C., Withers S.T., Shiba Y., Sarpong R., and Keasling J.D. 2006. Production of the antimalarial drug precursor artemisinic acid in engineered yeast. *Nature* 440: 940–943.

Romero M.R., Serrano M.A., Vallejo M., Efferth T., Alvarez M., and Marin J.J.G. 2006. Antiviral effect of artemisinin from *Artemisia annua* against a model member of the Flaviviridae family, the Bovine Viral Diarrhoea Virus (BVDV). *Planta Medica* 72: 1169–1174.

Roth R.J. and Acton N. 1989. A simple conversion of artemisinic acid into artemisinin. *Journal of Natural Products* 52: 1183–1185.

Sachs J. and Malaney P. 2002. The economic and social burden of malaria. *Nature* 415: 680–685.

Schmid G., and Hofheinz W. 1983. Total synthesis of qinghaosu. *Journal of the American Chemical Society* 105: 624–625.

Tellez M.R., Canel C., Rimando A.M., and Duke S.O. 1999. Differential accumulation of isoprenoids in glanded and gland-less *Artemisia annua* L. *Photochemistry* 52: 1035–1040.

Teoh K.H., Polichuk D.R., Reed D.W., Nowak G., and Covello P.S. 2006. *Artemisia annua* L. (Asteraceae) trichome-specific cDNA reveal CYP71AV1, a cytochrome P450 with a key role in the biosynthesis of the antimalarial sesquiterpene lactone artemisinin. *FEBS Letters* 580: 1411–1416.

Timbermont L., Lanckriey A., Nollet N., Schwarzer K., Haesebrouck A., Ducatelle R., and Van Immerseel F. 2010. Control of *Clostridium perfringens*-induced necrotic enteritis in broilers by target-released butyric acid, fatty acids and essential oils. *Avian Pathology* 39: 117–121.

Tilaoui M., Mouse H.A., Jaafari A., and Zyad A. 2014. Differential effect of artemisinin against cancer cell lines. *Natural Products and Bioprospecting* 4: 189–196. doi:10.1007/s13659-014-0024-4.

Turner K.E. and Ferreira J.F.S. 2005. Potential use of *Artemisia annua* in meat goat production systems. *The Conference of the American Forage and Grassland Council*, Bloomington, IL, June 11–15; Cassida, K., Ed. AFGC: Bloomington, IL; 221–225.

Van der Kooy F. and Verpoorte R. 2011 The content of artemisinin in the *Artemisia annua* tea infusion. *Planta Medica* 77: 1754–1756.

Van Herpen T.W.J.M., Cankar K., Nogueira M., Bosch D., Bouwmeester H.J., and Beekwilder J. 2010. *Nicotiana benthamiana* as a production platform for artemisinin precursors. *PLoS ONE* 5(12): e14222. doi:10.1371/journal.pone.0014222.

Wallaart T.E., Bouwmeester H.J., Hille J., Poppinga L., and Maijers N.C. 2001. Amorpha-4,11-diene synthase: Cloning and functional expression of a key enzyme in the biosynthetic pathway of the novel antimalarial drug artemisinin. *Planta* 212: 460–465.

Wang J.W. and Tan R.X. (2002). Artemisinin production in *Artemisia annua* hairy root cultures with improved growth by altering the nitrogen source in the medium. *Biotechnology Letters* 24: 1153–1156.

Wang J.X., Tang W., Zhou R., Wan J., Shi L-P., Zhang Y., Yang Y-F., Li Y., and Zuo J-P. 2008. The new water-soluble artemisinin derivative SM905 ameliorates collagen-induced arthritis by suppression of inflammatory and Th17 responses. *British Journal of Pharmacology* 153: 1303–1310.

Wang Y.C., Zhang H.X., Zhao B., and Yuan X.F. 2001. Improved growth of *Artemisia annua* L. hairy roots and artemisinin production under red light conditions. *Biotechnology Letters* 23: 1971–1973.

Wang Z.J., Qiu J., Guo T.B., Liu A., Wang Y., Li Y., and Zhang J.G. 2007. Anti-inflammatory properties and regulatory mechanism of a novel derivative of artemisinin in experimental autoimmune encephalomyelitis. *Journal of Immunology* 179: 5958–5965.

Weathers P.J., Arsenault P.R., Covello P.S., McMickle A., Teoh K.H., and Reed D.R. (2011) Artemisinin production in *Artemisia annua*: Studies *in planta* and results of a novel delivery method for treating malaria and other neglected diseases. *Phytochemistry Reviews* 10: 173–183. doi:10.1007/s11101-010-9166-0.

Weathers P.J., Reed K., Hassanali A., Lutgen P., and Engeu PO. 2014b. Whole plant approaches to therapeutic use of *Artemisia annua*, L. (Asteraceae). In: Aftab T., Ferreira J.F.S., Khan M.M.A. and Naeem M. (Eds.). *Artemisia annua – Pharmacology and Biotechnology*. Springer-Verlag, Berlin Heidelberg 2014. doi:10.1007/978-1-4614-5001-6_2.

Weathers P.J. and Towler M.J. 2014a. Changes in key constituents of clonally propagated *Artemisia annua* L. during preparation of compressed leaf tablets for possible therapeutic use.. *Industrial Crops and Products* 62: 173–178.

Weathers P.J., Towler M., Hassanali A., Lutgen P., Engeu P.O. 2014c. Dried-leaf *Artemisia annua*: A practical malaria therapeutic for developing countries? *World Journal of Pharmacology* 3: 39–55.

Willcox M. 2009. Artemisia species: From traditional medicines to modern antimalarials–and back again. *The Journal of Alternative and Complementary Medicine* 15: 101–109.

Willcox M., Falquet J., Ferreira J.F.S., Gilbert B., Hsu E., Melillo de Magalhães P, Plaizier-Vercammen J, Sharma VP and Wright CW. 2007. *Artemisia annua* as a herbal tea for malaria. *African Journal of Traditional, Complementary, and Alternative Medicines: AJTCAM* 4: 121–123.

Xu H., He Y., Yang X., Liang L., Zhan Z., et al. 2007. Antimalarial agent artesunate inhibits TNF-a-induced production of proinflammatory cytokines via inhibition of NF-kB and PI3 kinase/Akt signal pathway in human rheumatoid arthritis fibroblast-like synoviocytes. *Rheumatology* 46: 920–926.

Zhang Y. and Sun H. 2009. Immunosuppressive effect of ethanol extract of *Artemisia annua* on specific antibody and cellular responses of mice against ovalbumin. *Immunopharmacology and Immunotoxicology* 31: 625–630.

Zhou H.J., Wang W.Q., and Wu G.D. 2007. Artesunate inhibits angiogenesis and down-regulates vascular endothelial growth factor expression in chronic myeloid leukaemia K562 cells. *Vascular Pharmacology* 47: 131–138.

Zhu C. and Cook S.P. 2012. A concise synthesis of (+)-artemisinin. *Journal of the American Chemical Society* 134: 13577–13579.

CHAPTER 5

Artemisia annua and Its Bioactive Compounds as Anti-Inflammatory Agents

Bianca Ivanescu and Andreia Corciova

CONTENTS

5.1 INTRODUCTION

Inflammation is the body's immune response to harmful stimuli, such as pathogens, damaged cells, and irritants. It is crucial for the restoration of cellular homeostasis, by removal of foreign invaders or damaging molecules and by healing the damaged tissue. Inflammation is a generic immunovascular response and has been considered a mechanism of innate immunity, the first line of defense against any damaging condition (Medzhitov, 2008).

The process of inflammation starts with the recognition of specific molecular motifs associated with infection or tissue injury. The progress of inflammation involves several key mediators such as nitric oxide, leukotrienes, prostaglandins, adhesion molecules, cytokines, and chemokines that control vascular changes and inflammatory cell recruitment. The expression of pro-inflammatory molecules is regulated by activation of transcription factors, such as nuclear factor kappa B

(NF-κB), activator protein 1 (AP-1), or signal transducers and activators of transcription (STATs). Their ability to bind DNA is changed by different protein kinases involved in signal transduction, such as mitogen-activated protein kinases (MAPK) (Tunon et al. 2009).

Acute inflammation is a normal physiological response that includes a resolution phase dominated by anti-inflammatory mediators and tissue repair through generation of appropriate growth factors and cytokines. If the resolution is unsuccessful because of persistent stimuli or genetic variations, the acute inflammation develops into chronic inflammation. Alterations in the inflammatory responses account for various chronic inflammatory conditions, such as rheumatoid arthritis, atherosclerosis, type 2 diabetes, asthma, obesity, inflammatory bowel diseases, neurodegenerative diseases, and cancer. In chronic inflammatory diseases, the extensive injuries of the host are the result of the host inflammatory response itself and not of foreign attacks (Ahmed, 2011).

Considering the biological complexity of these conditions, developing effective anti-inflammatory agents with no severe side effects remains a challenge. Lately, numerous natural compounds and vegetal extracts are being tested for anti-inflammatory activity and among those *Artemisia annua* L. (Asteraceae) occupies a leading position. Not just traditional medicine recommends this species for the treatment of inflammatory diseases, but also its diverse chemical composition lends to this outlook, with over 600 phytochemicals identified to date, many of them with proven biological activities (van der Kooy and Sullivan, 2013). The existing research on *A. annua* anti-inflammatory activity refers generally to compounds from the following classes: sesquiterpene lactones, flavonoids, phenolic acids (mainly chlorogenic acids), lipids, and coumarins.

5.2 ANTI-INFLAMMATORY ACTIVITY OF *A. ANNUA* EXTRACTS

Extracts are complex mixtures of compounds from different chemical classes that contribute to the overall biological activity and, as numerous studies showed, the therapeutic effect of *A. annua* extracts is not solely due to artemisinin. With respect to the antimalarial action, some authors noted that the activity of artemisinin may be enhanced in whole plant extracts due to increased bioavailability and synergistic effect with other metabolites (Räth et al., 2004; Rasoanaivo et al., 2011). Also, aqueous extracts, with small amounts of artemisinin, may reduce the risk of *Plasmodium* resistance and even be more efficient in the treatment of malaria. These findings are in accordance with the traditional way of preparation of *A. annua* tea as described in the Chinese *materia medica*: soaking the fresh plant in water for 24 hours, wringing out the vegetal material, and drinking the resulting extract (van der Kooy and Sullivan 2013).

Moreover, Carbonara et al. (2012) proposed that phenolic constituents of the infusion, highly soluble in polar solvents, have a role in enhancing artemisinin solubility and extraction efficiency in water. In the case of anti-inflammatory activity, the few studies on *A. annua* extracts (described in Table 5.1) support the hypothesis of higher efficiency of extracts compared to individual isolated compounds.

Table 5.1 The effect of *A. annua* extracts in different inflammatory models

Models	Extract type	Concentration/exposure time	Results compared to control without *A. annua* treatment	References
Human intestinal Caco-2 cells + cocktail of inflammatory mediators	Lyophilized infusion (20 g plant in 1 L distilled boiling water for 15′)	Pretreatment with 3300 μg/mL for 1 h	↓ IL-6 ↓ IL-8	Magalhães et al. (2012)
LPS-activated RAW 264.7 murine macrophage cells	Water, 80% methanol, 80% ethanol, and 80% acetone extract	1, 5, 10, 50, and 100 μg/mL for 18 h	↓ NO and PGE$_2$ ↓ IL-6, IL-10, IL-1β	Kim et al. (2015)
LPS-activated neutrophils	Supercritical carbon dioxide extract (Arthrem®)	200, 100, and 50 μg/mL (20′ pretreatment) 400, 200, and 100 μg/mL (20′ pretreatment)	100% inhibition of TNF-α production >87% inhibition of PGE$_2$ production	Hunt et al. (2015)
LPS-activated RAW 264.7 murine macrophage cells	95% ethanol extract	12.5, 25, and 50 μg/mL	100% inhibition of NO production	Chougouo et al. (2016)
In vitro inhibition of COX-1 and 5-LO	70% ethanol extract	100, 10, 1, 0.1, and 0.01 μg/mL	No inhibition of COX-1 or 5-LO	Chagas-Paula et al. (2015)
Writhing test, formalin test, mechanical allodynia test, and, carrageenan-induced paw edema in mice	Sesquiterpene lactones enriched fraction	30, 100, and 300 mg/kg i.p. 30′/2 h prior to test	Up to 100% inhibition in writhing and formalin test Up to 64.18% inhibition in mechanical allodynia test	Favero et al. (2014)
Formalin- and ovalbumin-induced paw edema in rats	Lipid extract (hydrocarbon bp 75–80°C) containing essential oil	0.1, 0.5, and 5 mg/kg p.o.	18–28% inhibition of formalin-edema 8–31% inhibition of ovalbumin-edema	Ul'chenko et al. (2005)
High-fat diet–fed mice	80% aqueous ethanol (1 g: 20 mL agitated for 24 h at 25°C)	Chronic daily 400 mg/kg p.o. for 12 weeks	↓ TGF-β1, ↓ CTGF, ↓ HMGB1, ↓ COX-2	Kim et al. (2016)
Diet-induced obese mice	Aqueous extract (40 g plant: 1.8 L distilled water under 1.5 bar at 80°C for 30′)	Chronic daily 0.2 g/mL/kg p.o. for 4 weeks	↓ IL-6	Baek et al. (2016)

The anti-inflammatory activity of four extracts of *A. annua* (water, 80% methanol, 80% ethanol, and 80% acetone extract) was evaluated *in vitro* by assessing the inhibitory effect on inflammatory mediators produced by lipopolysaccharides(LPS)-activated RAW 264.7 murine macrophage cells. All extracts contained artemisinin and polyphenols in different amounts and suppressed the generation of nitric oxide (NO), prostaglandin E_2 (PGE_2), and interleukins IL-1β, IL-6, and IL-10. The acetone extract had the highest anti-inflammatory effect and the largest concentration of artemisinin compared to the other extracts (Kim et al., 2015).

The crude ethanol extract of *A. annua* inhibited NO production in LPS-activated RAW 264.7 murine macrophages even at the lowest concentration tested, without affecting the viability of the macrophage cells. Four compounds isolated from this extract—artemisinin, scopoletin, chrysosplenetin, and eupatin—also exhibited potent inhibitory activity on NO production, especially scopoletin which was efficient even at low concentrations (Chougouo et al., 2016).

Furthermore, *A. annua* infusion had a potent anti-inflammatory effect on inflamed intestinal Caco-2 (human epithelial colorectal adenocarcinoma) cells, manifested by the reduction of IL-6 and IL-8 cytokines secretion. Chlorogenic acid, rosmarinic acid, isoquercitrin, and scopoletin were identified in the aqueous extract, with rosmarinic acid being the main component. It is worth mentioning that although the infusion significantly lowered the production of pro-inflammatory mediators, artemisinin alone had no effect. The authors concluded that the anti-inflammatory effect was not related to the presence of artemisinin and could be attributable to the phenolic compounds, mainly to rosmarinic acid (Magalhães et al., 2012).

On the other hand, an ample *in vitro* study conducted on 57 *Asteraceae* plants species showed that *A. annua* 70% ethanol extract had no effect on the activity of two major enzymes involved in inflammatory processes: COX-1 (cyclooxygenase-1) and 5-LO (5-lipoxygenase). The chemical analysis showed that extracts were mainly composed of chlorogenic acids, flavonoids, and sesquiterpene lactones (Chagas-Paula et al., 2015).

A commercial product containing supercritical carbon dioxide extract of *A. annua*, Arthrem®, inhibited the production of tumor necrosis factor α (TNF-α) and PGE_2 in activated neutrophils in a concentration-dependent fashion. At the same concentrations, artemisinin was less efficient in inhibiting the inflammatory mediators. This suggests that other compounds present in the extract contributed to the anti-inflammatory activity (Hunt et al. 2015). Arthrem® is marketed as a dietary supplement for joint support in the form of gel capsules containing a 150 mg standardized extract of *A. annua* and grape seed oil. In a clinical trial conducted on 42 patients with osteoarthritis over 12 weeks, administration of one capsule twice daily showed significant improvement in physical function and reduction of pain and stiffness (Stebbings et al. 2016).

The anti-inflammatory effect of *A. annua* extracts was substantiated by some animal studies, too. An 80% ethanol extract from *A. annua* leaves reduced hepatic steatosis and inflammation in high-fat diet–fed mice and protected against fibrosis by decreasing the hepatic expression of profibrogenetic cytokines TGF-β1 (transforming growth factor beta 1), CTGF (connective tissue growth factor), and HMGB1 (high mobility group box 1 protein) and of COX-2 (cyclooxygenase-2). COX-2 mediated inflammation is critical for the evolution of hepatic steatosis (Kim et al., 2016).

Obesity is equally associated with chronic low-level inflammation. An aqueous extract of *A. annua* exhibited antiadipogenic activity in diet-induced obesity in mice through reduction of adipose-tissue accumulation and inhibition of chronic inflammatory state characterizing obesity (Baek et al., 2015). The same extract lowered IL-6 level in the blood and stimulated neuronal maturation in the brain of diet-induced obese mice. This latter effect was achieved without modifying COX-2 and microglial Iba-1 expression in the hippocampus, thus suggesting a different anti-inflammatory pathway. When tested in the *Caenorhabditis elegans* paralysis model of Alzheimer disease, the extract manifested a favorable effect, advancing the possibility of using *A. annua* as an anti-Alzheimer agent (Baek et al., 2016).

The sesquiterpene lactone enriched fraction of *A. annua* exhibited significant analgesic and anti-inflammatory activity in the writhing test, formalin test, mechanical allodynia test, carrageenan-induced paw edema, and tail flick test in mice. The fraction including artemisinin (1.72%) and deoxyartemisinin (0.31%) was administered by intraperitoneal injection. The results prove that the extract affected not only the perception of inflammatory stimuli, but also reduced the inflammatory mediators causing edema (Favero et al., 2014).

The antiphlogistic activity was also evaluated for the lipid extract obtained from leaves and flowers of *A. annua*. The lipid extract administered orally to rats inhibited formalin- and ovalbumin-induced paw edema (Ul'chenko et al., 2005).

5.3 ANTI-INFLAMMATORY ACTIVITY OF SESQUITERPENE LACTONES

Artemisinins have immunoregulatory properties on numerous components of the immune system. They decrease neutrophils count, suppress the secretion of cytokines, inhibit macrophage activation and their responses, and block lymphocytes proliferation. The anti-inflammatory effects of artemisinins are owed to the inhibition of different signaling pathways (Ho et al., 2014; Yao et al., 2016). In addition, artemisinins have an impact on autoimmunity, autoinflammation, and delayed-type hypersensitivity responses, through their immunosuppressive activity on T and B cells (Shakir et al. 2011).

The first report on the immunosuppressive effect of artemisinins dates from 1984 and shows that artesunate inhibited the proliferation of mouse splenocytes and human peripheral lymphocytes *in vitro* (Shen et al., 1984). Afterwards, artemisinin, dihydroartemisinin, and arteether were reported to exhibit potent suppression of humoral responses in mice (Tawfik et al., 1990).

Artesunate, artemether, dihydroartemisinin, and two water-soluble derivatives SM905 and SM934 have been shown to protect against collagen-induced arthritis, an experimental model of *rheumatoid arthritis* (Cuzzocrea et al., 2005; Mirshafiey et al., 2006; Wang et al., 2008; Lin et al., 2016). Rheumatoid arthritis is an autoimmune disorder characterized by persistence of circulating autoantibodies and uncontrolled lymphocyte activation, clinically manifested by synovial inflammation, cartilage damage, and bone destruction.

The remarkable antiarthritic effect of artesunate might be mediated by inhibition of pro-inflammatory cytokines and MMP-9 (metalloproteinase-9) activity through the suppression of NF-κB and MAPK signaling pathways (Li et al., 2013b). Artesunate also inhibited chemotactic factors CCL2 (chemokine [C-C motif] ligand 2) and CCL5 (chemokine [C-C motif] ligand 5) in the serum and the synoviocyte culture of collagen-induced arthritis in rats (Mo et al., 2012) and suppressed hypoxia-inducible factor-1α (HIF-1α) and secretion of vascular endothelial growth factor (VEGF) in human rheumatoid arthritis fibroblast-like synoviocytes by inhibition of the phosphoinositide 3-kinase/protein kinase B (PI3K/Akt) signaling pathway (He et al., 2011). Moreover, the high immunosuppressive activity of artesunate plays an important part in fighting this disease and other T cell-mediated immune disorders (Li et al., 2013a; Lee et al., 2015). It has also been reported that artemisinin analogues SM905 and SM934 suppressed T follicular helper cells and T helper 17 cells, hence ameliorating collagen-induced arthritis (Wang et al., 2008; Lin et al., 2016).

Artemisinins manifest beneficial effects in *systemic lupus erythematosus*, an autoimmune condition characterized by production of autoantibodies and abnormal accumulation of autoreactive T lymphocytes. Artesunate manifested immunosuppressive activity, comparable to cyclophosphamide in a murine model of systemic lupus erythematosus (Jin et al., 2009), and protected against lupus nephritis better than prednisone by reducing IL-6 and TNF-α serum levels and downregulating the expression of the NF-κB p65 subunit and TGF-β1 in renal tissue (Wu et al., 2010). Dihydroartemisinin was found to abolish LPS-induced cell activation in spleen cells from lupus-prone MRL/lpr mice by blocking the Toll-like receptor 4/interferon regulatory factor (TLR4/IRF) signaling pathway (Huang et al. 2014). A plausible anti-inflammatory mechanism of action of dihydroartemisinin involves downregulation of Akt and MAP kinases (ERKs—extracellular signal–regulated kinases, JNK—c-Jun N-terminal kinases, and p38 mitogen-activated protein kinases) signal transduction cascades (Kim et al., 2013).

The efficiency of the SM934 derivative in systemic lupus erythematosus is comparable to that of immunosuppressant drug rapamycin or corticosteroid prednisolone and is attributable to its immunomodulatory properties: SM934 inhibits memory/effector T cells and stimulates regulatory T cell development (Hou et al., 2011, 2012). Recent research demonstrated that SM934 also inhibited B-cell activation in lupus-prone MRL/lpr mice by suppression of TLR7/MyD88 (myeloid differentiation factor 88)/NF-κB pathway (Wu et al., 2016a). As a result of these compelling preclinical data, SM934 was recently approved by the China Food and Drug Administration for clinical trial as a novel drug candidate to treat systemic lupus erythematosus (Wu et al., 2016b).

In *inflammatory bowel disease*, artemisinin mitigated colonic inflammation by inducing cytochrome CYP3A expression through activation of the receptor pregnane X, an important component of the body's adaptive defense mechanism against toxic substances (Hu et al., 2014). Artesunate significantly ameliorated dextran sulfate sodium salt- and trinitrobenzene sulfonic acid-induced colitis in mice by lowering inflammatory mediators via NF-κB pathway inhibition and by suppressing T helper Th1/Th7 responses (Yang et al., 2012).

In spite of the well-known fact that sesquiterpene lactones can cause allergic contact dermatitis, it has been shown that artemisinins exhibit potent anti-inflammatory and immunomodulatory effects in *allergic inflammation*. Topical treatment with artemisinin notably reduced contact hypersensitivity, a T cell-mediated cutaneous inflammatory reaction. The proposed mechanism of action implies inhibition of Th17 development and stimulation of regulatory T cell production, related to a decrease in inflammatory mediators and of inflammation (Li et al., 2012). In the IgE-mediated passive cutaneous anaphylaxis mouse model, artesunate arrested IgE-mediated vascular hyperpermeability, whereas in the passive systemic anaphylaxis mouse model, artesunate blocked IgE-induced mast cell degranulation in the lungs, inhibited hypotermia and the rise of histamine levels. (Cheng et al., 2013). Furthermore, artesunate may constitute a potential therapeutic agent for the treatment of allergic asthma, as it ameliorated experimental allergic airway inflammation and house dust mite-induced asthma in mice. The anti-inflammatory effect was linked to downregulation of PI3/Akt signaling cascade and NF-κB activity (Cheng et al., 2011).

Pretreatment with artesunate potently inhibited TNF-α and IL-6 production in animal models of *septic inflammation* in a dose- and time-dependent manner, by reducing TLR4 and TLR9 expression and blocking NF-κB activation (Wang et al., 2006; Li et al., 2008). In addition, artesunate protects sepsis model mice against lethal doses of *Staphylococcus aureus* by decreasing TNF-α and IL-6 secretion via inhibition of TLR2 and NOD2 (nucleotide-binding oligomerization domain-containing protein) expression and NF-κB activation (Li et al., 2010). In particular, artesunate acts in synergy with oxacillin in protecting against mouse sepsis induced by methicillin-resistant *Staphylococcus aureus* (Jiang et al., 2011).

Artesunate treatment blocked the inflammation in endotoxin-induced *uveitis* in rats, by inhibiting the release of inflammatory mediators (TNF-α, PGE$_2$, NO, and CCL2) in the aqueous humor (Wang et al., 2011a,b).

Hepatic fibrosis represents the formation of scar tissue in response to persistent liver damage and can progress to cirrhosis. Oral artesunate alleviated hepatic fibrosis induced by bovine serum albumin in rats through modulation of metalloproteinases expression: reduction of MMP-2 and MMP-9 and the augmentation of MMP-13 (Xu et al., 2014). Another study showed that the antifibrotic activity of artesunate is linked to downregulation of α-smooth muscle actin (α-SMA), TLR4, MyD88, TGF-β1, and inhibition of NF-κB p65 subunit translocation into the nucleus (Lai et al., 2015).

Artesunate also generated favorable responses in HgCl$_2$-induced *nephrotic syndrome* in rats, by lowering levels of inflammatory markers TNF-α and C-reactive protein (Hassanin and Shata, 2014).

Considering the fact that neuroinflammation is an important component of *Alzheimer's disease*, the effect of artemisinin in this disease was investigated.

Researchers found that artemisinin diminished neuritic plaque burden, stopped β-secretase activity, and displayed anti-inflammatory activity via inhibition of NF-κB activity and NALP3 inflammasome activation in a mouse model of Alzheimer's disease (Shi et al., 2013a,b).

Atherosclerosis is an inflammatory disease of large- and medium-sized arteries, distinguished by the formation of atheromatous plaques, as a result of accumulation of white blood cells and proliferation of intimal smooth muscle cells. The effect of artemisinin on the inflammatory response in atherosclerosis-related differentiation of monocytes into macrophages was evaluated using phorbol myristate acetate-induced THP-1 monocytes. Artemisinin treatment attenuated the release of TNF-α, IL-1β, and IL-6 through inhibition of the NF-κB canonical pathway, therefore demonstrating a possible capacity for inhibition of inflammatory advancement of atherosclerosis (Wang et al., 2011a,b). Cao et al. (2015) found that artemisinin inhibits atherosclerosis via the ROS (reactive oxygen species)-mediated NF-κB signal pathway: artemisinin decreased NO, PGE_2, and ROS production and reduced the expression of MMP-2 and MMP-9, and consequently blocked the proliferation, migration, and inflammation of vascular smooth muscle cells (VSMC) induced by TNF-α. The same result was achieved in VSMC stimulated by platelet-derived growth factor BB, where artemisinin treatment reduced activation of ERK1/2 and attenuated MMP-9 expression (Lee et al., 2014).

5.4 ANTI-INFLAMMATORY ACTIVITY OF FLAVONOIDS

Flavonoids are well-known anti-inflammatory agents, but their mechanism of action is still unclear. They usually manifest antiphlogistic activity by modulating pro-inflammatory gene expression and intervening in multiple signaling pathways, chiefly NF-κB and MAPK. The anti-inflammatory response of flavonoids seems to be signal specific and dependent on the cell type. In addition, flavonoids have the ability to scavenge free radicals and manifest antioxidant activity, thus reducing intracellular oxidative stress and inflammation. However, they can act as a pro-oxidant at high concentrations and caution is necessary when administering flavonoids. Regarding structure-activity relation, it has been shown that the double bond between C2 and C3 and hydroxyl substitution on A- and B-ring are essential for the anti-inflammatory properties (Tunon et al., 2009).

Numerous phenolic compounds were identified in the *A. annua* plant, grouped into flavones, flavonols, coumarins, phenolic acids, and others. Ferreira et al. (2010) listed 11 prominent flavones and 29 flavonols reported in *A. annua*. Typically, the species contains high amounts of polymethoxylated flavonoids, such as cirsilineol, casticin, chrysosplenetin, chrysosplenol-D, eupatin, eupatorin, isorhamnetin, and artemetin. These compounds reach the greatest concentration in plants at full blooming stage (Baraldi et al., 2008). Lai et al. (2007) reported that the main flavonoids in *A. annua* are quercetin-glucoside, rhamnetin (quercetin-7-methyl ether), chrysosplenol-D (quercetagetin 3, 6, 7, 3′-tetramethyl ether), and pilloin (luteolin-7, 4′-dimethyl ether).

The flavonoids found in the acetone extract of *A. annua* are mainly glycosylated derivatives of apigenin, quercetin, luteolin, isorhamnetin, and kaempferol, but also luteolin and apigenin aglycones were identified (Gouveia and Castilho, 2013). Only small quantities of flavonoids are extracted in *A. annua* infusion: vitexin, isovitexin, patuletinglycoside, luteolin-7-O-glucoside, chrysoeriol rutinoside, jaceidin (1.1%), and cirsilineol (Carbonara et al. 2012). This fact was confirmed by Suberu et al. (2013a,b) who found casticin and isovitexin (apigenin-6-C-glucoside) in *A. annua* tea, only the latter being in notable amounts.

A. annua flavonoids have been tested for anti-inflammatory properties in different *in vitro* and *in vivo* models of inflammation. Artemetin and casticin manifested remarkable lipoxygenase inhibition *in vitro* (Choudhary et al., 2009).

Eupatorin attenuated carrageenan-induced paw inflammation in mice, by abrogating iNOS (*inducible nitric oxide synthase*) and COX-2 expression and the production of NO, PGE_2 and TNF-α in a dose-dependent manner. Eupatorin also blocked the activation of transcription factor STAT-1α induced by lipopolysaccharides (Laavola et al., 2012).

Cirsilineol and apigenin strongly inhibited lymphocyte proliferation induced by concanavalin A *in vitro*, while pectolinarigenin and quercetin exhibited a low level of inhibition. In addition, apigenin and cirsilineol potently inhibited T cell activation, manifesting immunosuppressive and anti-inflammatory activity (Yin et al. 2008).

Casticin and chrysosplenol-D inhibited in a dose-dependent manner the production of pro-inflammatory factors NO, PGE_2 and cytokines (VEGF, IL-1β, IL-6, and TNF-α) in LPS-induced rat peritoneal cells. They also suppressed PGE_2 and cytokines secretion in LPS-activated human peripheral blood mononuclear cells (Zhu et al., 2013). Given their anti-inflammatory activity, the two flavones were also evaluated on animal models of local and systemic inflammation: croton oil-induced ear dermatitis and edema and LPS-induced systemic inflammatory response syndrome in mice. Topically applied casticin and chrysosplenol-D reduced the edema, similar to indomethacin. Pretreatment with casticin and chrysosplenol-D protected the mice against the systemic immune response, a severe immune reaction characterized by overproduction of pro-inflammatory molecules that lead to serious organ damage and dysfunction (Li et al. 2015).

In vitro, casticin and chrysosplenol-D suppressed the LPS-induced release of inflammatory mediators (IL-1β, IL-6, and CCL2) by reducing IKK (inhibitor of kappa B protein kinase), phosphorylation, and NF-κB activation. Furthermore, the anti-inflammatory activity of chrysosplenol-D was mediated via JNK and the inhibition of c-JUN protein phosphorylation (Li et al., 2015). The anti-inflammatory effect of casticin was also investigated by Liou et al. (2014) who found that the molecular mechanism of action involved inhibition of the NF-κB, Akt, and MAPK signaling pathways. In this way, casticin reduced COX-2 and iNOS expression in murine macrophage cells through inhibition of IkB-α (inhibitor of kappa B protein α), ERK 1/2, p38 and Akt phosphorylation, and of translocation of NF-κB active subunit p65 into the nucleus. Also, heme oxygenase 1 (HO-1) and nuclear factor erythroid 2-related factor 2 (Nrf2) expression were promoted by casticin, leading to antioxidative and anti-inflammation responses.

Moreover, casticin and chrysosplenol-D were able to inhibit macrophages migration in LPS-activated RAW 264.7 cells (Li et al., 2015), while casticin decreased eosinophil migration in lung epithelial cells by reducing ICAM-1 (intercellular adhesion molecule 1) expression (Koh et al., 2011).

In an animal model of rheumatoid arthritis, casticin significantly reduced synovial hyperplasia and inflammatory cell infiltration by suppressing NF-κB expression *in vivo*. The underlying molecular mechanism was investigated in fibroblast-like synoviocytes activated by LPS. In this case, casticin induced the upregulation of anti-inflammatory cytokine IL-10 and downregulation of TNF-α, it also, decreased the expression of NF-κB and PKC (protein kinase C) and inhibited NF-κB transfer from cytoplasm to the nucleus (Li and Shen, 2016).

Isorhamnetin treatment significantly protected against carrageenan-induced acute inflammation in rats and decreased cells infiltration and inflammatory mediators in rats, through NF-κB inactivation, due to inhibition of JNK and Akt/IKK (Yang et al., 2013). In two animal models of inflammatory bowel disease (ulcerative colitis and the Crohn's disease model), isorhamnetin mitigated inflammation by inhibiting myeloperoxidase activity, the expression of inflammatory mediators and IkB-α and NF-κB activation. The authors also demonstrated that isorhamnetin is an activator of human pregnane X receptor, a target for reversing inflammation in inflammatory bowel disease (Dou et al., 2014).

It has been shown that quercetin intraperitoneal administration has prophylactic and therapeutic effects against LPS-induced systemic inflammation in mice, by stimulating the secretion of anti-inflammatory cytokine IL-10 (Liao and Lin, 2015). Furthermore, quercetin exhibits hepatoprotective and anti-inflammatory activity in CCl4-, perfluorooctanoic acid-, and Ni-induced liver injury through mitigating oxidative stress and decreasing production of pro-inflammatory markers (TNF-α, IL-1β, iNOS, COX-2, and NO). The underlying mechanism of action involves quercetin's ability to modulate TLR2/TLR4, MAPK/NF-κB, Nrf2/HO-1, and p38/STAT1/NF-κB signaling pathways (Liu et al., 2015; Ma et al., 2015; Zou et al., 2015). In a recent study, quercetin treatment protected against heat stroke-induced myocardial injury in rats; the cardioprotective effect may be due to the anti-inflammatory properties of quercetin which increased IL-10 levels and lowered TNF-α and IL-6 in heart tissue (Lin et al. 2017).

5.5 THE ANTI-INFLAMMATORY ACTIVITY OF CHLOROGENIC ACIDS

Chlorogenic acids are phenolic acids, a group of esters formed between quinic acid and trans-cinnamic acids, such as caffeic, ferulic, p-coumaric, sinapic, and dimethoxycinnamic. They are found in significant amounts in fruits and vegetables and their dietary presence is linked to a low-risk of chronic diseases. There is strong evidence for the antibacterial, antioxidant, and anti-inflammatory activity of chlorogenic acids (Liang and Kitts, 2016), as for neuroprotective (Han et al., 2010), hepatoprotective (dos Santos et al., 2005), and cardioprotective properties (Chiou et al., 2011; Zhang et al., 2016). The anti-inflammatory effect is traceable to their ability to relieve intracellular oxidative stress and to inhibit pro-inflammatory cytokines by regulation of key transcription factors (Liang and Kitts, 2016).

Gouveia and Castilho (2013) identified 18 chlorogenic acids in the acetone extract of *A. annua*, including mono-, di-, and tri-caffeoylquinic acids, feruloylquinic and p-coumaroylquinic acids. Also, chlorogenic acids which are highly soluble in polar solvents make up the bulk of *A. annua* infusion, along with some flavonoids. Thus, 16 hydroxycinnamic compounds were described and quantified by Carbonara et al. (132012) in the aqueous extract. Caffeic acid, 5-caffeoylquinic acid, 3,4-dicaffeoylquinic acid, 3,5-dicaffeoylquinic acid and 4,5-dicaffeoylquinic acid are the main components of *A. annua* tea. Suberu et al. (2013a,b) confirmed that hydroxycinnamic acids are the major components of *A. annua* tea and demonstrated their additive and synergistic (rosmarinic acid) interaction with artemisinin against *Plasmodium falciparum*.

Using an optimized method, Zhao et al. (2015) separated 36 chlorogenic acids from the 70% ethanol extract. The major compounds found in the leaves of *A. annua* were 5-caffeoylquinic acid, 1,5-dicaffeoylquinic acid, 3,5-dicaffeoylquinic acid, and 4,5-dicaffeoylquinic acid. They concluded that the content of monocaffeoylquinic acids and dicaffeoylquinic acids in the plant was much higher than the content of other compounds, therefore these compounds are suitable for extraction and further use in the pharmaceutical or food industry.

Orally administered caffeoylquinic acids showed a marked anti-inflammatory effect in acute airway inflammation induced by ammonia liquor in mice. Among the four tested compounds, 4,5-dicaffeoylquinic acid significantly inhibited leukocytosis in the bronchoalveolar lavage fluid (Wu et al., 2015).

Moreover, 3,5-dicaffeoylquinic acid inhibited the production of NO in LPS-stimulated RAW 264.7 murine macrophage cells through downregulation of iNOS, COX-2, and TNF-α gene expressions (Hong et al., 2015). Earlier, it has been shown that 3,5-dicaffeoylquinic acid limited LPS-mediated injury in human endothelial cells through inactivation of intracellular ROS and inhibition of caspase-3 activity (Zha et al., 2007).

Toyama et al. (2014) showed that 5-caffeoylquinic acid reduced the edema and myonecrosis induced by secretory phospholipase A2, a snake venom enzyme, similar in structure, function, and pharmacological effects with human secretory phospholipase A2, a known promoter of inflammation.

An extract of caffeoylquinic acids reduced the acute myocardial ischemia-reperfusion injury when administered intravenously to rats in a single bolus 1 min before reperfusion. The extract inhibited the cardiac inflammatory response by decreasing leukocytes infiltration and pro-inflammatory cytokines (TNF-α and IL-6) production in the ischemic tissue. The caffeoylquinic acids also suppressed activation of NF-κB and JNK in cardiomyocytes after reperfusion, thus exhibiting cardioprotective effect (Zhang et al., 2016).

At lower concentrations 3,5-dicaffeoylquinic acid and 4,5-dicaffeoylquinic acid, reduced PGE$_2$ levels, whereas at higher doses they stimulated PGE$_2$ and also TNF-α production by the LPS-stimulated U-937 cells. Both compounds inhibited the production of chemokine CCL7. The authors of the study suggest that stimulation of pro-inflammatory molecules by high concentrations of hydroxycinnamic acids may be due to the perturbation of cellular redox balance, traceable to the catechol group in the structure of caffeoylquinic acids able of exhibiting both anti and pro-oxidant activity (dos Santos et al., 2010).

The anti-inflammatory effect of caffeic and chlorogenic acid has been demonstrated in Caco-2 cells, where both compounds inhibited TNF-α and H_2O_2-induced production of IL-8 (Shin et al., 2015). Chlorogenic acid also reduced the secretion of pro-inflammatory mediators in LPS-activated RAW 264.7 murine macrophage and BV2 microglial cells by downregulating NF-κB and JNK/AP-1 signaling pathways (Shan et al., 2009; Hwang et al., 2014).

Numerous animal studies support the anti-inflammatory properties of chlorogenic acid. Treatment with 5-chlorogenic acid protected against trinitrobenzenesulfonic acid—and dextran sulfate sodium-induced colitis in mice (Shin et al., 2015; Zatorski et al., 2015) and in carrageenan-induced paw edema in rats (Chauhan et al., 2011). Furthermore, 5-chlorogenic acid accelerated wound healing in streptozotocin-induced diabetic rats by reducing the inflammation (Bagdas et al., 2015). The compound also displayed a marked anti-inflammatory effect in a LPS-induced rat knee joint inflammation model and a potent antiarthritic effect in adjuvant-induced arthritis in rats, an autoimmune model of rheumatoid arthritis (Chauhan et al., 2012). Chlorogenic acid significantly decreased CCl_4-induced liver damage and symptoms of liver fibrosis by inhibition of the TLR4/MyD88/NF-κB signaling pathway (Shi et al., 2013a,b).

5.6 OTHER COMPOUNDS WITH ANTI-INFLAMMATORY ACTIVITY

Scopoletin (7-hydroxy-6-methoxycoumarin) was one of the first anti-inflammatory compounds identified in *A. annua* (Huang et al., 1993) and the glucoside, scopolin, was also identified in *A. annua* infusion (van der Kooy and Sullivan, 2013). Additional coumarins found in *A. annua* are coumarin, aesculetin (6,7-dihydroxycoumarin), iso-fraxidin (7-hydroxy-6,8-dimethoxycoumarin), and tomentin (5-hydroxy-6,7-dimethoxycoumarin) (Ferreira et al., 2010).

Numerous studies certify the anti-inflammatory activity of scopoletin. Scopoletin treatment significantly ameliorates croton oil- and carrageenan-induced inflammatory models by mitigating neutrophil infiltration and PGE_2 and TNF-α overproduction (Ding et al., 2008). Scopoletin's mechanism of action relies on reduction of inflammatory mediators via inhibition of NF-κB activation and the MAPK signal pathway (Yao et al., 2012; Lee and Lee, 2015), in parallel with an augmentation of enzymatic and non-enzymatic antioxidant activity (Chang et al., 2012; Jamuna et al., 2015).

Further compounds from *A. annua* might have anti-inflammatory properties, as research carried out on other plants confirmed: phytosterols (Othman and Moghadasian, 2011; Aldini et al., 2014), polyacetylenes (Konovalov, 2014; Le et al., 2015), and components of volatile oil (Souza et al., 2003; Trinh et al., 2011).

5.7 METHODS OF ANALYSIS OF BIOACTIVE COMPOUNDS

The compounds discussed in the previous section belong to the following classes of compounds:

- Sesquiterpenes lactones: artemisinins and biosynthetic precursors—arteannuin B and dihydroartemisinic acid (considered in a previous paper [Ivanescu et al., 2015]).
- Flavones: apigenin, acacetin, artemetin, cirsiliol, cirsilineol, cirsimaritin, chrysoeriol, chrysosplenol, chrysosplenol-C, eupatorin, and luteolin.
- Flavonols: casticin, chrysin, chrysosplenetin, chrysosplenol-D, and eupatin.
- Phenolic acids: chlorogenic acid, quinic acid, and several derivatives.
- Coumarins.

5.8 EXTRACTION AND SEPARATION

In the analysis of active principles from plants, the steps prior to identification and quantification of compounds are important. Some of these steps are: pre-washing, drying directly or freeze-drying, grinding to bring the particles to a given size, and so on. But perhaps the most important step is extraction because constituents must be extracted without loss of weight or structural modifications. To extract the wanted components, the extraction method and suitable solvents must be carefully chosen (Sasidharan et al., 2011).

In the case of *A. annua*, the plant material used may be formed of leaves, flowers, or entire aerial parts. The vegetal products can be used as fresh plant, dried plant, or freeze-dried plant materials (Lai et al., 2007; Baraldi et al., 2008). The drying can be carried out in air at room temperature (Zhu et al., 2013), in an oven at 40°C for 2 days (Iqbal et al., 2012), or in a forced air oven at 45°C for 48 hours (Ferreira et al., 2011; Singh et al., 2011). The samples are then ground and the particle size is generally 2 mm (Ferreira et al., 2011; Singh et al., 2011).

For these compounds coming from *A. annua*, relevant literature provides the analysis after a preliminary extraction in water or in various solvents, with or without further processing, as it will be seen from the following examples. The purpose of extraction in water is obtaining infusions, by treating the dried plant with hot water in different proportions. The extracts are shaken for a certain period of time which varies from a few minutes to days, sometimes in the dark. Other types of aqueous extraction solutions prepared from *A. annua* are decoctions. Among infusions and decoctions, the best result for artemisinin extraction was obtained by preparing an infusion with boiling water for 15 minutes. On the other hand, the quantity of flavonoids extracted in these conditions was low (Bilia et al., 2006a,b).

The low quantity of flavonoids was confirmed by Carbonara et al. (2012) which showed that aqueous extraction produces an extract rich in phenolic acids, but with a low concentration of flavonoids. Extraction should not be approached by boiling, because artemisinin is thermally unstable and its concentration would decrease (Bilia et al., 2006a,b).

Ferreira et al. (2013) have shown that artemisinin drastically degrades during sample preparation by distillation, observing a reduction of 84% after 1.25 minutes, decreasing further and becoming undetectable after 40 minutes. Alternately, by Soxhlet extraction, a decrease of 30% was observed after 40 minutes.

Also, Weathers and Towler (2012) showed that by infusion, artemisinin extraction is efficient, while for polymethoxylated flavonoids, such as casticin, the extracted quantity is very small or even undetectable, as in the case of artemetin. Furthermore, the amount of casticin decreases after storage at room temperature (Weathers and Towler, 2012), while concentration of artemisinin does not significantly decrease for 24 hours (van der Kooy and Verpoorte, 2011).

Most often, the organic solvents are used for extraction. The methods used are various: maceration with various solvents, like dichloromethane or hexane for 72 hours (Bilia et al., 2006a,b; Baraldi et al., 2008). Also, Soxhlet extraction was used by treatment with 70% ethanol at 70°C for 72 hours (Owuna et al. 2013) or by sequentially extracting using petroleum ether, followed by ethanol, and then methanol (Engeu et al. 2015). Ethanol extracts presented higher concentrations of flavonoids than petroleum ether or hexane extracts (Singh et al., 2011). In the case of tinctures, the quantity of extracted compounds is higher for both artemisinin and flavonoids. The ethanol concentration of 60% was more efficient than 40% in extracting the bioactive compounds (Bilia et al., 2006a,b).

If, for flavonoids extraction, various solvents are used, such as 60%, 80%, 100% methanol, 99.7% ethanol, or 100% acetonitrile, it was observed that 60% and 80% methanol are more efficient for the extraction of hydrophilic compounds (such as quercetin and chlorogenic acid) than for the extraction of hydrophobic compounds (acacetin). 100% methanol is efficient for the extraction of all compounds, while ethanol and acetonitrile are less efficient (Lai et al., 2007).

The results obtained by Iqbal et al. (2012) suggest that the amount of extracted phenolic compounds depends on the solvent polarity, increasing with the polarity of the solvent. It was also observed that methanol is one of the best solvents. In addition, water showed a comparable extraction potential to that of methanol, regarding the extraction of phenolic compounds. The same thing can be seen in the case of flavonoids, polar solvents being the most efficient in their extraction, because flavonoids polarity increases by conjugating with glycosides having hydroxyl groups, which augments their solubility in polar solvents. The same remarks were supported by other authors who extracted compounds from different plant species (Spigno et al., 2007; Mohsen and Ammar, 2009).

Other extraction techniques considered promising, simple, and efficient are ultrasonication and microwave extraction. Microwave extraction has been used with good results in case of artemisinin (Bilia et al., 2006a,b). Ultrasonication can be performed for various periods of time, for example 30 minutes at room temperature using pentane for artemisinin extraction or methylene chloride for casticin and artemetin extraction (Weathers and Towler, 2012).

Other methods comprise two steps: a first step of extraction of dried and ground leaves with 70% ethanol, by stirring for 2 hours at 60°C, followed by a step of ultrasonication for 30 minutes. Using 70% ethanol, almost all compounds removable with nonpolar solvents can be extracted (Ferreira et al., 2011; Singh et al., 2011).

Ultrasonic extraction is an alternative for artemisinin extraction, because a temperature of 25°C and a 40 kHz frequency of ultrasound will tear the glands on the surface of the leaves, so the energy required for extraction will be lower, the extraction yield will be higher and the extracts purer (the quantity of co-extracted impurities will be smaller) (Briars and Paniwnyk, 2013).

Because plants contain a mixture of various structural compounds the separation remains a challenge. Thus, most often, after extraction it is necessary to purify the extract by using various techniques such as column chromatography on various stationary phases and eluents, such as thin layer chromatography (TLC) and high-speed counter-current chromatography (HSCCC).

Thus, after extraction in ethanol (100%) at room temperature and concentrating in vacuum at 50°C, Zhu et al. (2013) processed further the residue by way of fractionating with ethyl acetate. The obtained fractions were passed over silica gel eluted with petroleum ether, increasing amounts of ethyl acetate to 6:4 (petroleum ether:ethyl acetate, v/v). The following compounds were isolated: arteannuin B, artemisinic acid, artemisinin (sesquiterpene lactones), casticin, chrysosplenol-D (flavonoids), and 4-ethylesculetin (coumarin).

Another separation method for arteannuin B, artemetin, artemisinic acid, artemisinin, deoxyartemisinin, scopoletin, and identification of casticin and chrysosplenetin was conducted by Anshul et al. (2013). In order to isolate these compounds, the plant material was extracted with boiling n-hexane, and after removal of the solvent, more than 75% of the extract was partitioned with acetonitrile–water (4:1) to remove fatty material. For removal of water, sodium chloride was added to the system acetonitrile–water. After concentration and drying, the acetonitrile extract was passed through a silica gel column, which was eluted with n-hexane–ethyl acetate (95:5, 90:10, 85:15, 80:20, and 70:30) and ethyl acetate.

Bilia et al. (2002) isolated, by column chromatography in several stages, chrysosplenetin, casticin, eupatin, and chrysosplenol-D. Initially the plant was extracted with dichloromethane by percolation at room temperature for 24 hours. A portion of the extract was concentrated and passed over a column of Sephadex type using as eluent, initially, dichloromethane:methanol in 2:1 ratio. By repeating the elution with the same mixture, but at 1:1 ratio, eupatin and chrysosplenol-D were separated. To separate chrysosplenetin and casticin, the remaining part of the extract is passed over a column of silica gel using cyclohexane–dichloromethane:methanol mixtures with increasing polarity.

For the same purpose, the separation of components from *A. annua*, Chougouo et al. (2016) used multiple passes on different chromatographic columns and with different compositions of mobile phases. After obtaining the extract by maceration, this was passed through a silica gel column chromatography, eluted with hexane, hexane–ethyl acetate, ethyl acetate–methanol, and methanol to give 37 fractions. After TLC analysis, the four resulted fractions were purified successively on silica gel (Sephadex) and again on silica gel using an eluent system with hexane–acetone, dichloromethane–methanol, and dichloromethane–acetone. Finally, it separated the components: artemisinin, scopoletin chrysosplenetin, eupatin, and β-sitosterol-3-O-β-D-glucopyranoside.

HSCCC is a new dynamic separation technique. Advantages of this method include: efficient and rapid separation of compounds that, through usual methods, would be difficult to separate, separation of small quantities of compounds, separation of compounds for which the solubility is problematic, high selectivity, and obtaining pure compounds (Sethi et al., 2009).

Thus, for the isolation of high-purity casticin from A. annua, HSCCC can be used for an initial clean-up step. An extract obtained by maceration with 95% ethanol for 7 days was concentrated under reduced pressure. The resulting product was dissolved in water and extracted separately with ethyl acetate and heptane. Dried ethyl acetate extract was cleaned by passing through silica gel, eluted with light petroleum–ethyl acetate mixtures in different proportions: 55:45, 45:55, and 35:65. Next, HSCCC was applied, using the system of solvents n-hexane–ethyl acetate–methanol–water in a 7:10:7:10 (v/v) optimal ratio, at a flow rate of 2.0 ml/min. The quantification was realized by HPLC-DAD with a mobile phase consisting of acetonitrile:water (1% acetic acid) (45:44, v/v) and detection at 254 nm. Casticin recovery was 96.2% at a purity of over 99%, which demonstrated that the method can be successfully used for preparative separation (Han et al., 2007).

5.9 IDENTIFICATION AND QUANTIFICATION

For the identification and quantification of A. annua compounds, most literature data indicate spectroscopic and chromatographic methods.

UV-Vis spectroscopy is one of the traditional methods of analysis, it is simple, fast, and relatively inexpensive. To identify flavones, one of the methods requires the treatment of ethanol extract (70°C, 72 hours) with ammonia solution, followed by the addition of concentrated sulfuric acid to give a yellow color (Owuna et al., 2013). Another way to identify the presence of flavones requires adding metallic magnesium and concentrated hydrochloric acid which gives a red coloration (Engeu et al., 2015).

To determine the total phenolic content, the most used method is Folin–Ciocalteu. The extract is treated with Folin–Ciocalteu reagent (diluted 1:10 fold), followed by the addition of 20% sodium carbonate solution. After 60 minutes of being stored in the dark, absorbance at 650 nm is recorded. The results may be expressed in mg of gallic acid equivalent per g dry weight (Kaur and Kapoor, 2002). Other authors use 7.5% sodium carbonate solution and after 30 minutes at room temperature, the absorbance is recorded at 765 nm (Gouveia and Castilho, 2013).

To determine the total flavonoid content, the most used method implies treating the extract with 5% sodium nitrite, 10% aluminum chloride solution, and 1 mol/l sodium hydroxide solution. Absorbance is read at 510 nm after 15 minutes and the results can be expressed in mg per g dry weight equivalent rutin (Chang et al., 2002).

Starting from the method used by Chen et al. (2010), in order to determine the total flavonoid content expressed in rutin, Engeu et al. (2015) used, for A. annua, 5% sodium nitrite and 10% aluminum nitrate solution with the reaction period of 6 minutes, followed by neutralization of the acidity with a 4% sodium hydroxide solution. The absorbance was recorded after 15 minutes at 510 nm and casticin was used as standard.

Because of their high efficiency, *chromatographic methods* have been imposed widely as methods of separation and analysis. The chromatographic methods used in analyzing compounds in plants vary from simple, quick, and inexpensive methods which provide fast answers on the composition of a sample, such as Thin Layer

Chromatography (TLC) or High-Performance Thin Layer Chromatography (HPTLC) to more advanced methods like High-Performance Liquid Chromatography (HPLC) which are used in fingerprinting studies for the quality control of plants (Fan et al., 2006; Sasidharan et al., 2011). In chromatographic methods, the choice of the column, mobile phase, detector, and type of elution are important.

Data from the relevant literature show that, for the analysis of A. annua compounds, generally, the reversed-phase column is used. The preferred mobile phase systems contain acetonitrile, water, acetic acid, and formic acid, and for type of elution, both gradient and isocratic are used. As combinations HPLC/UV, HPLC/Diode Array Detector (DAD), HPLC/Evaporative Light Scattering Detector (ELSD), Liquid Chromatography coupled with Mass Spectrometry (LC/MS), Liquid Chromatography/Diode Array Detector-Atmospheric Pressure Chemical Ionization/Mass Spectrometry (LC/DADAPCI/MS) and Gas Chromatography coupled with Mass Spectrometry (GC/MS) are used.

When TLC was performed on silica gel using as eluent toluene: acetic acid (4:1), the Rf values of 0.12 for apigenin, 0.36 for acacetin, 0.21 for chrysoeriol, and 0.23 for cirsimaritin were obtained. Another system containing a microcrystalline cellulose layer as the stationary phase and acetic acid:water (3:10) as the mobile phase, gave the Rf values of 0.19 for apigenin and acacetin, 0.08 for chrysoeriol, and 0.46 for cirsimaritin. Using the same stationary phase but with a different proportion of the mobile phase components (acetic acid:water, 1:1) the following Rf values were obtained: 0.49 for apigenin, 0.61 for acacetin, 0.42 for chrysoeriol, and 0.72 for cirsimaritin (Greenham et al., 2003).

An HPTLC method was used for quantification of artemisinin. The mobile phase consisted of heptane–diethyl ether, 1:1 ratio. After derivatization, by placing the plate in acetic acid–sulfuric acid–anisaldehyde (10:0.02:0.1), the absorbance was measured at 366 nm and the Rf value was 0.5 (Engeu et al., 2015).

Greenham et al. (2003) combined chromatography with spectroscopy in order to identify the lipophilic flavones and flavonols. Thus, reverse phase HPLC/DAD was used, at 25°C, with a gradient elution formed of mobile phase A acetic acid:water (1:50) and mobile phase B methanol:acetic acid:water (18:1:1), starting at 2:3 (A:B) and changing to 0:100 in 20 minutes. Spectroscopic characterization was performed by measuring the absorbance in the first version at 365 nm, in the second version at 365 nm after treatment with ammonia vapors and in the third version after treatment with natural product reagent A. Also, UV spectra were recorded in methanol, or by treatment with various reagents such as sodium acetate, boric acid, aluminum chloride, aluminum chloride/hydrochloric acid, and sodium hydroxide. By using these methods several flavones and derivatives were identified, like methyl ethers, 5-deoxyflavones, 5-deoxyflavonols, 5-methoxyflavones, 6 and 8-hydroxyluteolin, and some derivatives of 6,8-dihydroxyapigenin and 6,8-dihydroxyluteolin. We noted the following compounds: apigenin, luteolin, tricetin, acacetin, chrysoeriol, cirsilineol, eupatorin, and chrysin.

Generally, for the identification of flavonoids from various plants, the most useful method is LC/MS. This is due to the fact that, in addition to providing m/z, the method has other advantages: it identifies compounds which are unstable in solution such as the acylated flavonoids, differentiates O-glycosides, C-glycosides, and

O,C-glycosides (LC/MS[n]) (Cuyckens and Claeys, 2004; Lai et al., 2007), and differentiates the position of functional groups, such as hydroxyl and carboxylic groups (Schram et al., 2004; Clifford et al., 2005; Lai et al., 2007).

By combining LC/DAD-APCI/MS both flavonoids and phenolic acids, as caffeoylquinic acids, have been identified. LC/MS analysis was performed on crude methanolic extracts, without pretreatment, using as solvents methanol, ethanol, and acetonitrile. A gradient elution was used, starting from 12% of mobile phase A (acetonitrile) and 88% of mobile phase B (0.1% formic acid at pH 4.0) for 10 minutes, increasing to 25% mobile phase A at 60 minutes, then to 60% mobile phase A at 80 minutes and finally to 100% mobile phase A at 85 minutes. For the MS analysis, an ion trapping mass spectrometer was used, coupled with ESI (Electrospray Ionization) and APCI (Atmospheric Pressure Chemical Ionization) in both positive and negative mode. Using LC/UV analysis at 335 nm, more than 40 components were detected in *A. annua* and from these, 27 have been identified by MS. Some of the identified compounds were: chlorogenic acid, chrysosplenol, chrysosplenol-C, chrysosplenol-D, chrysosplentin, cirsiliol, flaviolin, isorhamnetin, isorhamnetin-Glu, laricitin-Glu, mearnsetin-Glu, pilloin, quercetin, quercetin-Glu, quinic acid, rhamnetin, rhamnetin-Glu, 1,5-dicaffeoylquinic acid, 1,3-dicaffeoylquinic acid, and 3,5-dicaffeoylquinic acid (Lai et al., 2007).

Carbonara et al. (2012) investigated, by HPLC coupled with DAD and ELSD, the phytochemical profile of some *A. annua* infusions, obtained in different periods of time: 1, 24 and 40 hours. Both methods used the same solvent system: mobile phase A, 0.1% formic acid in water, mobile phase B, 0.1% formic acid in acetonitrile. In DAD detection, the pH of mobile phase A was adjusted with sodium hydroxide at pH 4. In ELSD detection, sodium hydroxide was not used due to its low vaporization. In both cases, a gradient elution was used: in ELSD detection 12% B increasing to 20% B at 30 minutes, 25% B at 46 minutes, and 100% B at 66 minutes, and in DAD detection from 12% B increasing to 25% B at 60 minutes, 60% B at 80 minutes, and 100% B at 85 min. DAD detection was made at 210, 270, 310, and 350 nm.

Thus, beside artemisinin other compounds were revealed, such as monocaffeoyl- and monoferuloyl-quinic acids, dicaffeoyl- and diferuloyl-quinic acids, and certain flavonoids. The following compounds were quantified: 3-caffeoylquinic acid, 5-caffeoylquinic acid, 4-caffeoylquinic acid, 3-feruloylquninic acid, tetramethoxyflavanone, 4-feruloylquinic acid, caffeic acid, 5-feruloylquininc acid, 6-C-arabinosyl-8-C-glucosyl apigenin, 6-C-glucosyl-8-C-arabinosyl apigenin, trimethoxycoumarin, chrysoeriol rutinoside, vitexin (8-C-glucosyl apigenin), patuletinglycoside, isovitexin (6-C-glucosyl apigenin), 3,4-dicaffeoylquinic acid, 3,5-dicaffeoylquinic acid, luteolin-7-O-glucoside, 4,5-dicaffeoylquinic acid, 3,4-diferuloylquinic acid, 3,5-diferuloylquinic acid, 4,5-diferuloylquinic acid, 3,5-caffeoylferuloylquinic acid, 4-caffeoyl-3,5-disuccinoylquinic acid, eriodictyol, jaceidin, and cirsilineol.

The content in phenolic compounds and flavonoids was different depending on the time of infusion, and in some cases increased with time of extraction (e.g., caffeoylquinic acids), while in other cases decreased (e.g., tetramethoxyflavanone). In some cases, the concentration decreased at 24 hours and then increased at 48 hours (e.g., chrysoeriol rutinoside), or increased at 24 hours and then decreased at 48 hours (e.g., 4-feruloylquinic acid).

Starting from Carbonara's method, Subero et al. (2013a,b) have made some modifications: the two mobile phases A and B contained acetic acid instead of formic acid and the gradient was: 0–60 minutes: 12–25% B, 60–80 minutes: 25–60% B, 80–85 minutes: 60–100% B. Thus, caffeic acid and its derivatives were quantified. Among them, those found in the greatest amounts were 3-caffeoylquinic acid, followed by 3,5-dicaffeoylquinic acid, 4,5-dicaffeoylquinic acid, 4-caffeoylquinic acid, and 5-caffeoylquinic acid. Isovitexin was found in large amounts and rosmarinic acid in small amounts.

In order to analyze the metabolites profile of *A. annua* tea, Suberu et al. (ibid) used HPLC/MS. Artemisinin was determined by MS/MS using ESI as an ionization source, in positive mode ionization. The chromatographic conditions were: gradient elution, mobile phase A consisting of 0.1% acetic acid in water and mobile phase B consisting of 0.1% acetic acid in acetonitrile.

In the case of simultaneous detection of artemisinin and flavonoids, HPLC/DAD/MS can be utilized. Extracts obtained by maceration in dichloromethane and hexane were separated by column, using isocratic elution, 50% mobile phase A (water adjusted to pH 3.2 with formic acid) and 50% mobile phase B (acetonitrile). For flavonoids, DAD detection was performed at 280 nm and for artemisinin at 210 nm. In case of MS detection, API (atmospheric pressure ionization) electrospray in the positive ion mode was used. Artemetin, casticin, cirsilineol, chrysosplenetin, chrysosplenol-D, and eupatin were revealed in this way. Hexane extracts showed a greater amount of artemisinin, while dichloromethane extracts had a higher content of flavonoids (Bilia et al., 2006a,b).

Also, isocratic elution with a mobile phase consisting of acetonitrile: 0.1% acetic acid (50:50) was used to determine artemisinin, artemisinic acid, and arteannuin B by HPLC/UV/ELSD, obtaining a good resolution (Zhang et al., 2007).

The same chromatographic conditions were used by Baraldi et al. (2008) for the analysis of hexane extracts. Flavonoids have been quantified in UV and artemisin by MS detection. The total amount of flavonoids (casticin, chrysosplenetin, eupatin, and artemetin) and artemisinin reached higher values in aerial parts during full blooming compared to pre- and post-flowering periods. Artemisinin was found in greater quantity in leaves than in the inflorescences. Flavones were found in equal amounts in both parts of the plant.

Another combination of methods included UPLC/DAD-MS/MS. Gradient elution used acetonitrile and ammonium acetate buffer, pH 2.6. Artemisinin, rosmarinic acid, chlorogenic acid, scopoletin, and isoquercitrin were identified by this method (Melillo de Magalhães et al., 2012).

To determine casticin and artemetin by GC/MS, Weathers and Towler (2012) used derivatization by adding pyridine, BSTFA-N,O-bis(trimethylsilyl)trifluoroacetamide, and TMCS-trimethylsilyl chloride, heating at 70°C for 4 hours. Chromatographic conditions included: Helium (He) as carrier gas at 1 mL/min, injection volumes in splitless mode, ion source temperature at 275°C, and oven temperature increased from 110°C at 280°C. In case of artemisinin, derivatization was not necessary.

In Table 5.2, data are presented obtained by mass spectrometry, LC/UV, and UV spectroscopy, which can be used to identify the previously mentioned compounds.

Table 5.2   Mass spectra, LC-UV, and UV spectra data of different compounds from *A. annua*

Compound	Chemical name	Mass spectramolecular ion/fragmental ions	LC-UVλ_{max} (nm)	UV spectraλ_{max} (nm)	Ref.
Acacetin	Apigenin-4'-methyl ether,5,7-dihydroxy-4-methoxy flavone	285.2 (APCI-MS)	267, 334	269, 325 (methanol)	Greenham et al., 2003; Lai et al., 2007; Ferreira et al., 2011
Apigenin	4',5,7-Trihydroxyflavone	271.2 (APCI-MS)	238, 298, 334[a] 267, 338[b]	269, 335 (methanol)	Greenham et al., 2003[b]; Lai et al., 2007[a]
Artemetin	Quercetagetin 3,6,7,3',4'-pentamethyl ether, 5-hydroxy-3,3',4',6,7-pentamethoxy-flavone	389 (API-electrospray)			Bilia et al., 2006a,b
Caffeoylquinic acids	3-Caffeoylquinic acid	353/191, 179, 173, 135 (ESI-MS/MS)	241, 300, 324		Carbonara et al., 2012; Gouveia and Castilho, 2013
	5-Caffeoylquinic acid	353/191, 179, 135 (ESI-MS/MS)	242, 300, 325		Carbonara et al., 2012; Gouveia and Castilho, 2013
	4-Caffeoylquinic acid	353/191, 179, 173, 135 (ESI-MS/MS)	240, 300, 323		Carbonara et al., 2012; Gouveia and Castilho, 2013
Feruloylquinic acids	3-Feruloylquinic acid	367/193, 191, 173 (ESI-MS/MS)	326		Carbonara et al., 2012; Gouveia and Castilho, 2013
	5-Feruloylquinic acid	367/193, 191, 173 (ESI-MS/MS)	327		Carbonara et al., 2012; Gouveia and Castilho, 2013
	4-Feruloylquinic acid	367/193, 191, 173 (ESI-MS/MS)			Carbonara et al., 2012
Dicaffeoylquinic acids	1,5-Dicaffeoylquinic acid	515/353	243, 300, 328		Gouveia and Castilho, 2013
	3,5-Dicaffeoylquinic acid	515/353, 191 (ESI-MS/MS)	218, 241, 328		Fraisse et al., 2011; Carbonara et al., 2012
	4,5-Dicaffeoylquinic acid	515/353, 191, 179, 173 (ESI-MS/MS)	217, 241, 326		Fraisse et al., 2011; Carbonara et al., 2012

(Continued)

Table 5.2 (Continued)   Mass spectra, LC-UV, and UV spectra data of different compounds from *A. annua*

Compound	Chemical name	Mass spectramolecular ion/fragmental ions	LC-UVλ_{max} (nm)	UV spectraλ_{max} (nm)	Ref.
	3,4-Dicaffeoylquinic acid	515/353, 191, 179, 173 (ESI-MS/MS)	246, 299, 325		Carbonara et al., 2012; Gouveia and Castilho, 2013
Diferuloylquinic acid	3,4-Diferuloylquinic acid	543/367, 349, 193, 173 (ESI-MS/MS)	543		Carbonara et al., 2012; Gouveia and Castilho, 2013
	3,5-Diferuloylquinic acid	543/367, 349, 193, 173 (ESI-MS/MS)	543		Carbonara et al., 2012; Gouveia and Castilho, 2013
	4,5-Diferuloylquinic acid	543/367, 349, 193, 173 (ESI-MS/MS)			Carbonara et al., 2012
Casticin	5,3'-Dihydroxy-3,6,7,4'-tetramethyl ether flavone, quercetagetin-3,6-7,4'-tetramethyl ether	375.1 (ESI-MS)[a] 375 (API-electrospray)[b]		257, 271, 350	Bilia et al., 2006a,b[b]; Han et al., 2007[a]; Ferreira et al., 2011
Chlorogenic acid		354.9/193.1 (APCI-MS)	226, 288, 336[a] 217, 238, 325[b]		Lai et al., 2007[a]; Fraisse et al., 2011[b]
Chrysin	5,7-Dihydroxy flavone	315.2, 300.1, 282.0, 254.1 (APCI-MS)			Lai et al., 2007
Chrysoeriol	Luteolin-3'-methyl ether, 5,7,4'-trihydroxy-3'-methoxy flavones		252, 267[sh], 348	250, 270, 346	Greenham et al., 2003; Ferreira et al., 2011
Chrysosplenetin	5,4'-Dihydroxy-3,6,7,3'-tetramethoxy flavone, quercetagetin-3,6-7,3'-tetramethyl ether	375.2/359.2, 342.1, 317.2 (APCI-MS)[a] 375 (API-electrospray)[b]	225, 278, 350		Bilia et al., 2006a,b[b]; Lai et al., 2007[a]; Ferreira et al., 2011
Chrysosplenol		345.2/331.2/315.1 (APCI-MS)	228, 280, 348		Lai et al., 2007
Chrysosplenol-C	Quercetagetin 3,7,3'-trimethyl ether	361.2/345.2, 303.1 (APCI-MS)	236, 258, 350		Lai et al., 2007
Chrysosplenol-D	Quercetagetin 3,6,7-trimethyl ether	361.2/346.2, 303.1 (APCI-MS)	230, 260, 345		Lai et al., 2007

(Continued)

Table 5.2 (Continued)   Mass spectra, LC-UV, and UV spectra data of different compounds from *A. annua*

Compound	Chemical name	Mass spectra molecular ion/fragmental ions	LC-UV λ_{max} (nm)	UV spectra λ_{max} (nm)	Ref.
Cirsilineol	6-Hydroxyluteolin-6,7,3'-trimethyl ether, 5,4'-dihydroxy-6,7,3'-trimethoxyflavone	345 (API-electrospray)	255, 270, 345		Greenham et al., 2003; Bilia et al., 2006a,b; Ferreira et al., 2011
Cirsiliol	3',4',5-Trihydroxy-6,7-dimethoxy-flavone	331.2/316.2, 301.2, 271.1 (APCI-MS)[a] 343/329, 313 (ESI-MS/MS)[b]	239, 296, 335		Lai et al., 2007; Carbonara et al., 2012
Cirsimaritin	6-Hydroxyapigenin-6,7-dimethyl ether, scutellarin-6,7-dimethyl ether, 5,4'-dihydroxy-6,7-dimethoxyflavone		275, 335	276, 333	Greenham et al., 2003; Ferreira et al., 2011
Eupatin	3,5,3'-Trihydroxy-6,7,4'-trimethoxyflavone, quercetagetin-3,6-dimethyl ether	361 (API-electrospray)			Bilia et al., 2006a,b; Ferreira et al., 2011
Eupatorin	6-Hydroxyluteolin-6,7,4'-trimethyl ether, 5,3'-dihydroxy-6,7,4'-trimethoxyflavone		252, 274, 344		Greenham et al., 2003; Ferreira et al., 2011
Luteolin	5,7,3',4'-Tetrahydroxy flavone	287.2 (APCI-MS)	244, 298, 328[a] 254, 267[sh], 350[b]	252, 260, 347	Greenham et al., 2003; Lai et al., 2007
Quinic acid		193.1/178.1[a] (APCI-MS) 191/173, 127 (ESI-MS/MS)[b]	204, 228, 344		Lai et al., 2007; Carbonara et al., 2012

[a] results obtained by Lai et al., 2007.
[b] results obtained by Carbonara et al., 2012.
[sh] shoulder.

5.10 CONCLUSIONS

Chronic inflammation is involved in many pathological conditions such as rheumatoid arthritis, atherosclerosis, type 2 diabetes, asthma, obesity, inflammatory bowel diseases, neurodegenerative diseases, and cancer. Currently, there is a demand for effective anti-inflammatory drugs with low toxicity and reduced side effects. In this context, *A. annua* appears to be a promising source of anti-inflammatory compounds, as numerous studies on different models of inflammation certify. Further research is needed in order to understand the underlying mechanism of action for each compound. The anti-inflammatory activity of extracts is also promising, considering the fact they have been demonstrated to be more efficient than isolated compounds, although they raise the problem of reproducibility and standardization. In addition, no clinical trials were yet undertaken to assert the anti-inflammatory effect of *A. annua* in the human body.

REFERENCES

Ahmed, A.U. 2011. An overview of inflammation: Mechanism and consequences. *Front Biol* 6:274–81.

Aldini R., Micucci M., Cevenini M., et al. 2014. Antiinflammatory effect of phytosterols in experimental murine colitis model: Prevention, induction, remission study. *PLoS ONE* 9:e108112. (doi:10.1371/journal.pone.0108112)

Anshul N., Bhakuni R.S., Gaur R., and Singh D. 2013. Isomeric flavonoids of *Artemisia annua* (Asterales: Asteraceae) as insect growth inhibitors against *Helicoverpa armigera* (Lepidoptera: Noctuidae). *Fla Entomol* 96:897–903.

Baek H.K., Kim P.S., Song J.A., et al. 2016. Neuronal maturation in the hippocampal dentate gyrus via chronic oral administration of *Artemisia annua* extract is independent of COX-2 signaling pathway in the diet-induced obesity (DIO) mice model. *J Vet Sci* [Epub ahead of print] 18:119–127.

Baek H.K., Shim H., Lim H., et al. 2015. Anti-adipogenic effect of *Artemisia annua* in diet-induced-obesity mice model. *J Vet Sci* 16:389–96.

Bagdas D., Etoz B.C., Gul Z., et al. 2015. In vivo systemic chlorogenic acid therapy under diabetic conditions: Wound healing effects and cytotoxicity/genotoxicity profile. *Food Chem Toxicol* 81:54–61.

Baraldi R., Isacchi B., Predieri S., Marconi G., Vincieri F.F., and Bilia A.R. 2008. Distribution of artemisinin and bioactive flavonoids from *Artemisia annua* L. during plant growth. *Biochem Syst Ecol* 36:340–8.

Bilia A.R., Gabriele C., Bergonzi M.C., Melillo de Malgalhaes P., and Vincieri F.F. 2006b. Variation in artemisinin and flavonoid content in different extracts of *Artemisia annua* L. *Nat Prod Commun* 1:1–5.

Bilia A.R., Lazari D., Messori L., Taglioli V., Temperini C., and Vincieri F.F. 2002. Simple and rapid physico-chemical methods to examine action of antimalarial drugs with hemin: Its application to *Artemisia annua* constituents. *Life Sci* 70:769–78.

Bilia A.R., Melillo de Malgalhaes P., Bergonzi M.C., and F.F. Vincieri. 2006a. Simultaneous analysis of artemisinin and flavonoids of several extracts of *Artemisia annua* L. obtained from a commercial sample and a selected cultivar. *Phytomedicine* 13:487–93.

Briars R., and Paniwnyk L. 2013. Effect of ultrasound on the extraction of artemisinin from *Artemisia annua*. *Ind Crops Prod* 42:595–600.

Cao Q., Jiang Y., Shi J., et al. 2015. Artemisinin inhibits the proliferation, migration, and inflammatory reaction induced by tumor necrosis factor-a in vascular smooth muscle cells through nuclear factor kappa B pathway. *J Surg Res* 194:667–78.

Carbonara T., Pascale R., Argentieri M.P., et al. 2012. Phytochemical analysis of a herbal tea from *Artemisia annua* L. *J Pharm Biomed Anal* 62:79–86.

Chagas-Paula D.A., Oliveira T.B., Faleiro D.P., Oliveira R.B., and Costa F.B. 2015. Outstanding anti-inflammatory potential of selected Asteraceae species through the potent dual inhibition of cyclooxygenase-1 and 5-lipoxygenase. *Planta Med* 81:1296–307.

Chang C., Yang M., Wen H., and Chern J. 2002. Estimation of total flavonoid content in propolis by two complementary colorimetric methods. *J Food Drug Anal* 10:178–82.

Chang T.N., Deng J.S., Chang Y.C., et al. 2012. Ameliorative effects of scopoletin from *Crossostephium chinensis* against inflammation pain and its mechanisms in mice. *Evid Based Complement Alternat Med* 595603. (doi:10.1155/2012/595603)

Chauhan P.S., Satti N.K., Sharma V.K., Dutt P., Avtar K., and Bani D. 2011. Amelioration of inflammatory responses by chlorogenic acid via suppression of pro-inflammatory mediators. *J Appl Pharm Sci* 1:67–75.

Chauhan P.S., Satti N.K., Sharma P., Sharma V.K., Suri K.A., and Bani S. 2012. Differential effects of chlorogenic acid on various immunological parameters relevant to rheumatoid arthritis. *Phytother Res* 26:1156–65.

Chen Y., Wang J., and Wan D. 2010. Determination of total flavonoids in three *Sedum* crude drugs by UV–Vis spectrophotometry. *Pharmacogn Mag* 6:259–63.

Cheng C., Ho W.E., Goh F.Y., et al. 2011. Anti-malarial drug artesunate attenuates experimental allergic asthma via inhibition of the phosphoinositide 3-kinase/Akt pathway. *PLoS ONE* 6:e20932. (doi:10.1371/journal.pone.0020932)

Cheng C., Ng D.S., Chan T.K., et al. 2013. Anti-allergic action of anti-malarial drug artesunate in experimental mast cell-mediated anaphylactic models. *Allergy* 68:195–203.

Chiou W.F., Chen C.C., and Wei B.L. 2011. 3,4-Di-O-caffeoylquinic acid inhibits angiotensin-II-induced vascular smooth muscle cell proliferation and migration by downregulating the JNK and PI3K/Akt signaling pathways. *Evid Based Complement Alternat Med* 634502. (doi:10.1093/ecam/nep140)

Choudhary M.I., Azizuddin J.S., Nawz S.A., Khan K.M., Tareen R.B., and Atta-ur-Rahman. 2009. Anti-inflammatory and lipoxygenase inhibitory compounds from *Vitex agnus-castus*. *Phytother Res* 23:1336–9.

Chougouo R.D.K., Nguekeu Y.M.M., Dzoyem J.P., et al. 2016. Anti-inflammatory and acetylcholinesterase activity of extract, fractions and five compounds isolated from the leaves and twigs of *Artemisia annua* growing in Cameroon. *SpringerPlus* 5:1525. (doi:10.1186/s40064-016-3199-9)

Clifford M.N., Knight S., and Kuhnert N. 2005. Discriminating between the six isomers of dicaffeoylquinic acid by LC-MS(n). *J Agric Food Chem* 53:3821–32.

Cuyckens F. and Claeys M. 2004. Mass spectrometry in the structural analysis of flavonoids. *J Mass Spectrom* 39:1–15.

Cuzzocrea S., Saadat F., Di Paola R., et al. 2005. Artemether: A new therapeutic strategy in experimental rheumatoid arthritis. *Immunopharmacol Immunotoxicol* 27:615–30.

Ding Z., Dai Y., Hao H., Pan R., Yao X., and Wang Z. 2008. Anti-inflammatory effects of scopoletin and underlying mechanisms. *Pharm Biol* 46:854–60.

dos Santos M.D., Martins P.R., Dos Santos P.A., Bortocan R., Iamamoto Y., and Lopes N.P. 2005. Oxidative metabolism of 5-o-caffeoylquinic acid (chlorogenic acid), a bioactive natural product, by metalloporphyrin and rat liver mitochondria. *Eur J Pharm Sci* 26:62–70.

dos Santos M.D., Chen G., Almeida M.C., et al. 2010. Effects of caffeoylquinic acid derivatives and *C*-flavonoid from Lychnophora ericoides on in vitro inflammatory mediator production. *Nat Prod Commun* 5:733–40.

Dou W., Zhang J., Li H., et al. 2014. Plant flavonol isorhamnetin attenuates chemically induced inflammatory bowel disease via a PXR-dependent pathway. *J Nutr Biochem* 25:923–33.

Engeu P.O., Omujal F., Agwaya M., Kyakulaga H., and Obua C. 2015. Variations in antimalarial components of *Artemisia annua* Linn from three regions of Uganda. *Afr Health Sci* 15:828–34.

Fan X.H., Cheng Y.Y., Ye Z.L., Lin R.C., and Qian Z. 2006. Multiple chromatographic fingerprinting and its application to the quality control of herbal medicines. *Anal Chim Acta* 555:217–24.

Favero F.F., Grando R., Nonato F.R., et al. 2014. *Artemisia annua* L.: Evidence of sesquiterpene lactones' fraction antinociceptive activity. *BMC Complement Altern Med J* 28:266. (http://www.biomedcentral.com/1472-6882/14/266)

Ferreira J.F.S., Luthria D.L., Sasaki T., and Heyerick A. 2010. Flavonoids from *Artemisia annua* L. as antioxidants and their potential synergism with artemisinin against malaria and cancer. *Molecules* 15:3135–70.

Ferreira J.F.S., Peaden P., and Keiser J. 2011. In vitro trematocidal effects of crude alcoholic extracts of *Artemisia annua*, *A. absinthium*, *Asimina triloba*, and *Fumaria officinalis*. *Parasitol Res* 109:1585–92.

Ferreira J.F.S., Zheljazkov, V.D., and Gonzalez J.M. 2013. Artemisinin concentration and antioxidant capacity of *Artemisia annua* distillation by product. *Ind Crops Prod* 41:294–8.

Fraisse D., Felgines C., Texier O., and Lamaison J.L. 2011. Caffeoyl derivatives: Major antioxidant compounds of some wild herbs of the Asteraceae family. *Food Nutr Sci* 2:181–92.

Gouveia S.C., and Castilho P.C. 2013. *Artemisia annua* L.: Essential oil and acetone extract composition and antioxidant capacity. *Ind Crops Prod* 45:170–81.

Greenham J., Harborne J.B., and Williams C.A. 2003. Identification of lipophilic flavones and flavonols by comparative HPLC, TLC and UV spectral analysis. *Phytochem Anal* 14:100–18.

Han J., Miyamae Y., Shigemori H., and Isoda H. 2010. Neuroprotective effect of 3,5-di-O-caffeoylquinic acid on SH-SY5Y cells and senescence-accelerated-prone mice 8 through the up-regulation of phosphoglycerate kinase-1. *Neuroscience* 169:1039–45.

Han X., Ma X., Zhang T., Zhang Y., Liu Q., and Ito Y. 2007. Isolation of high-purity casticin from *Artemisia annua* L. by high-speed counter-current chromatography. *J Chromatogr A* 1151:180–2.

Hassanin A.A., and Shata A. 2014. Effect of artesunate on atherosclerosis in experimentally induced nephrotic syndrome in rats. *Indian J Appl Res* 4:21–4.

He Y., Fan J., Lin H., et al. 2011. The anti-malaria agent artesunate inhibits expression of vascular endothelial growth factor and hypoxia-inducible factor-1α in human rheumatoid arthritis fibroblast-like synoviocyte. *Rheumatol Int* 31:53–60.

Ho W.E., Peh H.Y., Chan T.K., and Wong W.S. 2014. Artemisinins: Pharmacological actions beyond anti-malarial. *Pharmacol Ther* 142:126–39.

Hong S., Joo T., and Jhoo J.W. 2015. Antioxidant and anti-inflammatory activities of 3,5-dicaffeoylquinic acid isolated from *Ligularia fischeri* leaves. *Food Sci Biotechnol* 24:257–63.

Hou L.F., He S.J., Li X., et al. 2011. Oral administration of artemisinin analog SM934 ameliorates lupus syndromes in MRL/lpr mice by inhibiting Th1 and Th17 cell responses. *Arthritis Rheum* 63:2445–55.

Hou L.F., He S.J., Li X., et al. 2012. SM934 treated lupus-prone NZB x NZW F1 mice by enhancing macrophage interleukin-10 production and suppressing pathogenic T cell development. *PLoS ONE* 7:e32424. (doi:10.1371/journal.pone.0032424)

Hu D., Wang Y., Chen Z., et al. 2014. Artemisinin protects against dextran sulfate-sodium-induced inflammatory bowel disease, which is associated with activation of the pregnane X receptor. *Eur J Pharmacol* 5:273–84.

Huang C., Xie Z., Liu F., et al. 2014. Dihydroartemisinin inhibits activation of the Toll-like receptor 4 signaling pathway and production of type I interferon in spleen cells from lupus-prone MRL/lpr mice. *Int Immunopharmacol* 22:266–72.

Huang L., Liu J.F., Liu L.X., et al. 1993. Antipyretic and anti-inflammatory effects of *Artemisia annua* L. *Zhongguo Zhong Yao Za Zhi* 18:44–8.

Hunt S., Yoshida M., Davis C.E.J., Greenhill N.S., and Davis P.F. 2015. An extract of the medicinal plant *Artemisia annua* modulates production of inflammatory markers in activated neutrophils. *J Inflamm Res* 8:9–14.

Hwang S.J., Kim Y.W., Park Y., Lee H.J., and Kim K.W. 2014. Anti-inflammatory effects of chlorogenic acid in lipopolysaccharide-stimulated RAW264.7 cells. *Inflamm Res* 63:81–90.

Iqbal S., Younas U., Wei Chan K., Zia-Ul-Haq M., and Ismail M. 2012. Chemical composition of *Artemisia annua* L. leaves and antioxidant potential of extracts as a function of extraction solvents. *Molecules* 17:6020–32.

Ivanescu B., Miron A., and Corciova A. 2015. Sesquiterpene lactones from *Artemisia* genus: Biological activities and methods of analysis. *J Anal Methods Chem.* (doi:10.1155/2015/247685)

Jamuna S., Karthika K., Paulsamy S., Thenmozhi K., Kathiravan S., and Venkatesh R. 2015. Confertin and scopoletin from leaf and root extracts of *Hypochaeris radicata* have anti-inflammatory and antioxidant activities. *Ind Crops Prod* 70:221–230.

Jiang W., Li B., Zheng X., et al. 2011. Artesunate in combination with oxacillin protect sepsis model mice challenged with lethal live methicillin-resistant *Staphylococcus aureus* (MRSA) via its inhibition on proinflammatory cytokines release and enhancement on antibacterial activity of oxacillin. *Int Immunopharmacol* 11:1065–73.

Jin O., Zhang H., Gu Z., et al. 2009. A pilot study of the therapeutic efficacy and mechanism of artesunate in the MRL/lpr murine model of systemic lupus erythematosus. *Cell Mol Immunol* 6:461–7.

Kaur C., and Kapoor H.C. 2002. Antioxidant activity and total phenolic content of some Asian vegetables. *Int J Food Sci Technol* 37:153–61.

Kim H.G., Yang J.H., Han E.H., et al. 2013. Inhibitory effect of dihydroartemisinin against phorbol ester-induced cyclooxygenase-2 expression in macrophages. *Food Chem Toxicol* 56:93–9.

Kim K.E., Ko K.H., Heo R.W., et al. 2016. *Artemisia annua* leaf extract attenuates hepatic steatosis and inflammation in high-fat diet-fed mice. *J Med Food* 19:290–9.

Kim W.S., Choi W.J., Lee S., et al. 2015. Anti-inflammatory, antioxidant and antimicrobial effects of artemisinin extracts from *Artemisia annua* L. *Korean J Physiol Pharmacol* 19:21–7.

Koh D.J., Ahn H.S., Chung H.S., et al. 2011. Inhibitory effects of casticin on migration of eosinophil and expression of chemokines and adhesion molecules in A549 lung epithelial cells via NF-kappaB inactivation. *J Ethnopharmacol* 136:399–405.

Konovalov D.A. 2014. Polyacetylene compounds of plants of the *Asteraceae* family (review). *Pharm Chem J* 48:613–31.

Laavola M., Nieminen R., Yam M.F., et al. 2012. Flavonoids eupatorin and sinensetin present in *Orthosiphon stamineus* leaves inhibit inflammatory gene expression and STAT1 activation. *Planta Med* 78:779–86.

Lai J.P., Lim Y.H., Su J., Shen H.M., and Ong C.N. 2007. Identification and characterization of major flavonoids and caffeoylquinic acids in three Compositae plants by LC/DAD-APCI/MS. *J Chromatogr B Analyt Technol Biomed Life Sci* 848:215–25.

Lai L., Chen Y., Tian X., et al. 2015. Artesunate alleviates hepatic fibrosis induced by multiple pathogenic factors and inflammation through the inhibition of LPS/TLR4/NF-κB signaling pathway in rats. *Eur J Pharmacol* 765:234–41.

Le J., Lu W., Xiong X., et al. 2015. Anti-inflammatory constituents from *Bidens frondosa*. *Molecules* 20:18496–510.

Lee H.I., and Lee M.K. 2015. Coordinated regulation of scopoletin at adipose tissue-liver axis improved alcohol-induced lipid dysmetabolism and inflammation in rats. *Toxicol Lett* 237:210–8.

Lee K.P., Park E.S., Kim D.E., Park I.S., Kim J.T., and Hong H. 2014. Artemisinin attenuates platelet-derived growth factor BB-induced migration of vascular smooth muscle cells. *Nutr Res Pract* 8:521–25.

Lee S.H., Cho Y.C., Kim K.H., Lee I.S., Choi H.J., and Kang B.Y. 2015. Artesunate inhibits proliferation of naïve CD4(+) T cells but enhances function of effector T cells. *Arch Pharm Res* 38:1195–203.

Li B., Li J., Pan X., et al. 2010. Artesunate protects sepsis model mice challenged with *Staphylococcus aureus* by decreasing TNF-α release via inhibition TLR2 and Nod2 mRNA expressions and transcription factor NF-κB activation. *Int Immunopharmacol* 10:344–50.

Li B., Zhang R., Li J., et al. 2008. Antimalarial artesunate protects sepsis model mice against heat-killed *Escherichia coli* challenge by decreasing TLR4, TLR9 mRNA expressions and transcription factor NF-κB activation. *Int Immunopharmacol* 8:379–89.

Li T., Chen H., Wei N., et al. 2012. Anti-inflammatory and immunomodulatory mechanisms of artemisinin on contact hypersensitivity. *Int Immunopharmacol* 12:144–50.

Li T., Chen H., Yang Z., Liu X.G., Zhang L.M., and Wang H. 2013a. Evaluation of the immunosuppressive activity of artesunate in vitro and in vivo. *Int Immunopharmacol* 16:306–12.

Li Y., and Shen Y. 2016. Casticin attenuates rheumatoid arthritis through PKC-NF-κB signaling in vitro and in vivo. *Int J Clin Exp Pathol* 9:2879–87.

Li Y., Wang S., Wang Y., et al. 2013b. Inhibitory effect of the antimalarial agent artesunate on collagen-induced arthritis in rats through nuclear factor kappa B and mitogen-activated protein kinase signaling pathway. *Transl Res* 161:89–98.

Li Y.J., Guo Y., Yang Q., et al. 2015. Flavonoids casticin and chrysosplenol D from *Artemisia annua* L. inhibit inflammation in vitro and in vivo. *Toxicol Appl Pharmacol* 286:151–8.

Liang N., and Kitts D.D. 2016. Role of chlorogenic acids in controlling oxidative and inflammatory stress conditions. *Nutrients* 8:16. (doi:10.3390/nu8010016)

Liao Y.R., and Lin J.Y. 2015. Quercetin intraperitoneal administration ameliorates lipopolysaccharide-induced systemic inflammation in mice. *Life Sci* 137:89–97.

Lin S., Zhang H., Han T., et al. 2007. In vivo effect of casticin on acute inflammation. *J Integr Med* 5:573–6.

Lin X., Lin C.H., Zhao T., et al. 2017. Quercetin protects against heat stroke-induced myocardial injury in male rats: Antioxidative and antiinflammatory mechanisms. *Chem Biol Interact*, In Press, Accepted Manuscript. 265: 47–54. (doi:10.1016/j.cbi.2017.01.006)

Lin Z.M., Yang X.Q., Zhu F.H., He S.J., Tang W., and Zuo J.P. 2016. Artemisinin analogue SM934 attenuates collagen-induced arthritis by suppressing T follicular helper cells and T helper 17 cells. *Sci Rep* 6:38115. (http://www.nature.com/articles/srep38115)

Liou C.J., Len W.B., Wu S.J., Lin C.F., Wu X.L., and Huang W.C. 2014. Casticin inhibits COX-2 and iNOS expression via suppression of NF-κB and MAPK signaling in lipopolysaccharide-stimulated mouse macrophages. *J Ethnopharmacol* 158:310–6.

Liu C.M., Ma J.Q., Xie W.R., et al. 2015. Quercetin protects mouse liver against nickel-induced DNA methylation and inflammation associated with the Nrf2/HO-1 and p38/STAT1/NF-κB pathway. *Food Chem Toxicol* 82:19–26.

Ma J.Q., Li Z., Xie W.R., Liu C.M., and Liu S.S. 2015. Quercetin protects mouse liver against CCl$_4$-induced inflammation by the TLR2/4 and MAPK/NF-κB pathway. *Int Immunopharmacol* 28:531–9.

Magalhães P.M., Dupont I., Hendrickx A., et al. 2012. Anti-inflammatory effect and modulation of cytochrome P450 activities by *Artemisia annua* tea infusions in human intestinal Caco-2 cells. *Food Chem* 134:864–71.

Medzhitov R. 2008. Origin and physiological roles of inflammation. *Nature* 454:428–35.

Mirshafiey A., Saadat F., Attar M., et al. 2006. Design of a new line in treatment of experimental rheumatoid arthritis by artesunate. *Immunopharmacol Immunotoxicol* 28:397–410.

Mo H.Y., Wang L.F., and Zhang L.H. 2012. Effects of artesunate on tumor necrosis factor alpha and chemotactic factors in the serum and the synoviocyte culture supemate of collagen-induced arthritis rats. *Zhongguo Zhong Xi Yi Jie He Za Zhi* 32:253–6.

Mohsen S.M., and Ammar A.S.M. 2009. Total phenolic contents and antioxidant activity of corn tassel extracts. *Food Chem* 112:595–8.

Othman R.A., and Moghadasian M.H. 2011. Beyond cholesterol-lowering effects of plant sterols: Clinical and experimental evidence of anti-inflammatory properties. *Nutr Rev* 69:371–82.

Owuna G., Mustapha A.A., Ogbonna C.I.C., and Kaladi P.H. 2013. Antimicrobial effects and phytoconstituents of ethanolic extract of leaves of *Artemisia annua* L. *J Med Plants Stud* 1:97–101.

Rasoanaivo P., Wright C.W., Willcox M.L., and Gilbert B. 2011. Whole plant extracts versus single compounds for the treatment of malaria: Synergy and positive interactions. *Malaria J* 10(Suppl 1):S4.

Rasul A., Zhao B.J., Liu J., et al. 2014. Molecular mechanisms of casticin action: An update on its antitumor functions. *Asian Pac J Cancer Prev* 15:9049–58.

Räth K., Taxis K., Walz G., Gleiter C.H., Li S.M., and Heide L. 2004. Pharmacokinetic study of artemisinin after oral intake of a traditional preparation of *Artemisia annua* L. (annual wormwood). *Am J Trop Med Hyg* 70:128–32.

Sasidharan S., Chen Y., Saravanan D., Sundram K.M., and Yoga Latha L. 2011. Extraction, isolation and characterization of bioactive compounds from plants' extracts. *Afr J Tradit Complement Altern Med* 8:1–10.

Schram K., Miketova P., Slanina J., Humpa O., and Taborska E. 2004. Mass spectrometry of 1,3- and 1,5-dicaffeoylquinic acids. *J Mass Spectrom* 39:384–95.

Sethi N., Anand A., Sharma A., Chandrul K.K., Jain G., and Srinivasa K.S. 2009. High speed counter current chromatography: A support-free LC technique. *J Pharm Bioall Sci* 1:8–15.

Shakir L., Hussain M., Javeed A., Ashraf M., and Riaz A. 2011. Artemisinins and immune system. *Eur J Pharmacol* 668:6–14.

Shan J., Fu J., Zhao Z., et al. 2009. Chlorogenic acid inhibits lipopolysaccharide induced cyclooxygenase-2 expression in RAW264.7 cells through suppressing NF-kB and JNK/AP-1 activation. *Int Immunopharmacol* 9:1042–8.

Shen M., Ge H.L., He Y.X., Song Q.L., and Zhang H.Z. 1984. Immunosuppressive action of Qinghaosu. *Sci Sin B* 27:398–406.

Shi H., Dong L., Jiang J., et al. 2013a. Chlorogenic acid reduces liver inflammation and fibrosis through inhibition of Toll-like receptor 4 signaling pathway. *Toxicology* 303:107–14.

Shi J.Q., Zhang C.C., Sun X.L., et al. 2013b. Antimalarial drug artemisinin extenuates amyloidogenesis and neuroinflammation in APPswe/PS1dE9 transgenic mice via inhibition of nuclear factor-κB and NLRP3 inflammasome activation. *CNS Neurosci Ther* 19:262–8.

Shin H.S., Satsu H., Bae M.J., et al. 2015. Anti-inflammatory effect of chlorogenic acid on the IL-8 production in Caco-2 cells and the dextran sulphate sodium-induced colitis symptoms in C57BL/6 mice. *Food Chem* 168:167–75.

Singh N.P., Ferreira J.F.S., Park J.S., and Lai H.C. 2011. Cytotoxicity of ethanolic extracts of *Artemisia annua* to molt-4 human leukemia cells. *Planta Med* 77:1788–93.

Souza M.C., Siani A.C., Ramos M.F., Menezes-de-Lima O.J., and Henriques M.G. 2003. Evaluation of anti-inflammatory activity of essential oils from two Asteraceae species. *Pharmazie* 58:582–6.

Spigno G., Tramelli L., and De Faveri D.M. 2007. Effects of extraction time, temperature and solvent on concentration and antioxidant activity of grape marc phenolics. *J Food Eng* 81:200–8.

Stebbings S., Beattie E., McNamara D., and Hunt S. 2016. A pilot randomized, placebo-controlled clinical trial to investigate the efficacy and safety of an extract of *Artemisia annua* administered over 12 weeks, for managing pain, stiffness, and functional limitation associated with osteoarthritis of the hip and knee. *Clin Rheumatol* 35:1829–36.

Suberu J., Song L., Slade S., Sullivan N., Barker G., and Lapkin A.A. 2013b. A rapid method for the determination of artemisinin and its biosynthetic precursors in *Artemisia annua* L. crude extracts. *J Pharm Biomed Anal* 84:269–77.

Suberu, J.O., Gorka, A.P., Jacobs L., et al. 2013a. Anti-plasmodial polyvalent interactions in *Artemisia annua* L. aqueous extract—Possible synergistic and resistance mechanisms. *PLoS ONE* 8. (doi:10.1371/journal.pone.0080790)

Tawfik A.F., Bishop S.J., Ayalp A., and El-Feraly F.S. 1990. Effects of artemisinin, dihydroartemisinin and arteether on immune responses of normal mice. *Int J Immunopharmacol* 12:385–9.

Toyama D.O., Ferreira M.J.P., Romoff P., et al. 2014. Effect of chlorogenic acid (5-caffeoylquinic acid) isolated from *Baccharis oxyodonta* on the structure and pharmacological activities of secretory phospholipase A2 from *Crotalus durissus* terrificus. *BioMed Res Int* 726585. (doi:10.1155/2014/726585)

Trinh H.T., Lee I.A., Hyun Y.J., and Kim D.H. 2011. Artemisia princeps Pamp essential oil and its constituents eucalyptol and α-terpineol ameliorate bacterial vaginosis and vulvovaginal candidiasis in mice by inhibiting bacterial growth and NF-κB activation. *Planta Med* 77:1996–2002.

Tunon M.J., Garcia-Mediavilla M.V., Sanchez-Campos S., et al. 2009. Potential of flavonoids as anti-inflammatory agents: Modulation of pro-inflammatory gene expression and signal transduction pathways. *Curr Drug Metab* 10:256–71.

Ul'chenko N.T., Khushbaktova Z.A., Bekker N.P., Kidisyuk E.N., Syrov V.N., and Glushenkova A.I. 2005. Lipids from flowers and leaves of *Artemisia annua* and their biological activity. *Chem Nat Comp* 3:280–4.

van der Kooy F., and Sullivan S.E. 2013. The complexity of medicinal plants: The traditional *Artemisia annua* formulation, current status and future perspectives. *J Ethnopharmacol* 150:1–13.

van der Kooy F., and Verpoorte R. 2011. The content of artemisinin in the *Artemisia annua* tea infusion. *Planta Med* 77:1754–6.

Wang J., Zhou H., Zheng J., et al. 2006. The antimalarial artemisinin synergizes with antibiotics to protect against lethal live *Escherichia coli* challenge by decreasing proinflammatory cytokine release. *Antimicrob Agents Chemother* 50:2420–7.

Wang J.X., Tang W., Zhou R., et al. 2008. The new water-soluble artemisinin derivative SM905 ameliorates collagen-induced arthritis by suppression of inflammatory and Th17 responses. *Br J Pharmacol* 153:1303–10.

Wang X.Q., Liu H.L., Wang G.B., et al. 2011b. Effect of artesunate on endotoxin-induced uveitis in rats. *Invest Ophthalmol Vis Sci* 52:916–9.

Wang Y., Huang Z., Wang L., et al. 2011a. The anti-malarial artemisinin inhibits pro-inflammatory cytokines via the NF-κB canonical signaling pathway in PMA-induced THP-1 monocytes. *Int J Mol Med* 27:233–41.

Weathers P.J., and Towler M.J. 2012. The flavonoids casticin and artemetin are poorly extracted and are unstable in an *Artemisia annua* tea infusion. *Planta Med* 78:1024–6.

Wu Q.Z., Zhao D.X., Xiang J., Zhang M., Zhang C.F., and Xu X.H. 2015. Antitussive, expectorant, and anti-inflammatory activities of four caffeoylquinic acids isolated from *Tussilago farfara*. *Pharm Biol* 54:1117–24.

Wu X., Zhang W., Shi, X., An, P., Sun W., and Wang Z. 2010. Therapeutic effect of artemisinin on lupus nephritis mice and its mechanisms. *Acta Biochim Biophys Sin* 42:916–23.

Wu Y., He S., Bai B., et al. 2016a. Therapeutic effects of the artemisinin analog SM934 on lupus-prone MRL/lpr mice via inhibition of TLR-triggered B-cell activation and plasma cell formation. *Cell Mol Immunol* 13:379–90.

Wu Y., Tang W., and Zuo J. 2016b. Development of artemisinin drugs in the treatment of autoimmune diseases. *Sci Bull* 61:37–41.

Xu Y., Liu W., Fang B., Gao S., and Yan J. 2014. Artesunate ameliorates hepatic fibrosis induced by bovine serum albumin in rats through regulating matrix metalloproteinases. *Eur J Pharmacol* 744:1–9.

Yang, Z., Ding J., Yang C., et al. 2012. Immunomodulatory and anti-inflammatory properties of artesunate in experimental colitis. *Curr Med Chem* 19:4541–51.

Yang J.H., Kim S.C., Shin B.Y., et al. 2013. O-methylated flavonol isorhamnetin prevents acute inflammation through blocking of NF-κB activation. *Food Chem Toxicol* 59:362–72.

Yao X., Ding Z., Xia Y., et al. 2012. Inhibition of monosodium urate crystal-induced inflammation by scopoletin and underlying mechanisms. *Int Immunopharmacol* 14:454–62.

Yao W., Wang F. and Wang H. 2016. Immunomodulation of artemisinin and its derivatives. *Sci Bull* 61:1399–1406.

Yin Y., Gong F.Y., Wu X.X., et al. 2008. Anti-inflammatory and immunosuppressive effect of flavones isolated from *Artemisia vestita*. *J Ethnopharmacol* 120:1–6.

Zatorski H., Sałaga M., Zielińska M., et al. 2015. Experimental colitis in mice is attenuated by topical administration of chlorogenic acid. *Naunyn Schmiedebergs Arch Pharmacol* 388:643–51.

Zha R.P., Xu W., Wang W.Y., Dong L., and Wang Y.P. 2007. Prevention of lipopolysaccharide-induced injury by 3,5-dicaffeoylquinic acid in endothelial cells. *Acta Pharmacol Sin* 28:1143–8.

Zhang D., Yang L., Yang, L.X., Wang, M.Y., and Tu Y.Y. 2007. Determination of artemisinin, arteannuin B and artemisinic acid in herba *Artemisiae annuae* by HPLC-UV-ELSD. *Acta Pharm Sin* 42:978–81.

Zhang Z., Liu Y., Ren X., et al. 2016. Caffeoylquinic acid derivatives extract of *Erigeron multiradiatus* alleviated acute myocardial ischemia reperfusion injury in rats through inhibiting NF-KappaB and JNK activations. *Mediators Inflamm* 7961940. (doi:10.1155/2016/7961940)

Zhao W., Zhang W., Chen Y., et al. 2015. Identification and purification of novel chlorogenic acids in *Artemisia annua* L. *J Exp Biol Agric Sci* 3:415–22.

Zhu X.X., Yang L., Li Y.J., et al. 2013. Effects of sesquiterpene, flavonoid and coumarin types of compounds from *Artemisia annua* L. on production of mediators of angiogenesis. *Pharmacol Rep* 65:410–20.

Zou W., Liu W., Yang B., et al. 2015. Quercetin protects against perfluorooctanoic acid-induced liver injury by attenuating oxidative stress and inflammatory response in mice. *Int Immunopharmacol* 28:129–35.

CHAPTER 6

Biosynthetic Pathway of Artemisinin

Bushra Hafeez Kiani

CONTENTS

6.1　ARTEMISININ

Artemisinin is an endoperoxide sesquiterpene lactone that is produced by aerial parts of *Artemisia annua* L. and is effective against multidrug-resistant strains of the malarial parasite. The isolation and characterization of artemisinin from *A. annua* is considered one of the most novel discoveries in recent medicinal plant research.

It was isolated from the plant in 1972 (Roth and Acton, 1989) and its structure was determined in 1979 by x-ray analysis (Brown, 1993). It has an empirical formula of C (carbon), H (hydrogen), and O (oxygen). Artemisinin has a peroxide bridge to which its antimalarial properties are attributed. It has a unique structure and lacks a nitrogen-containing heterocyclic ring, which is found in most antimalarial compounds, as explained in Figure 6.1. Artemisinin is an odorless, colorless compound that forms crystals with a melting point of $156°C-157°C$. The molecular weight as determined by high-resolution mass spectroscopy is m/e 282.1742 m+ (Figure 6.1) (Brown, 1993).

Artemisinin has proved to be one of the most promising drugs. It also possesses considerable antitumor, antimicrobial, and antioxidant activities (Dhingra et al., 2000). Although the complete process of organic synthesis has been established, chemical synthesis of artemisinin is not yet economically feasible because of its complexity and low yield. Currently the leaves, roots, and flowers of *Artemisia* species, especially *A. annua*, form the only source of this drug. Artemisinin is found in very low quantities (0.05%–1.1%) in different cultivars of *A. annua* (1%–4%). High-artemisinin-yielding clones are being isolated by selection and other nonconventional approaches; however, these have their own limitations. Therefore, in the recent past, *in vitro* systems of *A. annua* culture have been exploited for the enhanced production of artemisinin (Delabays et al., 1993).

6.2 ARTEMISININ SYNTHESIS AND STORAGE

Artemisinin is found in many parts of plants. Artemisinin has been reported to be present in higher concentrations at the top of the plant (Charles and Simon 1990; Laughlin, 1995), while other scientists suggest it is equally distributed throughout the plant (Laughlin, 1995). Artemisinin has also been produced by differentiated (shoots + roots) shoot cultures (Martinez and Staba, 1988; Fulzele et al., 1991; Whipkey et al., 1992; Ferreira and Janick, 1996), but only trace levels are found in shoots without roots, suggesting a regulatory effect (Martinez and Staba, 1988; Jha et al., 1988; Fulzele et al., 1991; Woerdenbag et al., 1993; Paniego and Giuliette, 1994) of roots on the storage and synthesis of artemisinin. Very small amounts of artemisinin are also found in the roots and seeds of *A. annua*. Most workers (Martinez

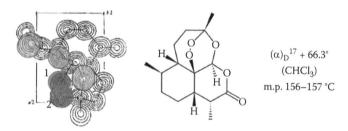

Figure 6.1   Structure of artemisinin, $C_{15}O_5H_{22}$ (MW = 282).

and Staba, 1988; Tawfiq et al., 1989; Fulzele et al., 1991; Kim and Kim 1992) were unable to detect artemisinin in roots, although Nair et al. (1986) and Jha et al. (1988) reported trace amounts. Artemisinin is also found in negligible amounts in seeds due to the presence of floral debris (Ferreira et al., 1995b). The highest concentration of artemisinin is found in the inflorescences, which contain more than ten times as much artemisinin as the leaves (Ferreira and Janick, 1996).

Artemisinin is produced in ten-celled glandular trichomes that are located on leaves, floral buds, and flowers (Ferreira et al., 1995a; Tellez et al., 1999; Olsson et al., 2009) and sequestered in the epicuticular sac at the apex of the trichome (Olsson et al., 2009). Artemisinin concentrations are higher in leaves that are formed later in development than in leaves formed early in the plant's development; this difference has been attributed to a higher trichome density and a higher capacity per trichome in the upper leaves (Lommen et al., 2006). Artemisinin can also be produced in green roots and hairy roots produced by infection with *Agrobacterium rhizogenes* (Jaziri et al., 1995; Liu et al., 1999). This suggests that the plant is capable of producing artemisinin in the absence of trichomes. However, to date, the regulatory mechanisms that control artemisinin biosynthesis and its formation outside the trichomes are poorly characterized.

6.3 DERIVATIVES OF ARTEMISININ

Derivatives of artemisinin such as the ether derivatives artemether and arteether, its primary metabolite artenimol, and the ester derivative artesunate (Figure 6.2) exhibit superior activity against both *Plasmodium falciparum* and

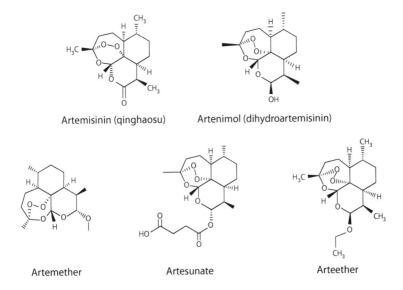

Artemisinin (qinghaosu)   Artenimol (dihydroartemisinin)

Artemether   Artesunate   Arteether

Figure 6.2 Chemical structure of artemisinin and its derivatives.

Plasmodium vivax. The reactivity of the endoperoxide bridge, which is the common structural feature of artemisinin and all its derivatives, is the key to their biological activity.

Artemisinin derivatives are fast-acting substances that can rapidly remove the malarial parasites from blood. Their short biological half-life makes artemisinin derivatives less suitable for monotherapy, but they can be used in combination with longer-acting drugs that have a slower onset of activity (Davis et al., 2005).

The WHO Roll Back Malaria program has advocated a strategy of using artemisinin-containing therapies (ACTs) in areas of emerging, high resistance to the most commonly used antimalarial drugs (WHO, 2004). Another benefit of ACTs is their apparently excellent safety and tolerability in humans (Price et al., 1999).

6.4 MODE OF ACTION OF ARTEMISININ

Artemisinin and its derivatives are the only group of compounds against which malarial parasites have not yet developed resistance. They are effective even against chloroquine-resistant and chloroquine-sensitive strains of *P. falciparum* as well as against cerebral malaria (Anon, 1979; Li et al., 1982; Roth and Acton, 1989; Pras et al., 1991). The unusual structure of artemisinin is the basis of its potent antimalarial activity, and it has a distinctive mode of action from other antimalarial drugs. There is an assumption that death of the parasite is due to the interaction of artemisinin with heme, because malarial parasites contain large amounts of heme iron (Meshnick et al., 1993; Kamchonwongpaisan and Meshnick, 1996; Meshnick, 1996). Furthermore, artemisinin is activated by ferrous iron, independent of heme, inside the parasite. The source of this iron is not yet known (Eckstein et al., 2003).

Artemisinin's mode of action has two steps. The first step involves the production of an oxygen free radical through cleavage of the endoperoxide bridge by the heme iron; therefore, this step is called the *activation step*. The oxygen free radical is subsequently rearranged to give a carbon free radical. The second step involves alkylation of specific malarial proteins by the carbon free radical, which causes lethal damage to the malarial parasite. This step is called the *alkylation step* (Figure 6.3) (Kamchonwongpaisan and Meshnick, 1996).

6.5 BIOSYNTHETIC PATHWAY OF ARTEMISININ

The biosynthetic pathway of artemisinin involves two independent and differently localized mechanisms. Both pathways converge to produce a terpenoid precursor, isopentenyl diphosphate (IPP) (Croteau et al., 2000; Lange et al., 2000) (Figure 6.4).

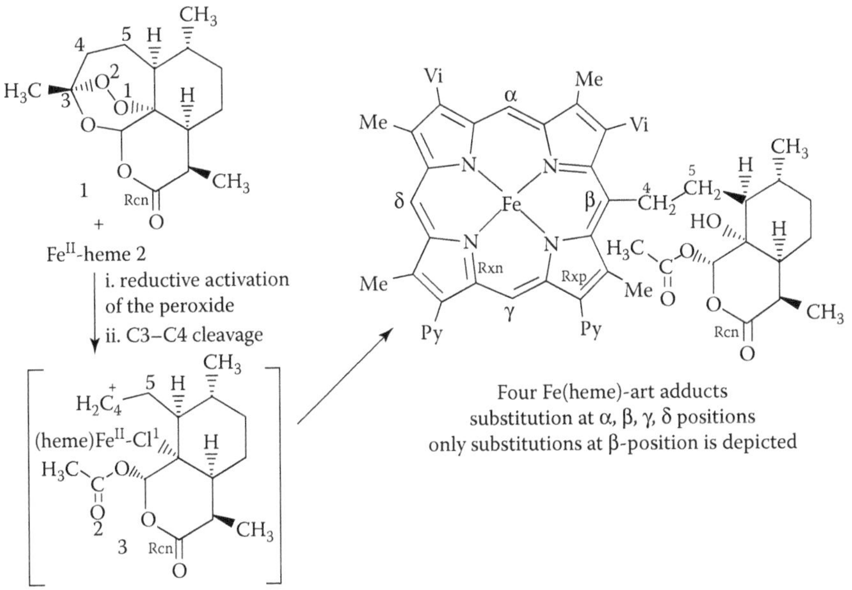

Figure 6.3 Mechanism of the reductive activation of artemisinin by iron (II)-heme, leading to covalent heme-artemisinin adducts.

6.5.1 Early Steps to Isopentenyl Diphosphate in Isoprenoid Biosynthesis

6.5.1.1 Mevalonic Acid Pathway

The mevalonic acid pathway (MVA) initiates from acetyl-CoA and is located in the cytosol. The key, regulatory, step in this pathway is the conversion of hydroxymethylglutaryl-CoA (HMG) to mevalonate via a regulatory enzyme, hydroxymethylglutaryl-CoA reductase (HMGR). Several subsequent steps lead to formation of the cytosolic-localized pool of IPP, which is used as a precursor for the synthesis of artemisinin (Towler and Weathers, 2007).

6.5.1.2 Non-Mevalonate Pathway

The other pathway, the non-mevalonate pathway (MEP), to IPP begins with pyruvate and occurs in the plastid with no mevalonate intermediate. The first key, regulatory, step toward the synthesis of terpenes is the synthesis of 1-deoxy-D-xylulose-5-phosphate (DXP) via 1-deoxy-D-xylulose-5-phosphate synthase (DXS). DXP is then converted to 2-C-methyl-D-erythritol-4-phosphate via

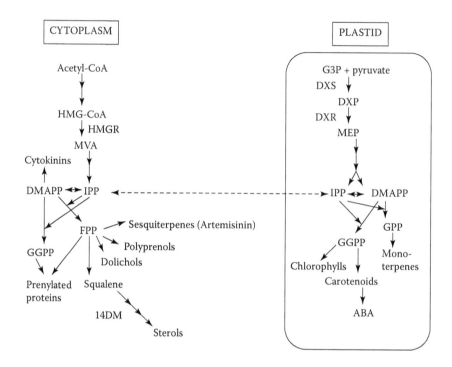

Figure 6.4   Early biosynthetic pathway for artemisinin: Steps in the MVA and MEP and post-IPP steps.

1-deoxy-D-xylulose-5-phosphate reductoisomerase (DXR). Several subsequent steps synthesize the plastid pool of IPP (Rohmer et al., 1996).

IPP in the cytosol is generally used for the biosynthesis of sesquiterpenes, triterpenes, polyterpenes, and sterols, and the IPP present in the plastid is used for the biosynthesis of monoterpenes, diterpenes, and carotenoids (Towler and Weathers, 2007). Translocation of IPP from the cytosol to the plastid or vice versa may occur depending on the needs of the plant (Adam et al., 1998; Lange et al., 2000).

6.5.1.3 Cytosolic and Plastidic Metabolic Crosstalk

These two compartmentalized pathways of terpene biosynthesis communicate with each other to regulate metabolic intermediate availability. There are no absolute restrictions on the compartmentalization of intermediates in the pathways, and the degree of separation depends on the species and physiological conditions (Hampel et al., 2005). In *A. annua,* both the MVA and MEP seem to play a role in artemisinin production. When the cytosolic MVA is disrupted by inhibiting HMGR with mevinolin, artemisinin levels drop by about 80%. When DXR in the MEP is inhibited by fosmidomycin, artemisinin levels drop by about 70%. Use of both inhibitors leads to negligible artemisinin production in *A. annua,* suggesting that the two pathways interlink with each other to enhance this production (Towler and Weathers, 2007).

6.5.2 Post-IPP Terpene Biosynthesis

Once IPP is formed and available in the cytosol, the next step toward artemisinin biosynthesis is the production of farnesyl diphosphate (FPP) via the enzyme farnesyl diphosphate synthase (FPS). Sequence analysis of FPS from *A. annua* has shown a very close similarity to FPS from other plants (Matsushita et al., 1996). Indeed, transgenic *A. annua* plants expressing FPS under a constitutive promoter produced three to four times more artemisinin than control lines, suggesting that FPS is one of the regulatory points in artemisinin biosynthesis (Chen et al., 2000a). FPP is a branch point, leading to the biosynthesis of triterpenes, polyterpenes, sterols, and sesquiterpenes (Figure 6.4).

There is evidence suggesting that the option to branch toward either sterols or sesquiterpenes is under coordinate control in plants. To produce sesquiterpenes, a sesquiterpene cyclase (SQC) is the first required catalyst; to produce sterols, the first required catalyst is a specific squalene synthase (SQS). In plants, when sterol production is upregulated, sesquiterpene production is often downregulated, and vice versa. Vogeli and Chappell (1988) demonstrated that fungal elicitors in cell suspension cultures lead to rapid increases in sesquiterpenoid production paralleled by a rapid decrease in sterol production. This coordinate control of SQS and SQC genes has been observed in many plant species (Yoshioka et al., 1999; Krits et al., 2007). Furthermore, upon wounding of the plant, metabolic flow is directed toward sterols by the upregulation of SQS and downregulation of SQC.

On the other hand, exposure of the wound to fungal pathogens or elicitors causes redirection of metabolic flow toward sesquiterpenoid phytoalexins, and SQC is upregulated while SQS is downregulated. The coordinate control between these two pathways may be at play in the production of artemisinin in *Artemisia* species (Towler and Weathers, 2007). Inhibition of SQS with miconazole causes a significant increase in artemisinin, suggesting that carbon is channeled toward sesquiterpene synthesis once sterol biosynthesis is inhibited.

6.5.3 Committed Steps in Artemisinin Biosynthesis

The first committed step in artemisinin biosynthesis (Figure 6.5) is the cyclization of FPP to generate amorpha-4,11-diene, catalyzed by amorpha-4,11-diene synthase (ADS) (Mercke et al., 2000; Wallaart et al., 2001). Subsequent oxidation at the C_{12} position, mediated by the cytochrome P450 enzyme CYP71AV1, produces artemisinic alcohol (Ro et al., 2006; Teoh et al., 2006). While arteannuin B has been suggested as a late precursor in artemisinin biosynthesis (Sangwan et al., 1993; Zeng et al., 2008), evidence now favors a route from artemisinic alcohol via dihydroartemisinic acid (Bertea et al., 2005; Covello et al., 2007; Covello, 2008).

This route is supported by the cloning and characterization of double bond reductase 2 (DBR2), which reduces the D11 (13) double bond of artemisinic aldehyde, but not of arteannuin B (Zhang et al., 2008), and the cloning of aldehyde dehydrogenase 1 (ALDH1), which catalyzes the oxidation of artemisinic and dihydroartemisinic aldehyde (Teoh et al., 2009). The conversion of dihydroartemisinic acid to artemisinin and

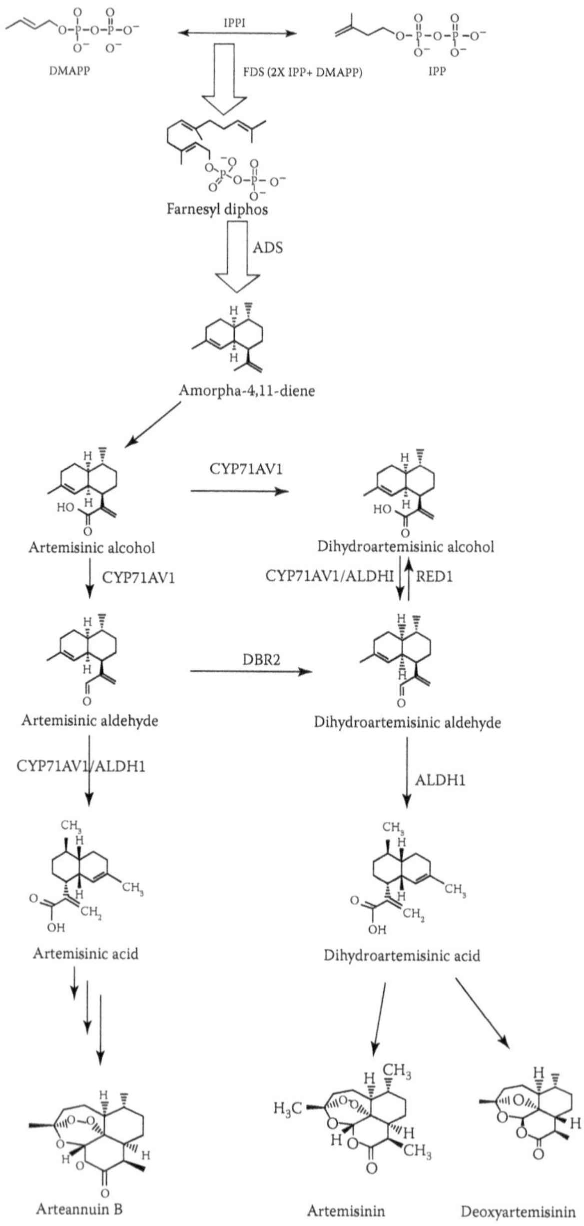

Figure 6.5 Biosynthetic pathway of artemisinin. Figure showing pathway and key genes and enzymes involved in artemisinin biosynthesis. ADS: amorpha-4,11-diene synthase; ALDH: aldehyde dehydrogenase; CYP: CYP71AV1 (p450 enzymes); Dbr2: artemisinic aldehyde D11 (13) double bond reductase; DMAPP: dimethylallyl diphosphate; IPP: isopentenyl diphosphate (Kiani et al., 2016).

of artemisinic acid to arteannuin B occurs via enzyme-independent reactions (Sy and Brown, 2002; Brown and Sy, 2004, 2007). Recently, a broad substrate oxidoreductase (RED1) with high affinity for dihydroartemisinic aldehyde and monoterpenes that may have a negative impact on the flux to artemisinin biosynthesis has been identified (Rydén et al., 2010).

6.5.4 Regulation of the Artemisinin Biosynthetic Pathway

Although many of the genes involved in artemisinin biosynthesis have been isolated and cloned from *A. annua*, little is known about their regulation. Most of the data relate to the effects of light, culture age, and tissue location on the expression of these genes, with most results to date measured in hairy root cultures (Souret et al., 2003; Teoh et al., 2006).

The shift of *A. annua* from vegetative growth to reproductive growth is another factor that has been shown to influence artemisinin production (Ferreira et al., 1995b). Since many terpenoids have floral fragrances, upregulation of terpene biosynthesis during the shift from vegetative to flowering states is not surprising. Previous studies had suggested a link between flowering and artemisinin biosynthesis, although these studies differed in relation to the stage during which peak artemisinin production was reached. One study, for example, showed that peak artemisinin production occurs in the budding stage just before flowering (Liersch et al., 1986; Woerdenbag et al., 1994; Chan et al., 1995), while others reported that peak production was only reached when the flowers were in full bloom (Ferreira et al., 1995b). Some genes related to terpene synthesis in other plants have also been shown to be transcriptionally activated as the shift to flowering occurs (Dudareva et al., 2003).

More recently, two separate studies investigated the link between flowering and artemisinin. The flowering promoter factor gene from *Arabidopsis*, fpf1, and the early flowering gene from *Arabidopsis*, CONSTANS, were constitutively expressed in *A. annua*, and although flowering was induced approximately two to three weeks earlier in transgenic lines, there was no corresponding increase in artemisinin biosynthesis. These data suggested that there was no direct regulatory link between flowering and artemisinin synthesis, and that some other factor is likely contributing to the observed increase in artemisinin content as the shift to the reproductive stage progresses (Wang et al., 2004, 2007).

6.6 KEY ENZYMES INVOLVED IN ARTEMISININ BIOSYNTHETIC PATHWAY

6.6.1 Amorpha-4,11-Diene Synthase

ADS is a key regulatory enzyme, catalyzing the rate-limiting step in the biosynthesis of artemisinin. It is involved in the first committed step in artemisinin biosynthesis. ADS is most abundant in young leaves and flower buds. Minor amounts of ADS are found in mature leaves and glandular trichomes (Zeng et al., 2008).

6.6.2 Aldehyde Dehydrogenase 1

$ALDH_1$ catalyzes the oxidation of artemisinic and dihydroartemisinic aldehydes into artemisinin. Cytochrome P_{450} CYP71AV1 is capable of oxidizing artemisinic alcohol to the corresponding aldehyde and acid. However, the ability of CYP71AV1 to oxidize dihydroartemisinic aldehyde is not detectable. A full-length cDNA encoding an aldehyde dehydrogenase homolog has been isolated and named $ALDH_1$, which is capable of catalyzing the oxidation of artemisinic aldehyde, dihydroartemisinic aldehyde, and a limited range of other aldehydes. $ALDH_1$ is expressed in the glandular trichomes, with moderate expression in the flower buds and low expression in the leaves (Zhang et al., 2008).

6.6.3 Cytochrome P_{450}

Cytochrome P_{450} enzymes participate in the oxidation of C_{12} in amorpha-4,11-diene through dehydrogenase and reductase activities. Moreover, it catalyzes the oxidation of amorpha-4,11-diene, artemisinic alcohol, and artemisinic aldehyde (Teoh et al., 2006). Cytochrome P_{450} is highly expressed in the glandular trichomes, moderately expressed in the flower buds and expressed at low levels in leaves, roots, and stems. Co-expression of cytochrome P_{450} reductase and CYP71AV1 leads to the production of high levels of artemisinic acid (Covello et al., 2007).

6.6.4 Artemisinic Aldehyde Δ11 (13) Reductase

Artemisinic Aldehyde Δ11 (13) Reductase (DBR2) is the checkpoint enzyme catalyzing artemisinic aldehyde to form dihydroartemisinic aldehyde, and is thus directly involved in the artemisinin biosynthetic pathway. DBR2 is highly expressed in the glandular trichomes, with low expression in other parts of the plant. It is relatively specific for artemisinic aldehyde and has some activity on small α,β-unsaturated carbonyl compounds (Wallaart et al., 2000; Covello, 2008; Teoh et al., 2009).

6.6.5 Other Possible Intermediates and Enzymes

The entire biosynthetic pathway from amorpha-4,11-diene to artemisinin remains uncertain. The main controversial issue in the latter stages of artemisinin biosynthesis is whether artemisinic or dihydroartemisinic acid serves as the later precursor (Dhingra and Narasu, 2001).

Artemisinic acid is a common precursor of arteannuin B and artemisinin. Artemisinin has been formed by incubating artemisinic acid and arteannuin B with *A. annua* leaf extract. Scientists have purified and characterized an enzyme from *A. annua* that is involved in the biochemical transformation of arteannuin B to artemisinin (Zhang et al., 2004). However, the genes encoding this enzyme have not been reported. This evidence suggests the possibility of artemisinin biosynthesis from artemisinic acid (Zhang et al., 2003). It has been reported that the level of artemisinin obtained from artemisinic acid was elevated approximately doubled when horseradish peroxidase was included in cell-free extracts of *A. annua* in a phosphate buffer system (Brown and Sy, 2007).

Recently, it was reported that artemisinic acid is converted *in vivo*, possibly non-enzymatically, into several compounds including arteannuin B, but not artemisinin (Brown and Sy, 2004). In addition, *in vivo*, dihydroartemisinic acid undergoes rapid plant pigment photosensitized oxidation, followed by subsequent spontaneous oxidation to form artemisinin. Dihydroartemisinic acid acts as a quencher of the singlet oxygen to produce dihydroartemisinic acid hydroperoxide, which is later transformed into artemisinin (Bertea et al., 2005).

6.7 SUMMARY

The study of the artemisinin biosynthesis pathway has made remarkable progress, especially since the discovery of important regulating enzymes. Since 1995, partial mRNA sequences of about 12 genes involved in artemisinin biosynthesis have been cloned from *A. annua*, and their complete data is now accessible in GenBank. As plant materials are affected by the natural environment, genetic regulation is implemented all over the world to increase artemisinin production with the development of genetic technology. Certainly, the explication of the artemisinin biosynthesis pathway is much needed. However, there are still some problems that must be resolved in the future. Advances or new discoveries in the study of this pathway are expected to lower the price of artemisinin and bring great benefits to millions around the world.

REFERENCES

Adam K.P., Thiel R., Zapp J., and Becker H. 1998. Involvement of the mevalonic acid pathway and the glyceraldehydes-pyruvate pathway in terpenoid biosynthesis of the liverworts *Ricciocarpos natans* and *Conocephalum conicum*. *Archives of Biochemistry and Biophysics* 354(1): 181–187.

Anon. 1979. Qinghaosu antimalarial coordinating research group. Antimalarial study on qinghaosu. *Chinese Medical Journal* 92(12): 811–816.

Bertea C.M., Freije J.R., Van der Woude H., Verstappen F.W., Perk L., and Marquez V. 2005. Identification of intermediates and enzymes involved in the early steps of artemisinin biosynthesis in *Artemisia annua*. *Planta Medica* 71(1): 40–47.

Brown G.D. and Sy L.-K. 2004. *In vivo* transformations of dihydroartemisinic acid in *Artemisia annua* plants. *Tetrahedron* 60(5): 1139–1159.

Brown G.D. and Sy L.-K. 2007. *In vivo* transformations of artemisinic acid in *Artemisia annua* plants. *Tetrahedron* 63(38): 9548–9566.

Brown T.A. 1993. Gene cloning and DNA analysis and introduction. How to obtain a clone of a specific gene Brown, T. A. (Ed.). Blackwell Science, 139–144.

Chan K.L., Teo C.K., Jinadasa S., and Yuen K.H. 1995. Selection of high artemisinin yielding *Artemisia annua*. *Planta Medica* 61(3): 285–287.

Charles D.J. and Simon J.E. 1990. Germplasm variation in artemisinin content of *Artemisia annua* using an alternative method of artemisinin analysis from crude plant extracts. *Journal of Natural Products* 53(1): 157–160.

Chen D., Ye H., and Li G. 2000. Expression of a chimeric farnesyl diphosphate synthase gene in *Artemisia annua* L. transgenic plants via *Agrobacterium tumefaciens*-mediated transformation. *Plant Science* 155(2): 179–185.

Covello P.S. 2008. Making artemisinin. *Phytochemistry* 69(17): 2881–2885.

Covello P.S., Teoh K.H., Polichuk D.R., Reed D.W., and Nowak G. 2007. Functional genomics and the biosynthesis of artemisinin. *Phytochemistry* 68(14): 1864–1871.

Croteau R., Kutchan T., and Lewis N. 2000. Natural products (secondary metabolites). In: B. Buchanan, W. Gruissem, and R. Jones (eds), *Biochemistry and Molecular Biology of Plants*. American Society of Plant Biologists, Rockville, MD, 1250–1318.

Davis T.M.E., Karunajeewa H.A., and Illett K.F. 2005. Artemisinin-based combination therapies for uncomplicated malaria. *The Medical Journal of Australia* 182 (4): 181–185.

Delabays N., Benakis A., and Collet G. 1993. Selection and breeding for high artemisinin (Qinghaosu) yielding strains of *Artemisia annua*. *Acta Horticulturae* 330: 203–207.

Delabays N., Simonnet X., and Gaudin M. 2001. The genetics of artemisinin content in *Artemisia annua* L. and the breeding of high yielding cultivars. *Current Medicinal Chemistry* 8(15): 1795–1801.

Dhingra V. and Narasu M.L. 2001. Purification and characterization of an enzyme involved in biochemical transformation of arteannuin B to artemisinin from *Artemisia annua*. *Biochemical and Biophysical Research Communications* 281(2): 558–561.

Dhingra V., Rajoli C., and Narasu M.L. 2000. Partial purification of proteins involved in the bioconversion of arteannuin B to artemisinin. *Bioresource Technology* 73(3): 279–282.

Dudareva N., Martin D., Kish C.M., Kolosova N., Gorenstein N., Faldt J., et al. 2003. E-beta-ocimene and myrcene synthase genes of floral scent biosynthesis in snapdragon: Function and expression of three terpene synthase genes of a new terpene synthase subfamily. *Plant Cell* 15(5): 1227–1241.

Eckstein L.U., Webb R.J., Van G.I., East J.M., Lee A.G., Kimura M., O'Neill P.M., et al. 2003. Artemisinin target the SERCA of Plasmodium falciparum. *Nature* 424: 957–961.

Ferreira J.F.S. and Janick J. 1996. Roots as an enhancing factor for the production of artemisinin in shoot cultures of *Artemisia annua*. *Plant Cell Tissue Organ Cult.* 44(3): 211–217.

Ferreira J.F.S., Simon J.E., and Janick J. 1995. Developmental studies of *Artemisia annua*: Flowering and artemisinin production under greenhouse and field conditions. *Planta Medica* 61(2): 167–170.

Ferreira J.F.S., Simon J.E., and Janick J. 1995b. Relationship of artemisinin content of tissue-cultured, greenhouse-grown, and field-grown plants of *Artemisia annua*. *Planta Medica* 61(4): 351–355.

Fulzele D.P., Sipahimalani A.T., and Heble M.R. 1991. Tissue cultures of *Artemisia annua*: Organogenesis and artemisinin production. *Phytotherapy Research* 5(4): 149–153.

Hampel D., Mosandl A., and Wust M. 2005. Biosynthesis of mono- and sesquiterpenes in carrot roots and leaves (*Daucus carota* L.): Metabolic cross talk of cytosolic mevalonate and plastidial methylerythritol phosphate pathways. *Phytochemistry* 66(3): 305–311.

Jaziri M., Shimonura K., Yoshimatsu K., Fauconnier M.L., Marlier M., and Homes J. 1995. Establishment of normal and transformed root cultures of *Artemisia annua* L. for artemisinin production. *Journal of Plant Physiology* 145(1–2): 175–177.

Jha J., Jha T.B., and Mahato S.B. 1988. Tissue culture of *Artemisia annua* L.—a potential source of an antimalarial drug. *Current Science* 57(6): 344–346.

Kamchonwongpaisan S. and Meshnick S.R. 1996. The mode of action of antimalarial artemisinin and its derivatives. *General Pharmacology* 27(4): 587–592.

Kiani B.H., Suberu J., Mirza B. 2016. Cellular engineering of Artemisia annua and Artemisia dubia with the rol ABC genes for enhanced production of potent anti-malarial drug artemisinin. *Malar J* 15: 252. DOI 10.1186/s12936-016-1312-8

Kim N.C. and Kim S.U. 1992. Biosynthesis of artemisinin from 11,12-dihydroarteannuic acid. *Journal of the Korean Society of Agricultural Chemistry and Biotechnology* 35(2): 106–109.

Krits P., Fogelman E., and Ginzberg I. 2007. Potato steroidal glycoalkaloid levels and the expression of key isoprenoid metabolic genes. *Planta* 227(1): 143–150.

Lange B.M., Rujan T., Martin W. and Croteau R. 2000. Isoprenoid biosynthesis: The evolution of two ancient and distinct pathways across genomes. *Proceedings of the National Academy of Sciences of the United States of America* 97(24): 13172–13177.

Laughlin J.C. 1995. The influence of distribution of antimalarial constituents in *Artemisia annua* L. on time and method of harvest. *Acta Horticulturae* 390: 67–73.

Li G., Guo X., Jin R., Wang Z., Jian H., and Li Z. 1982. Clinical studies on treatment of cerebral malaria with qinghaosu and its derivatives. *Journal of Traditional Chinese medicine* 2: 125–130.

Liersch R., Soicke H., Stehr C., and Tullner H.U. 1986. Formation of artemisinin in *Artemisia annua* during one vegetation period. *Planta Medica* 52(5): 387–390.

Liu C.Z., Wang X.C., Zhao B., Guo C., Ye H.C., and Li G.F. 1999. Development of a nutrient mist bioreactor for growth of hairy roots. *In Vitro Cellular and Developmental Biology Plant* 35(3): 271–274.

Lommen W.J.M., Schenk E., Bouwmeester H.J., and Verstappen F.W.A. 2006. Trichome dynamics and artemisinin accumulation during development and senescence of *Artemisia annua* leaves. *Planta Medica* 72(4): 336–345.

Martinez B.C. and Staba J. 1988. The production of artemisinin in *Artemisia annua* L. tissue cultures. *Advances in Cell Culture* 6: 69–87.

Matsushita Y., Kang W., and Charlwood B.V. 1996. Cloning and analysis of a cDNA encoding farnesyl diphosphate synthase from *Artemisia annua*. *Genes* 172(2): 207–209.

Mercke P., Bengtsson M., Bouwmeester H.J., Posthumus M.A., and Brodelius P.E. 2000. Molecular cloning, expression, and characterization of amorpha-4,11-diene synthase, a key enzyme of artemisinin biosynthesis in *Artemisia annua* L. *Archives of Biochemistry and Biophysics* 381(2): 173–180.

Meshnick S.R. 1996. Is haemozoin a target for antimalarial drugs? *Annals of Tropical Medicine and Parasitology* 90(4): 367–372.

Meshnick S.R., Yang Y.Z., Lima V., Kuypers F., Kamchonwongpaisan S., and Yuthavong Y. 1993. Iron-dependent free radical generation from antimalarial agent artemisinin (qinghaosu). *Antimicrobial Agents and Chemotherapy* 37(5): 1108–1114.

Meshnick S., Taylor T., and Kamchonwongpaisian S. 1996. Artemisinin and the antimalarial endoperoxides: From herbal remedy to targeted chemotherapy. *Microbiological Reviews* 60(2): 301–315.

Nair M.S.R., Acton N., Klayman D.L., Kendrick K., Basile D.V., and Mante S. 1986. Production of artemisinin in tissue cultures of *Artemisia annua*. *Journal of Natural Products* 49(3): 504–507.

Olsson M.E., Olofsson L.M., Lindahl A.L., Lundgren A., Brodelius M., and Brodelius P.E. 2009. Localization of enzymes of artemisinin biosynthesis to the apical cells of glandular trichomes of *Artemisia annua* L. *Phytochemistry* 70(9): 1123–1128.

Paniego N.B. and Giuliette A.M. 1994. *Artemisia annua* L.: Dedifferentiated and differentiated cultures. *Plant Cell Tissue and Organ Culture* 36(2): 163–168.

Pras N., Visser J.F., Batterman S., Woerdenbag H.J., Malingre T.M., and Lugt C.B. 1991. Laboratory selection of *Artemisia annua* L. for high artemisinin yielding types. *Phytochemical Analysis* 2(2): 80–83.

Price R., Van Vugt M., Phaipun L., Luxemburger C., Simpson J., McGready R., et al. 1999. Adverse effects in patients with acute falciparum malaria treated with artemisinin derivatives. *American Journal of Tropical Medicine and Hygiene* 60(4): 547–555.

Ro D.K., Paradise E.M., Ouellet M., Fisher K.J., Newman K.L., Ndungu J.M., et al. 2006. Production of the antimalarial drug precursor artemisinic acid in engineered yeast. *Nature* 440(7086): 940–943.

Rohmer M., Seemann M., Horbach S., Bringermeyer S., and Sahm H. 1996. Glyceraldehyde 3-phosphate and pyruvate as precursors of isoprenic units in an alternative non-mevalonate pathway for terpenoid biosynthesis. *Journal of the American Chemical Society* 118(11): 2654–2566.

Roth R.J. and Acton N. (1989). The isolation of sesquiterpines from *Artemisia annua. Journal of Chemical Education* 66(4): 349–350.

Rydén A.-M., Ruyter-Spira C., Quax W.J., Osada H., Muranaka T., Kayser O., and Bouwmeester H. 2010. The molecular cloning of dihydroartemisinic aldehyde reductase and its implication in artemisinin biosynthesis in *Artemisia annua. Planta Medica* 76(15): 1778–1783.

Sangwan R.S., Agarwal K., Luthra R., Thakur R.S., and Singh-Sangwan N. 1993. Biotransformation of arteannuic acid into arteannuin-B and artemisinin in *Artemisia annua. Phytochemistry* 34(5): 1301–1302.

Souret F.F., Kim Y., Wyslouzil B.E., Wobbe K.K., and Weathers P.J. 2003. Scale-up of *Artemisia annua* L. hairy root cultures produces complex patterns of gene expression. *Biotechnology and Bioengineering* 83(6): 653–657.

Sy L.-K. and Brown G.D. 2002. The mechanism of the spontaneous autoxidation of dihydroartemisinic acid. *Tetrahedron* 58(5): 897–908.

Tawfiq N.K., Anderzson L.A., Roberts M.F., Phillipson J.D., Bray D.H., and Warhurst D.C. 1989. Antiplasmodial activity of *Artemisia annua* plant cell culture. *Plant Cell Reports* 8(7): 425–428.

Tellez M.R., Canel C., Rimando A.M., and Duke S.O. 1999. Differential accumulation of isoprenoids in glanded and glandless *Artemisia annua.* L. *Photochemistry* 52(6): 1035–1040.

Teoh K.H., Polichuk D.R., Reed D.W., and Covello P.S. 2009. Molecular cloning of an aldehyde dehydrogenase implicated in artemisinin biosynthesis in *Artemisia annua. Botany* 87(6): 635–642.

Teoh K.H., Polichuk D.R., Reed D.W., Nowak G., and Covello P.S. 2006. *Artemisia annua* L. (Asteraceae) trichome-specific cDNAs reveal CYP71AV1, a cytochrome P450 with a key role in the biosynthesis of the antimalarial sesquiterpene lactone artemisinin. *FEBS Letters* 580(5): 1411–1416.

Towler M.J. and Weathers P.J. 2007. Evidence of artemisinin production from IPP stemming from both the mevalonate and the nonmevalonate pathways. *Plant Cell Reports* 26(12): 2129–2136.

Vogeli U. and Chappell J. 1988. Induction of sesquiterpene cyclase and suppression of squalene synthetase activities in plant cell cultures treated with fungal elicitor. *Plant Physiology* 88(4): 1291–1296.

Wallaart T.E., Bougtsson J., Hille L., and Poppingats N.C.A. 2001. Hormonal content and sensitivity of transgenic tobacco and potato plants expressing single *rol* genes of *A. rhizogenes* T-DNA. *Plant Journal* 212: 460–468.

Wallaart T.E., Pras N., Beekman A.C., and Quax W.J. 2000. Seasonal variation of artemisinin and its biosynthetic precursors in plants of *Artemisia annua* of different geographical origin: Proof for the existence of chemotypes. *Planta Medica* 66(1): 57–62.

Wang H., Ge L., Ye H.C., Chong K., Liu B.Y., and Li G.F. 2004. Studies on the effects of fpf1 gene on *Artemisia annua* flowering time and on the linkage between flowering and artemisinin biosynthesis. *Planta Medica* 70(4): 347–352.

Wang H., Liu Y., Chong K., Liu B.Y., Ye H.C., Li Z.Q., Yan F., and Li G.F. 2007. Earlier flowering induced by over-expression of CO gene does not accompany increase of arte-misinin biosynthesis in *Artemisia annua*. *Plant Biology* 9(3): 442–446.

Weathers P.J., Cheetham R.D., Follansbee E., and Theoharides K. 1994. Artemisinin pro-duction by transformed roots of *Artemisia annua*. *Biotechnology Letters* 16(12): 1281–1286.

Whipkey A., Simon J.E., Charles D.J., Janick J. 1992. In vitro production of artemisinin from *Artemisia annua* L. *J. Herbs Spices Med. Plants* 1: 15–25.

Woerdenbag H.J., Moskal T.A., Pras N., and Malingre T.M. 1993. Cytotoxicity of artemis-inin-related endoperoxides to Ehrlich ascites tumor cells. *Journal of Natural Products* 56(6): 849–856.

Woerdenbag H.J., Pras N., Chan N.G., Bang B.T., Bos R., Von U.W., et al. 1994. Artemisinin, related sesquiterpenes, and essential oil in *Artemisia annua* during a vegetation period in Vietnam. *Planta Medica* 60(3): 272–275.

World Health Organization (WHO). 2004. Malaria Medicines and Supplies Service (MMSS). Accessed September 21, 2006. http://www.rollbackmalaria.org.

Yoshioka H., Yamada N., and Doke N. 1999. cDNA cloning of sesquiterpene cyclase and squa-lene synthase, and expression of the genes in potato tuber infected with *Phytophthora infestans*. *Plant and Cell Physiology* 40(9): 993–998.

Zeng Q., Qui F., and Yuan L. 2008. Production of artemisinin by genetically-modified microbes. *Biotechnology Letters* 30(4): 581–592.

Zhang Y., Teoh K.H., Reed D.W., Maes L., Goossens A., Olson D.J.H., et al. 2008. The molecular cloning of artemisinic aldehyde D11 (13) reductase and its role in glandu-lar trichome-dependent biosynthesis of artemisinin in *Artemisia annua*. *Journal of Biological Chemistry* 283(31): 21501–21508.

Zhang Y.S., Liu B.Y., Li Z.Q., Ye H.C., Wang H., and Li G.F. (2004). Molecular cloning of a classical plant peroxidase from *Artemisia annua* and its effect on the biosynthesis of artemisinin *in vitro*. *Acta Botanica Sinica* 46(11): 1338–1346.

Zhang Y.S., Ye H.C., and Li G.F. 2003. Effect of horseradish peroxidase on the biosyn-thesis of artemisinin in *Artemisia annua in vitro*. *Chinese Journal of Applied and Environmental Biology* 9(6): 616–618.

CHAPTER 7

Cultivation of *Artemisia annua*—The Environmental Perspective

Karina Knudsmark Sjøholm (nee Jessing),
Bjarne W. Strobel, and Nina Cedergreen

CONTENTS

7.1 INTRODUCTION

Bioactive secondary metabolites represent a wide range of different chemical compounds, but the three major classes are nitrogen-containing compounds, phenols, and terpenes. Artemisinin, produced by *Artemisia annua* L. (sweet wormwood, annual wormwood, sweet annie, sweet sagewort) belongs to the terpenes group. Artemisinin is a sesquiterpene lactone with an endoperoxide bridge (Liu et al., 1979), and is very bioactive against chloroquine-resistant strains of the malaria parasite *Plasmodium falciparum* (Klayman, 1985). Today, combination therapies containing artemisinin are recommended as the first-line treatment for malaria in 77 out of 80 countries and amounted to 311 million treatments worldwide in 2015 (WHO, 2015).

The bioactivity of a plant-produced compound is two-sided: on one side, it has pharmaceutical benefits or can be useful in pest management; on the other side, its bioactivity can cause unwanted toxicity to beneficial non-target organisms. Just as anthropogenic pesticides and pharmaceuticals, plant-produced compounds can be toxic to both target and non-target organisms in the environment. In several cases, loss of bioactive secondary metabolites from plants to soil has been documented (Strobel et al., 2005; Gimsing and Kirkegaard, 2006; Hoerger et al., 2009; Zhang et al., 2011). When plants grow naturally in mixed vegetation, the load of individual secondary metabolites will usually be relatively low. This changes in intensely cultivated monocultures as the density of the toxin-producing plant species increases, and the plant variety may have been bred to produce even more of the desired bioactive secondary metabolite. In these dense monocultures and crops, large amounts of bioactive substances may potentially be released to the soil and water environments (Strobel et al., 2005; Gimsing and Kirkegaard, 2006; Hoerger et al., 2009; Carlsen et al., 2012). At present, total chemical synthesis (Abdin et al., 2003) or *in vitro* production of artemisinin is not economically feasible (Peplow, 2016), and cultivation of the plant is still the only cost-effective source. The cultivation of high-producing *A. annua* on a field scale, *Pharmland* (Figure 7.1), poses a risk of losing bioactive compound to the environment, where it might affect vulnerable organisms in the soil and/or leach into ground or surface waters. Hence, sustainable growing of *A. annua* requires careful assessment of the potential effect to the environment. Therefore, in this chapter, we perform a risk assessment of artemisinin, just as if it was a commercial chemical.

As *A. annua* is cultivated in fields, and artemisinin released directly to the environment, we use the risk assessment paradigm that exist for pesticides. In this type of risk assessment, the first tier is hazard identification. In hazard identification, the physicochemical properties of the chemical are evaluated in relation to environmental fate, and exposure scenarios are considered. Ecotoxicity of the compound of interest is also evaluated, and the most vulnerable species in exposed environmental compartments are identified. The environmental fate evaluation yields a predicted environmental concentration (PEC), and the effect evaluation produces a predicted no-effect concentration (PNEC) value. The PNEC is the lowest available ecotoxicity value divided by a safety factor. The magnitude of the safety factor (10–1000) depends on the level of certainty on which the data are obtained. From these, the hazard quotient (HQ) is calculated:

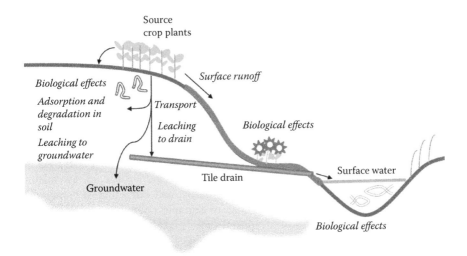

Figure 7.1   Schematic presentation of "Pharmland." A crop is the source of bioactive chemical substances that are released to the surrounding soil and water environments,
where non-target organisms might be exposed to them.

$$HQ = \frac{PEC}{PNEC} \qquad (7.1)$$

If HQ > 1, there will be a risk to the environment, as the exposure concentration
exceeds the estimated effect level.

Before we go into the details of the environmental fate analysis, we need to
explain how artemisinin is extracted and quantified from environmental matrices.
Most reported techniques and studies of extraction efficiencies of artemisinin from
the leaves of *A. annua* have the medical use of artemisinin as their purpose. This
chapter provides new, additional knowledge on methods to determine artemisinin in
soil and water samples.

7.2 EXTRACTION AND QUANTIFICATION OF ARTEMISININ

To reveal the fate of artemisinin in the environment and establish knowledge
on the concentrations of the compound under natural growing conditions, robust
and sensitive methods for monitoring of environmental artemisinin were needed.
Such information is of foremost importance for risk evaluation and assessment of
possible side effects of the cultivation of *A. annua*. Measurements include those
obtained by both extraction from the environmental matrices and analytical quantification. Extraction of artemisinin from soil can be performed using supercritical fluid extraction, with recovery in spiked soils being 74%–84% (Jessing et al.,
2009a). Simple solvent shake extraction with ethanol gives recoveries of 71%–88%

(Jessing et al., 2009b). The latter method was chosen due to its simplicity for quantification of artemisinin in soil. In both cases, the recoveries were best in sandy soil, followed by clayey, loamy, and humic soils. In very humic soil, the recovery using this method was only 55% (Jessing et al., 2013), due to sorption of artemisinin to the very organic soil. In a Danish study, the loamy soil recovery was 84% (Herrmann et al., 2013). Despite the low recoveries, the measured concentrations are proportional to the fraction considered readily bioavailable and hence the interesting fraction in relation to assessment of potential hazard. Extraction from plant material can be performed with recoveries >90% by dipping the leaves in chloroform (Tellez et al., 1999), or methanol (Jessing et al., 2011, 2013; Herrmann et al., 2013). As the artemisinin is located in the trichomes present on the leaf surface, all artemisinin diffuses into the solvent phase almost immediately, and extraction of compounds that would interfere with the quantification, for example, chlorophyll, is minimized.

The hydrophobic behavior of artemisinin also makes *in situ* passive sampling with silicone tubes possible and facilitates simple liquid:liquid extraction (LLE) from water samples by partitioning it into an organic solvent (Jessing et al., 2013). Solid phase extraction (SPE) has been applied to environmental samples, when needed, to purify and concentrate artemisinin in extracts from biological matrices (Jessing et al., 2011).

Quantification in the early studies of artemisinin in soil was done using the original HPLC–UV method (Liu et al., 1979) in an optimized version (Jessing et al., 2009a,b). The limit of detection (LOD) of this method was 0.18 mg/L, corresponding to 0.36 mg/kg soil (Jessing et al., 2009b). The more selective and sensitive mass spectrometry detection is superior to UV detection, especially when derivation is needed to obtain a measureable UV absorption, as it is the case with artemisinin. Hence, mass spectrometry is preferred for quantifying artemisinin in soil matrices. The LOD of the current LC–MS/MS method is 0.03 mg/L and 0.025 mg/kg soil (Jessing et al., 2011; Herrmann et al., 2013).

7.3 ENVIRONMENTAL FATE OF ARTEMISININ

7.3.1 Physicochemical Properties of Artemisinin

The physicochemical properties of the chemical substance are the first to consider in risk assessment, as this reveals the potential toxicity, mobility, and degradation in the environment. The physicochemical properties of artemisinin are summarized in Table 7.1. Artemisinin is not volatile and has relatively low solubility in water. The estimated octanol and water partitioning coefficient (log K_{OW}) suggests hydrophobic behavior, and, for example, partitioning to organic matter in soil.

7.3.2 Field Production of Artemisinin

The total amount of bioactive chemical substance potentially released from the plant cover is very important to the evaluation of risk to the environment. The

Table 7.1 Physicochemical Properties and Numbered Chemical Structure of Artemisinin

CAS no.	63968-64-9
IUPAC name	(3*R*,5a*S*,6*R*,8a*S*,9*R*,12*S*,12a*R*)-octahydro-3,6,9-trimethyl-3,12-epoxy-12*H*-pyrano[4,3-j]-1,2-benzodioxepin-10(3*H*)-one

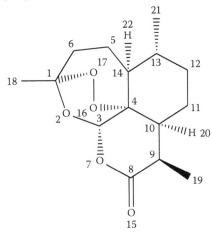

Molecular formula	$C_{15}H_{22}O_5$
Molar mass	282.34 g mol^{-1}
Solubility in water	49.7 ± 3.7 mg L^{-1}
Log K_{OW}	2.90
Henrys law constant:	4.92·10^{-9} atm m^3 mol^{-1}

Source: Modified after Jessing, K. et al., *Int. J. Environ. Anal. Chem.*, 89, 1–10, 2009.

annual yield of artemisinin varies from 1.5 to 10 kg/hectare in Europe and Africa to almost 90 kg/hectare in Asia, where multi-harvesting is common. In this harvesting practice, plants are cut 40–50 cm above ground, and a few twigs are left intact, four times per growing season (Kumar et al., 2004). Commercial *A. annua* is cultivated in monocultures to produce as much artemisinin as possible, and much effort is invested into enhancing the artemisinin yield (Abdin et al., 2003), both through breeding high-yield varieties and improved agricultural practice. The artemisinin content in *A. annua* is mainly governed by genetic potential, and high-artemisinin-producing cultivars have been selected through breeding schemes (Ferreira et al., 2005). However, various environmental factors also play a role in determining the artemisinin content, and exposure to stress such as frost at night (Wallaart et al., 2000), potassium deficiency (Ferreira, 2007), or moderate drought (Marchese et al., 2010) can trigger a higher production of artemisinin. *A. annua* is a native herb of temperate climate zones of Asia (Dhingra et al., 2000), but it has spread worldwide and can be cultivated under various climatic conditions. Clones adapted to tropical latitudes contain high levels of artemisinin, so even though the biomass production is less than at temperate latitudes a high yield of artemisinin can be achieved (Ferreira et al., 2005).

In the plant, artemisinin is stored in glandular trichomes on the surface of leaves and stems (Duke and Paul, 1993), on the corolla, and on receptacles of the florets (Janick and Ferreira, 1996). The glandular trichomes (Figure 7.2) consist of ten cells, where the two most apical cells form a bilobed sac by filling and expanding the space between the external cell walls and cuticle with secretory products (Ferreira and Janick, 1995). The subcuticular space in this sac stores secondary metabolites, including artemisinin. The trichomes burst as they mature, and artemisinin is leaked to the leaf surface (Duke and Paul, 1993). Younger leaves have a higher density of trichomes and produce more artemisinin per unit of fresh or dry weight plant biomass than older leaves (Arsenault et al., 2008). However, in mature plants, dead leaves are very important artemisinin stores and contain nearly half of the total yield of artemisinin at the time of harvest (Lommen et al., 2007).

From an environmental fate perspective, the production of artemisinin in the field without any process to control or reduce the exposure (e.g., human metabolism, or wastewater treatment) is high. In addition, there is a possibility that it can be higher at local hotspots (piles of harvested plant material) or when certain stressors trigger higher production in groups of plants. In contrast to spraying commercial pesticides in low and regulated doses once or twice in a growing season, the plants produce artemisinin continuously throughout the growing season, and potentially release this bioactive compound to the environment over a long period (Figure 7.3).

7.3.3 Artemisinin Concentrations in the Environment

The spread of artemisinin to the surrounding soil environment was examined in detail in a Danish study (Taastrup), where the plant and soil artemisinin contents were measured over a full growing season (Herrmann et al., 2013). As artemisinin theoretically sorbs quite strongly to the topsoil (Table 7.1), the initial conceptual idea was

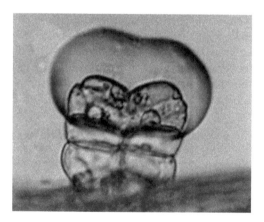

Figure 7.2 Glandular trichome from *Artemisia annua* at the mature stage where the subcuticular space is expanded. (Reprinted with permission from Springer, "Potential ecological roles of artemisinin produced by *Artemisia annua* L." *Journal of Chemical Ecology* 40, no. 2 (2014): 100–117.)

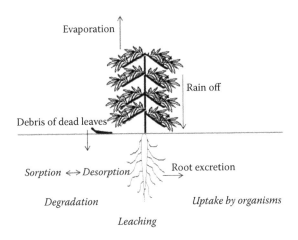

Figure 7.3   Artemisinin release routes to, and pathways in, the soil environment.

that artemisinin distribution was limited to the topsoil just below the *A. annua* plants. As a consequence of this idea, the occurrence of artemisinin was expected to spread in larger and larger circles below the plant as it expanded during growth. From a previous study at Aarslev in Denmark, where only the soil content of artemisinin was measured once a month, it was known that artemisinin could accumulate in topsoil over the growing season with concentrations up to 11 mg/kg (Jessing et al., 2009b). The highest soil concentration measured in the Taastrup study was 0.44 mg/kg. The 25-fold difference in the soil concentrations between the two Danish studies can partly be explained by the different *A. annua* cultivars used in each. In Aarslev, it was a high-yielding Vietnamese cultivar, whereas in Taastrup it was a garden variant. Apart from the first soil measurement at a plant age of 105 days, the soil concentration increased with the artemisinin content of the plants (Figure 7.4). The distribution pattern of artemisinin in the soil appeared patchy, as no trends in soil concentrations relative to distance from the plant were found within a two-meter distance. For further investigation, soil samples (0–10 cm layer) were collected up to 20 m from the plants on day 212. The highest concentration of 0.05 mg/kg was found at the margin of the plot, while lower, fluctuating concentrations of 0.001–0.024 mg/kg occurred at distances up to 15 m, beyond which no artemisinin was detected.

To investigate the magnitude of artemisinin exposure of the environment in areas where *A. annua* is cultivated on a larger scale for medical purposes, the soil artemisinin content was measured in soils from *A. annua* fields in Kenya, Italy, and West Virginia (Table 7.2). Measurements from Denmark at the time of harvest are included in Table 7.2 for comparison. Measurements from Kenya showed that all soils exposed to *A. annua* contained artemisinin, in concentrations ranging from 0.01 to more than 2 mg/kg dry weight. In many locations, artemisinin was also detected in soils not exposed directly to *A. annua*, but at significantly lower levels than those observed in the *A. annua* crop field samples ($p = 0.0130$). Artemisinin was still measurable in soils one year after cultivation of *A. annua*, as seen in Egerton (F), which

Table 7.2 Magnitude and Occurrence of Artemisinin in Soil and Plant Material from Different Climatic Locations and Cultivation Systems

Sample location	Plant density, (plants/m²)	Type of plantation	Depth	Artemisinin in soil (mg/kg DW)	Artemisinin in leaves (mg/kg DW)
			Kenya[a]		
			Tropical climate, yearly precipitation: 1060 mm, average temperature: 17.9°C		
Naivasha	8	Monoculture, 4 years in row	Not cul.	0.04	
			0–2 cm	0.32	1.48
			2–5 cm	0.21	
			0–2 cm	0.17	
			2–5 cm	0.15	
Kirinyaga	4	Intercropping, melons and beans, first year	Not cul.	0.07	
			0–2 cm	0.07	0.54
			2–5 cm	0.05	
			0–2 cm	0.03	
			2–5 cm	0.01	
Mweiga	8	Monoculture, first year	Not cul.	0.003	
			0–2 cm	1.05	2.00
			2–5 cm	0.74	
			0–2 cm	1.14	
			2–5 cm	2.03	
			Not cul.	0.06	
Kakuzi	5	Self-seeded, 4 years in row	0–2 cm, young	0.141	0.52
			2–5 cm, young	0.12	
			0–2 cm, old	0.46	
			2–5 cm, old	0.06	
Kajulu	6	Monoculture, first year	Not cul.	0.09	
			0–2 cm	0.15	0.76
			2–5 cm	0.07	
			2–5 cm	0.07	
Vihiga	3	Intercropping with maize, 3 year in row	0–2 cm	0.05	0.77
			0–2 cm	0.02	
Butere	4	Intercropping with maize, 3 year in row	0–2 cm, young	0.4	0.53
			0–2 cm, old	23	0.97
Egerton (A)	8	Monoculture, first year	Not cul.	0.13±0.04	NA
			0–5 cm	0.49±0.18	NA
Egerton (F)	0	Fallow field the year after *A. annua* cultivation	0–5 cm	0.1±0.03	NA

(Continued)

Table 7.2 (Continued)   Magnitude and Occurrence of Artemisinin in Soil and Plant Material from Different Climatic Locations and Cultivation Systems

Sample location	Plant density, (plants/m²)	Type of plantation	Depth	Artemisinin in soil (mg/ kg DW)	Artemisinin in leaves (mg/kg DW)
		West Virginia, USAb			
	Inland, cool summer climate, yearly precipitation: 1049 mm, average temperature day: 11.1°C				
Beaver reference	0		0–5 cm	0.16 ± 0.005	
			5–15 cm	0.15 ± 0.003	
Beaver 1	6	Experimental field	0–5 cm	0.16 ± 0.02	NA
			5–15 cm	0.18	
Beaver 2	6		0–5 cm	0.23 ± 0.03	
			5–15 cm	0.14 ± 0.001	
Beaver 3	6		0–5 cm	0.23 ± 0.07	
			5–15 cm	0.14	
Beaver 4	6		0–5 cm	0.3 ± 0.03	
			5–15 cm	0.15	
Beaver 5	6		0–5 cm	0.43 ± 0.06	
			5–15 cm	0.22 ± 0.09	
		South Italyb			
	Semi-arid mediterranean climate, yearly precipitation: 586 mm, average temperature day: 15.3°C				
Italy, Bari	6	Monoculture, 2005	0–10 cm	0.35 ± 0.13	NA
Italy, Bari	0	Green barley, 2006	0–10 cm	0.14 ± 0.021	
Italy, Bari	0	Yellow barley, after *A. annua* cultivation 2006	0–10 cm	0.14 ± 0.006	
		Denmark, Taastrup			
	Marine coast climate, yearly precipitation: 712 mm, average temperature day: 7.7°C				
Taastrup, *A. annua*	4	Experimental field, 1 year, at time of harvest	0–5 cm	0.44	0.35
			5–10 cm	0.11	

Source: Data from Herrmann, S., Jessing, K.K., Jørgensen, N.O.G., Cedergreen, N., Kandeler, E., and Strobel, B.W., *Soil Biol. Biochem.*, 57, 164–172, 2013; Jessing, K.K., Juhler, R.K., and Strobel, B.W., *J. Agric. Food Chem.*, 59(21), 11735, 2011.

a Soil artemisinin is analysed in true duplicates, why two measurements are given.
b Soil artemisinin is an average of duplicates. When standard derivation is left out the number only represents one analysis, as the duplicate gave 0.
NA Not analysed.

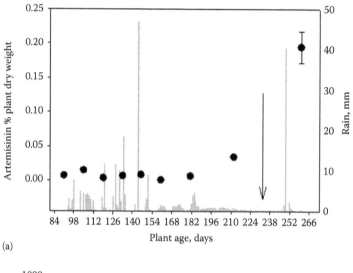

(a)

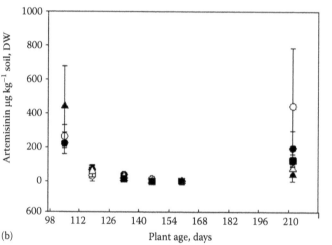

(b)

Figure 7.4 (a) Artemisinin content in leaves (% DW) (left *y*-axis, black circles) and daily precipitation (right axis, gray bars), both as a function of plant age in days. Data given as average ± standard error. Arrow indicates time of flowering. (b) Artemisinin content in soil at distances of 0.1 m (black circle), 0.25 m (open circle), 0.5 m (black triangle), 1 m (open triangle), and 2 m (black square) from the stem of the plant as a function of plant age at 0–5 cm depth. (Reprinted with permission from Elsevier, "Distribution and ecological impact of artemisinin derived from *Artemisia annua* L. in an agricultural ecosystem," *Soil Biology & Biochemistry* no. 57 (2013): 164–172.)

was a fallow field were *A. annua* had been cultivated the previous year (Jessing et al., 2011). The samples from Italy were taken from the topsoil of an *A. annua* culture in 2005. In this culture, high-yielding *A. annua* was grown for experimental purposes in rows, with grass in between the rows. The following year, barley was sown in the whole field. As discoloration of the barley was observed in the rows were *A. annua* had been cultivated, soil samples from both this and the healthy barley in between the rows were sampled from the upper 25 cm and analyzed for artemisinin. The average artemisinin concentration in the soil that had been cultivated with *A. annua* the previous year was 0.26 mg/kg. There was no significant difference in the artemisinin content of the yellow and healthy barley, which was 0.14 mg/kg on average. These data do, however, support the finding in the Egerton site in Kenya, showing that artemisinin is still measurable in the soil one year after cultivation of *A. annua*. The cultivation of *A. annua* in Beaver, West Virginia was for experimental purposes, and *A. annua* had been cultivated there continuously for six years. The *A. annua* was of high-yielding cultivars and sampling was performed at the end of the growing season, when the artemisinin content in the plants was expected to be high. On average, the artemisinin concentration in the 0–5 cm soil layer was 0.25 mg/kg, and in the 5–15 cm layer was 0.12 mg/kg. There was no correlation between concentrations at depths of 0–5 and 5–15 cm at the same spot. The measured concentrations presented in Table 7.2 are, as mentioned, "snapshots"; the result of dynamics involving continuous input from the plant and outputs in the form of processes taking place in the soil. From an exposure evaluation perspective, it is surprising that artemisinin could be measured several meters away from the plants, especially one year after cultivation. This tells us that the outputs of artemisinin from the soil environment are exceeded by the inputs.

7.3.4 Inputs and Outputs of Artemisinin to and from the Soil Environment

The inputs of artemisinin from the *A. annua* culture are of course important in determining the soil content of artemisinin. The distribution and routes of emission are also important in order to understand the potential risk of *A. annua* cultivation. A schematic overview of the proposed inputs and outputs of artemisinin in the soil environment is given in Figure 7.3.

7.4 ROUTES OF ARTEMISININ RELEASE TO THE SOIL

Examining the magnitude of the different routes of emission of artemisinin from *A. annua* to the soil aids in understanding environmental loading in various exposure scenarios, and also in quantifying and reducing the loss of valuable artemisinin compound to the environment. The potential release routes were divided into (1) debris from dead leaves, (2) rain wash-off from leaves, and (3) root exudation below ground as in Figure 7.3 (Jessing et al., 2013). Possible excretion of artemisinin from roots was determined without interference from aboveground plant sources. Root hairs are considered to be tubular outgrowths of trichoblasts, a specialized type of trichome (Peterson and Farquhar, 1996), and as artemisinin accumulates in the leaf trichomes, the same could be anticipated to also happen in root hairs. The accumulated amounts extrapolated to hectare scale are shown in Figure 7.5, assuming 100,000 *A. annua* plants per hectare, and even distribution of all excreted artemisinin in the upper 5 cm of soil. Artemisinin released from decaying leaves could explain 86%–108% of the total input. Rain wash-off contributed less than 0.5%, and scanning electron microscopy did not reveal any difference in the number of broken trichomes on leaves exposed to rain and leaves not exposed to rain (Jessing et al., 2013). The rain probably washed off artemisinin leaked to the leaf surface from already burst trichomes (Duke and Paul, 1993). The soil used in this experiment had a very high content of humic material, and relatively strong adsorption resulted in a relatively low recovery of 55% in spiked soil samples. Without correction for low recovery, soil concentrations of artemisinin up to 1 mg/kg were measured in the top 5 cm of soil. The highest measured leaf content of artemisinin in this experiment was 0.06%, which was low compared with a content of up to 1.6% reported elsewhere (Simonnet et al., 2011).

Root excretion of artemisinin *in situ* was determined with passive sampling silicone tubes. The tubes were placed in the soil at the beginning of the growing season and left there to facilitate regular, non-disturbing sampling throughout the growing season. The soil surface was covered with foil to avoid artemisinin contribution

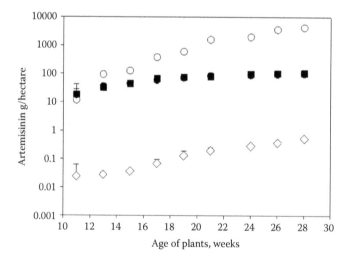

Figure 7.5 Accumulated artemisinin content (extrapolated to g/hectare) as a function of time in leaves above ground (open circles) ($n = 3$); in top 5 cm of soil with dead leaves as the only source (closed circles) ($n = 5$), in top 5 cm of soil with root, rain runoff, and dead leaf material as sources (closed squares) ($n = 3$), and in rain runoff with 10 mm precipitation events (open diamonds) ($n = 5$). Data given as average ± standard error. *Note*: Logarithmic *y*-axis. (Reprinted with permission from Elsevier, "Loss of artemisinin produced by *Artemisia annua* L. to the soil environment," *Industrial Crops and Products*, no. 43 (2013): 132–140.)

from sources other than the roots. A small amount of artemisinin was measured by *in situ* silicone tube-facilitated micro-extraction of the soil where root excretion was the only possible source of artemisinin. The diffusion of artemisinin, and other hydrophobic compounds, is expected to be very limited in soil (Schmidt and Ley, 1999) and, thus, only artemisinin in the close vicinity of the silicone tube would be sampled.

Dead leaf debris was identified as the major contributor to soil artemisinin input. The soil environment is exposed to continuous release of artemisinin, not single doses, as is the case when spraying with pesticides. As the soil concentration increases over the growing season (Figure 7.5), the inputs must be larger than the outputs, supporting the field observation that artemisinin is present in the soil environment even after the inputs stop at the time of harvesting the crop.

7.5 ADSORPTION, DEGRADATION, AND LEACHING

Artemisinin's occurrence and fate in soil is much dependent on interactions and processes within the soil. The main processes are adsorption, desorption, leaching, chemical and biological degradation, and uptake by organisms (Figure 7.3). The low vapor pressure of artemisinin almost eliminates evaporation from both the plants and soil (Table 7.1). It is expected that the proportion of artemisinin uptake by higher organisms such as plants and soil biota is relatively small compared with that being adsorbed, leached, and degraded by microorganisms in the soil.

7.6 SORPTION OF ARTEMISININ TO SOIL

Sorption of artemisinin to soil can occur either by partitioning or by surface complexation. Partitioning with soil organic matter is considered dominant, as artemisinin lacks specific functional groups that bind strongly to soil mineral surfaces. Based on K_{oc}, and assuming 5% organic matter in the soil, the partitioning coefficient, K_d, in soil is estimated to 16 L/kg. This suggests that the majority of artemisinin (86% in a fugacity calculation with only soil and water present) will be adsorbed to soil, when partitioning to soil organic matter is considered the only sorption mechanism. In degradation studies of artemisinin in different soil types, a large fraction of artemisinin disappeared within the first 24 h. This initial disappearance is believed to reflect partitioning into organic matter (Jessing et al., 2009a,b). Adsorption isotherms (Figure 7.6) were obtained on the previously mentioned Taastrup loamy soil and could be described by linear regressions. The slopes of the linear regressions were used as an estimate of K_d. They were 6.6 L/kg for the topsoil (0–10 cm), 3.3 L/kg for the subsurface (10–50 cm), and 0.92 L/kg for the subsoil (50–90 cm). Adsorption affinity decreased with increasing soil depth at the Taastrup site due to the decrease in soil organic matter content (Herrmann et al., 2013). If partitioning to organic matter was the only adsorption mechanism, K_d, based on the estimated K_{oc},

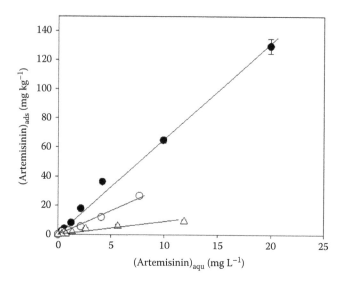

Figure 7.6 Adsorption isotherms of artemisinin to Taastrup soil. Adsorped artemisinin determined as function of artemisinin in aqueous phase at equilibrium. Linear regressions were used to estimate K_d (solid lines). Symbols: 0–10 cm depth (top-soil), closed circles; 10–50 cm depth (subsurface soil), open circles; and 50–90 cm depth (subsoil), open triangles. (Reprinted with permission from Elsevier, "Distribution and ecological impact of artemisinin derived from *Artemisia annua* L. in an agricultural ecosystem," *Soil Biology and Biochemistry* no. 57 (2013): 164–172.)

would have been 4.3 L/kg in the topsoil given the actual content of organic matter. As the measured K_d was higher, artemisinin probably adsorbs to clay material as well. Partitioning to organic matter may be reversible and desorption may occur.

7.7 DEGRADATION OF ARTEMISININ IN SOIL AND WATER

Degradation of specific compounds in soil is described by their half-life, when half of the initial amount is still present. The half-lives of artemisinin in soil have been calculated to be 0.9 and 4.22 days in loamy and sandy soil, respectively (Jessing et al., 2009b), when modeled as a sum of two first-order processes, where adsorption is considered to be the first fast process. When modeled as a first-order decay function with two parameters, the half-lives were 2.1, 8.3, and 13.5 days in sandy, clayey, and humic soil, respectively (Jessing et al., 2009a). The half-life of artemisinin in aqueous media was found to be 2.2–2.4 days in nutrient media at 22°C–23°C, and the degradation rate constant increased with increasing temperature (Jessing et al., 2014a). This is also expected to take place in soil matrices. The polycyclic structure of artemisinin suggests that the compound is relatively inert to hydrolysis. Biotic degradation of artemisinin in environmental matrices most likely occurs as well; as it does for most other organic compounds, but will not be discussed further. Chemical degradation, that is, reactions involving artemisinin, leading to transformed products not analyzed, is considered the dominant removal process in both water and soil.

The fact that artemisinin can be found one year after cultivation of *A. annua*, as was the case at the Egerton and the Italian sites, suggests that artemisinin can be sorbed in a way that protects the compound from degradation.

7.8 LEACHING

Artemisinin adsorbs strongest to the topsoil, with a K_d of 6.6 L/kg (Figure 7.6), but heavy rain events might provide conditions for preferential flow, causing transport of artemisinin further down through the soil where it adsorbs less due to the lower organic matter content. This was the case in the Danish study (Herrmann et al. 2013), where artemisinin was present along a vertical concentration gradient in the middle of the growing season after the soil had been exposed to artemisinin inputs for 17 weeks (Figure 7.7). Artemisinin concentration increased with depth and had the highest concentration at a depth of 90 cm, where the drainage pipes where located. As shown in Figure 7.1, artemisinin can reach surface waters via drainage pipes. Biotic degradation is expected to decrease with increasing depth as the conditions become more and more anaerobic. Despite its hydrophobic properties, artemisinin seems to have great potential to leach into ground water and nearby surface waters.

In Egerton, Kenya, the concentration was 0.1 mg/kg, and in Italy it was 0.14 mg/kg, one year after *A. annua* cultivation; hence, disappearance from the soil

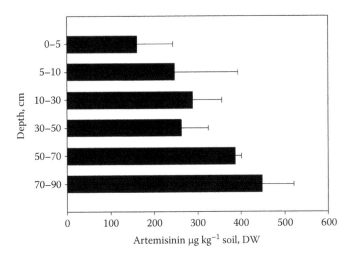

Figure 7.7 Artemisinin concentration by depth in the Danish culture at a plant age of 212 days (17 weeks after transplant to the field). Data given as average ± standard error (*n* = 3). (Reprinted with permission from Elsevier, "Distribution and ecological impact of artemisinin derived from *Artemisia annua* L. in an agricultural ecosystem," *Soil Biology and Biochemistry* no. 57 (2013): 164–172.)

environment is a relatively slow process. Artemisinin can be measured at a distance of up to 15 m away from the plantation, and has a rather heterogeneous and patchy distribution. This heterogeneous pattern correlates well with the observation that artemisinin is mainly released to the environment through plant debris, consisting largely of dead leaves, which can be distributed long distances with the wind. The long persistence of artemisinin, despite relative short laboratory-measured half-lives, suggests that desorption plays a role that is still not understood in making artemisinin bioavailable under field conditions.

7.9 ECOTOXICITY OF ARTEMISININ

As the environmental fate evaluation revealed exposure of the soil environment to artemisinin in the areas surrounding *A. annua* fields and potential leaching to surface waters through drains, the known ecotoxicities to soil- and water-living organisms are evaluated herein.

7.9.1 Ecotoxicity of Artemisinin in Soil

Most toxicity studies of natural bioactive compounds on invertebrates in the literature have the aim of finding potential new insecticides for use in pest management. Also, the focus has been on leaf and seed pests, whereas potential toxic effects on non-target soil organisms are studied much less. In soil spiked with artemisinin, Jessing et al. (2009b) did not see any effects on the reproduction of soil-living springtails, *Folsomia candida*, in a soil microcosm setup testing up to a concentration of 100 mg/kg soil. Fifty percent of earthworms of the species *Eisenia fetida* avoided artemisinin at a soil concentration of 21.6 mg/kg (EC_{50}) and 10% avoided 5.3 mg/kg soil (EC_{10}) in an experiment where the earthworms had the choice between uncontaminated soil and soil spiked with artemisinin. Earthworm growth, reproduction, or survival were not monitored, hence, the correlation between the avoidance behavior and toxicity could not be established. There was no observed effect on reproduction for soil-living potworms, *Enchytraeus crypticus*, in the tested concentration range (7.8–500 mg/kg) or on the soil-living nematode *Caenorhabditis elegans* in the tested range of soil (1.56–975 mg/kg) and water (3.175–50 mg/L) concentrations. Tests on agar did, however, show a significant response to artemisinin, with *C. elegans* showing an EC_{50} of 12.4 ± 7.2 mg/L for reproduction (Jessing et al., 2014b).

The quest to evaluate artemisinin as new potential herbicide has revealed it to have a well-established high level of phytotoxicity, which indicates that allelopathy is one of the main benefits of artemisinin production to *A. annua*. Therefore, it follows that the most artemisinin-sensitive species are plants. Examples of sensitive plants include several crops, for example, soybean and corn, which were inhibited by 25% and 9%, respectively, at a soil artemisinin concentration of 3.3 mg/kg. At the same concentration, growth of the weeds redroot pigweed and common lambsquarters were inhibited by 82% and 49%, respectively. The toxicity seems to be selective, as sorghum, for example, was not affected at the tested concentrations (Duke et al.,

1987). Jessing et al. (2009b) found that lettuce germination was not inhibited up to 100 mg/kg in a Petri dish study with sandy soil, but growth, measured as dry weight of aboveground plant parts after 21 days, was reduced, with an EC_{50} of 2.5 mg/kg sandy soil. Chen et al. (1991) conducted a comparable study, and although it was on filter paper and not on soil, it is relevant to mention here, as artemisinin was observed to have comparable toxicity to glyphosate and 2,4-D, two widely used herbicides, when compared on a molar basis.

To sum up the available present knowledge of ecotoxicity on soil-living organisms and plants, the most sensitive invertebrate species is the earthworm, *E. fetida*, as it avoids artemisinin at an EC_{10} of 5.3 mg/kg soil. Terrestrial plants are more sensitive, as germination and growth of several plants are also affected by artemisinin, and lettuce growth is reduced by 50% at a soil concentration of 2.5 mg/kg.

7.9.2 Ecotoxicity of Artemisinin in Surface Water

The ecotoxicity of artemisinin in water has been studied at three trophic levels. To be able to compare the aquatic ecotoxicological effect of artemisinin to that of conventional pesticides, artemisinin was tested on the freshwater crustacean *Daphnia magna*, which is the most frequently used test organism for freshwater ecotoxicology. The mobility of *D. magna* was not affected significantly by artemisinin at concentrations up to the solubility limit of the compound (50 mg/L). The tendency was, however, for artemisinin to slightly inhibit mobility at the highest concentration of 50 mg/L, both after 24 and 48 hours (Jessing et al., 2014b). At the two other trophic levels, Jessing et al. (2009b) found EC_{50} values for the freshwater algae *Pseudokirchneriella subcapitata* and *Lemna minor* of 0.08 and 1.3 mg/L, respectively, depending on test temperature and using relative growth rate as an endpoint. Similar effective concentrations for *L. minor* were found by Chen and Leather (1990) and Stiles et al. (1994), who also found a 44% reduction in chlorophyll content at a concentration of 0.7 g/L artemisinin. So, looking at aquatic plant species, there is little variance in toxicity between the different studies, and the aquatic phytotoxicity of artemisinin is comparable to that of commercial pesticides such as atrazine, which has EC_{50} values for *L. minor* and *P. subcapitata* of 0.08 and 0.29 mg/L, respectively (Giddings et al. 2005).

7.10 RISK ASSESSMENT OF *A. ANNUA* CULTIVATION

Current regulations do not require risk assessment for either the use or production of naturally produced bioactive chemical substances. But, especially with the *Pharmland* situation (Figure 7.1), where monocultures of plants producing bioactive compounds are grown, as is the case with *A. annua* for artemisinin production, future regulation may be a possibility. Even if regulations are not established, it is in everybody's interest, particularly the farmers, to be able to grow *A. annua* in a way that does not damage beneficial organisms and the crop of the following years.

Based on the environmental fate evaluation, it is known that artemisinin is spread from the plants and enters the soil environment, from where it may leach to surface waters. The soil environment exposure is continuous over the growing season and artemisinin is present in the soil at least one year after cultivation. The latter is either due to continuous release from dead leaves left in the field after harvest or desorption from soil organic matter, or both. Some soil and aquatic organisms are sensitive to artemisinin, and the remaining question to be answered is whether artemisinin production on a field scale is environmentally safe. The risk assessment is performed here as a calculation of the HQ, as described in the Introduction, using Eq. 7.1.

7.10.1 Considerations on PEC Values in Soil

The actual measured concentrations will be represented as PEC_{meas}. But, as these are a function of various processes, worst-case scenarios in various soil types will be provided as well. The scenario is as follows: According to the findings in Jessing et al. (2013), in an *A. annua* field with a yield of 30 kg artemisinin per hectare, 4.5 kg will be lost to the soil environment. In this scenario, the entire artemisinin load is released at once. Three soil types are considered: sandy, loamy, and humic, with respective soil densities assumed to be 2.3, 2, and 1.8×10^3 kg/m³. Both the sandy and loamy soils are quite representative of the Kenyan soil types in terms of content of organic matter and clay. Assuming artemisinin is distributed in the top 5 cm of soil, $PEC_{initial}$ will be 3.9, 4.5, and 5.0 mg/kg, respectively, for the three scenarios. As discussed earlier, there is an initial fast disappearance of extractable artemisinin within the first 24 h after its application to the soil. Using the disappearance parameters found in Jessing et al. (2009a,b), $PEC_{24\ hours}$ will then be 2.27, 2.03, and 1.4 mg/kg for sandy, loamy, and humic soil, respectively. Thirty days after the exposure, $PEC_{30\ days}$ would correspondingly be 0.46 and 0.17 mg/kg, in the sandy and loamy soil, respectively, whereas none would be left in the humic soil. This is assuming a dry period without rain events that would remove more artemisinin by leaching. These values correspond well with the ranges measured in cultivated soils. The last PEC value to be introduced is $PEC_{15\ m}$, a concentration of 0.014 mg/kg measured at a 15 m distance from the Danish cultivation (Herrmann et al., 2013). $PEC_{15\ m}$ will be used to evaluate risk in the near surroundings of *A. annua* plantations.

7.10.2 Considerations on PEC Values in Surface Water

As it was observed that artemisinin is leachable, it is relevant to evaluate the risk to aquatic organisms in surface waters close to *A. annua* cultivations. A worst-case PEC_{water} was calculated in the following way: Again, we are looking at a field with an artemisinin production of 30 kg per hectare, where 4.5 kg is lost to the surroundings (at once). It is then assumed that one-thousandth of the lost amount ends up in a pond nearby, either by leaching or as wind spread of dead leaves. The pond has a surface area of 100 m² and a depth of 1 m, resulting in an artemisinin concentration, PEC_{water}, of 0.005 mg/L. This is without considering loss of artemisinin, for example, sorption in the drainage pipes or degradation, on the way.

7.11 RISK ASSESSMENT FOR SOIL ENVIRONMENT

In Jessing et al. (2009b, 2014b), we did not see any effects on the soil-living invertebrates, springtails, nematodes, and potworms at nominal concentrations below 100, 975, and 500 mg/kg. These concentrations are at least 50 times higher that the highest measured artemisinin soil concentration. Earthworms, which seem to be the most sensitive of the tested soil invertebrates, avoid artemisinin-containing soil with EC_{10} and EC_{50} values of 5.24 and 21.6 mg/kg, respectively (Jessing et al., 2009b). The HQ is not calculated on this scarce information, but the ecological relevance of worms avoiding artemisinin-containing soil cannot be neglected. Soil-living worms are responsible to a large degree for decomposition of organic material, as well as aeration of the soil. There is a risk of decreased soil quality as a consequence of *A. annua* cultivation, especially in large-scale monocultures, if earthworms avoid the field and surface soil, where artemisinin exposure is massive. Lack of suitable food, if *A. annua* leaves are avoided (Maggi et al., 2005), could also be a problem for earthworms, but this would have to be tested. The avoidance EC_{10} reported is higher than PEC_{meas}, but only by a factor of 2.7 compared with the highest measured concentration in Kenya. The estimated $PEC_{initial}$ values are within the same range as this EC_{10}, indicating that the concentration can be higher in spots where leaves fall. In addition, the exposure in the field will be long-term, whereas the reported effect concentration is from only 48 hours of exposure.

The reported growth-inhibiting effects of artemisinin in soil on the germination and growth of *A. annua* itself (autotoxicity) (Duke et al., 1987) may explain why the bioactive compound is produced and stored in the non-living trichomes, where it is harmless to *A. annua* itself. A decrease in *A. annua* and artemisinin yield has been reported (Lenardis et al., 2011) when cultivating *A. annua* in the same field consecutively. This could be due to the phytotoxic effects of artemisinin on *A. annua* itself if the crop is intensively grown at the same location several years in a row.

In the study by Jessing et al. (2009b), the EC_{10} and EC_{50} were reported to be 0.54 and 2.5 mg/kg, respectively, for lettuce growth. Due to the wide selectivity of artemisinin phytotoxicity reported by Lydon et al. (1997), a safety factor of 1000 is initially used. As a 10% reduction in crop yield is considered relevant, the actual EC_{10} of lettuce is used as the PNEC value. Both the lowest and the highest PEC_{meas} values are used to cover the broad range of concentrations found. This results in the following calculation:

$$HQ = \frac{0.0004 \text{ mg/kg}}{0.54 \text{ mg/kg} \times 1000^{-1}} \text{ to } \frac{2.03 \text{ mg/kg}}{0.54 \text{ mg/kg} \times 1000^{-1}} = 0.74 \text{ to } 3760$$

Even without a safety factor, the HQ is 3.75 at the highest measured soil concentration. Using $PEC_{30 \text{ days}}$ instead to stimulate the situation in a cropping system where another crop follows *A. annua* 30 days after harvest, using a safety factor of 100, results in HQs of 7 and 18 in sandy and loamy soil, respectively. In Egerton, the concentration was 0.1 mg/kg, and in Italy it was 0.14 mg/kg, one year after

A. annua cultivation. These are realistic concentrations to expect one year after *A. annua* cultivation. If the safety factor is reduced to 10, the HQs will in this case be 1.8–2.6, and hence there is a risk for artemisinin-sensitive terrestrial plants being growth-inhibited after cultivation of *A. annua*. Negative effects on weeds may be appreciated in a cultivation system, but negative effects in the form of reduced yield of following crops is a problem. Plant diversity around the *A. annua* field is also at risk of being reduced, as the HQ using $PEC_{15\,m}$ is 26. As the $PEC_{15\,m}$ value is a result from measurements in the surroundings of a field with low artemisinin production compared with the reported and observed levels from Kenya and Asia, a safety factor of 1000 is used in this scenario.

7.12 RISK ASSESSMENT FOR AQUATIC ORGANISMS

There is a risk that artemisinin can reach surface waters through drainage pipes. The observed patchy distribution of artemisinin in the soil indicates that spread of dead leaf material by wind occurs as well. Surface waters close to *A. annua* cultivations could then also be contaminated by dead leaves blowing into the water. The phytotoxicity of artemisinin to aquatic organisms is well documented, and average EC_{50} values of 0.18 mg/L for *P. subcapitata* and 0.43 mg/L for *L. minor* (Jessing et al., 2009a,b) will be used as the PNEC. A safety factor of 1000 will be used here since long-term exposure concentrations and effects are unknown for other aquatic plants and algal species as well. Using this safety factor, the HQs will be 25 and 10, indicating that there is a risk to the algae and macrophytes in surface waters close to *A. annua* cultivations. The mobility of *D. magna*, the only animal tested, was not affected significantly by artemisinin at concentrations up to 50 mg/L, the solubility limit of the compound (Jessing et al., 2014b). Long-term effects on reproduction were, however, not investigated, and it has to be underlined that long-term exposure to low concentrations can be worse than short-term exposure to higher concentrations. Also, other aquatic heterotrophic organisms may be more sensitive than daphnids to artemisinin.

7.13 CONCLUSION ON RISK ASSESSMENT
OF *A. ANNUA* CULTIVATION

The HQs and data used to produce them for the most sensitive species are summarized in Table 7.3.

The data show that cultivating *A. annua* poses a risk of reduced soil quality in terms of reduced number of worms aiding decomposition in the soil. The most vulnerable species are, however, plants, and there is a risk of reduced floral biodiversity in terms of less species and fewer of each species in the near surroundings of *A. annua* fields. A change in the flora will also indirectly affect the faunal composition and diversity. The risk assessment did also show a risk of reduced yield of the following crop, a

Table 7.3 Predicted Effect Concentration (PEC), Predicted No Effect Concentration (PNEC), the Applied Safety Factor and the Calculated Hazard Quotient (HQ) for the Species Identified as Being Most Sensitive Towards Artemisinin. If HQ > 1 There Is a Risk

Organism	PEC	PNEC	Safety Factor	HQ
Lettuce	PEC_{meas}	EC10	1000	3760
	PEC_{meas}	EC50	100	81
	$PEC_{nextyear}$	EC10	10	2.6
	PEC_{15m}	EC10	1000	26
Duckweed	PEC_{water}	EC50	1000	10

problem in third-world countries in particular, where the largest areas of *A. annua* are cultivated and food sources are generally scarce. Artemisinin can leach, and cultivation of *A. annua* may pose a risk to aquatic plants in surface waters near the cultivations.

Hence, even though plant-produced active compounds have not yet been subjected to regulation and risk assessment, either when used against pests or, as here, produced in the environment, the present study shows that it could be very relevant to do so. To reduce the risk of adverse environmental effects, crop rotation with resistant crops and intercropping are recommended in farming systems with *A. annua* to minimize the artemisinin exposure of both the field and the environment. As decay of dead leaves was the largest contributor to soil artemisinin, careful collection of as much plant material as possible and not drying the plants directly on the ground during the harvesting process are strongly recommended. Minimizing leaf loss during harvest is also beneficial to the farmers as it will increase the artemisinin yield.

7.14 PERSPECTIVES

One of the problems with regulating artemisinin biomedicine production is that the production takes place in third-world countries where environmental regulation is limited. As a representative case study for biomedicine production, the knowledge from this work should, however, lead to considerations on whether *in situ* production of bioactive compounds, for example, ginseng in Europe, should be risk-assessed and possibly regulated.

No economically feasible alternatives to artemisinin production by *A. annua* cultivation currently exist. Therefore, if we want to fight malaria, recommendations on better cultivation practice are worth considering. Pruning during the growing season, along with the removal of withering leaves, would increase yield, as seen in studies from India. In addition, this would reduce the biomass liberated to the soil surface. Harvesting must be carried out before the plants dry out and massive loss of biomass occurs. Situations where very dry plants are cut and left on the ground in the field should be avoided, as these generate artemisinin hotspots. In general, drying of the plant material should be carried out on plastic or concrete and not directly on the soil. Cultivation of *A. annua* year after year in the same field results in long-term exposure to artemisinin and should be avoided, in order to maintain the same level of yield year after year. Intercropping is recommended, as this will provide soil-living organisms with "refugee spots." In both crop rotation and intercropping, the phytotoxic properties of *A. annua* might even be utilized if we can obtain more knowledge on its selectivity. More knowledge is also needed on the long-term effects of artemisinin on soil-living and aquatic organisms as well as on insects and microbial communities. Sorption, desorption, and leaching of artemisinin in various soil types under different climatic conditions, as well as measurements of actual artemisinin concentrations in drainage outlets, streams, and lakes in the surroundings of *A. annua* cultures require further investigations. In addition, a full risk assessment of the mixture of bioactive secondary substances that *A. annua* produces and releases is worth considering, as artemisinin is most likely not the only secondary metabolite from *A. annua* with bioactive properties.

REFERENCES

Abdin M.Z., Israr M., Rehman R.U., and Jain S.K. 2003. Artemisinin, a novel antimalarial drug: Biochemical and molecular approaches for enhanced production. *Planta Medica* 69(4): 289–299.

Arsenault P.R., Wobbe K.K., and Weathers P.J. 2008. Recent advances in artemisinin production through heterologous expression. *Current Medicinal Chemistry* 15(27): 2886–2896.

Carlsen S.C.K., Pedersen H.A., Spliid N.H., and Fomsgaard I.S. 2012. Fate in soil of flavonoids released from white clover (*Trifolium repens* L.). *Applied and Environmental Soil Science* 2010: 1–10.

Chen P.K. and Leather G.R. 1990. Plant growth regulatory activities of artemisinin and its related compounds. *Journal of Chemical Ecology* 16(6): 1867–1876.

Chen P.K., Polatnick M., and Leather G. 1991. Comparative study on artemisinin, 2,4-D, and glyphosate. *Journal of Agriculture and Food Chemistry* 39(5): 991–994.

Dhingra V., Rao K.V., and Narasu M.L. 2000. Current status of artemisinin and its derivatives as antimalarial drugs. *Life Sciences* 66(4): 279–300.

Duke S.O. and Paul R.N. 1993. Development and fine structure of the glandular trichomes of *Artemisia-annua* L. *International Journal of Plant Sciences* 154(1): 107–118.

Duke S.O., Vaughn K.C., Croom E.M., and El Sohly H.N. 1987. Artemisinin, a constituent of annual wormwood (*Artemisia annua*), is a selective phytotoxin. *Weed Science* 35(4): 499–505.

Ferreira J.F.S. 2007. Nutrient deficiency in the production of artemisinin, dihydroartemisinic acid, and artemisinic acid in *Artemisia annua* L. *Journal of Agriculture and Food Chemistry* 55(5): 1686–1694.

Ferreira J.F.S. and Janick J. 1995. Floral morphology of *Artemisia annua* with special reference to trichomes. *International Journal of Plant Sciences* 156(6): 807–815.

Ferreira J.F.S., Laughlin J.C., Delabays N., and Magalhães P.M. 2005. Cultivation and genetics of *Artemisia annua* L. for increased production of the antimalarial artemisinin. *Plant Genetic Resources* 3(2):206–229.

Giddings J.M., Anderson T.A., Hall Jr. L.W., Hosmer A.J., Kendall R.J., Richards R.P., Solomon, K.R., and Williams W.M. 2005. Tier 3 risk assessment. In A. Green (ed) *Atrazine in North American Surface Waters—A Probalistic Aquatic Ecological Risk Assessment.* SETAC, Pensacola, FL, 85–140.

Gimsing A.L. and Kirkegaard J.A. 2006. Glucosinolate and isothiocyanate concentration in soil following incorporation of Brassica biofumigants. *Soil Biology and Biochemistry* 38(8): 2255–2264.

Herrmann S., Jessing K.K., Jørgensen N.O.G., Cedergreen N., Kandeler E., and Strobel B.W. 2013. Distribution and ecological impact of artemisinin derived from *Artemisia Annua* L. in an agricultural ecosystem. *Soil Biology and Biochemistry* 57: 164–172.

Hoerger C.C., Schenzel J., Strobel B.W., and Bucheli T.D. 2009. Analysis of selected phytotoxins and mycotoxins in environmental samples. *Analytical and Bioanalytical Chemistry* 395(5): 1261–1289.

Janick J. and Ferreira J.F.S. 1996. Distribution of artemisnin in *Artemisia annua*. In J. Janick (ed) *Progress in New Crops*. ASHS Press, Alexandria, VA, 579–584.

Jessing K., Bowers T., Strobel B.W., Svensmark B., and Hansen H.C. 2009a. Artemisinin determination and degradation in soil using supercritical fluid extraction and HPLC–UV. *International Journal of Environmental and Analytical Chemistry* 89(1): 1–10.

Jessing K.K., Andresen M., and Cedergreen N. 2014a. Temperature-dependent toxicity of artemisinin toward the macrophyte lemna minor and the algae *Pseudokirchneriella subcapiata*. *Water, Air, and Soil Pollution* 225(6): 2010.

Jessing K.K., Cedergreen N., Jensen J., and Hansen H.C. 2009b. Degradation and ecotoxicity of the biomedical drug artemisinin in soil. *Environmental Toxicology and Chemistry* 28(4): 701–710.

Jessing K.K., Cedergreen N., Mayer P., Libous-Bailey L., Strobel B.W., Rimando A., and Duke S.O. 2013. Loss of artemisinin produced by *Artemisia annua* L. to the soil environment. *Industrial Crops and Products* 43:132–140.

Jessing K.K., Duke S.O., and Cedergreeen N. 2014b. Potential ecological roles of artemisinin produced by *Artemisia annua* L. *Journal of Chemical Ecology* 40(2): 100–117.

Jessing K.K., Juhler R.K., and Strobel B.W. 2011. Monitoring of artemisinin, dihydroartemisinin, and artemether in environmental matrices using high-performance liquid chromatography-tandem mass spectrometry (LC–MS/MS). *Journal of Agriculture and Food Chemistry* 59(21): 11735–11743.

Klayman D.L. 1985. Qinghaosu (artemisinin)—An antimalarial drug from China. *Science* 228: 1049–1055.

Kumar S., Gupta S.K., Singh P., Bajpai P., Gupta M.M., Singh D., Gupta A.K., et al. 2004. High yields of artemisinin by multi-harvest of *Artemisia annua* crops. *Industrial Crops and Products* 19(1): 77–90.

Lenardis A.E., Morvillo C.M., Gil A., and de la Fuente E.B. 2011. Arthropod communities related to different mixtures of oil (*Glycine max* L. Merr.) and essential oil (*Artemisia annua* L.) crops. *Industrial Crops and Products* 34(2): 1340–1347.

Liu J.M., Ni M.Y., Fan J.F., Tu Y.Y., Wu Z.H., Qu Y.L., and Chou M.S. 1979. Structure and reaction of arteannuin. *Acta Chim. Sin.* 37: 129–143.

Lommen W.J.M., Elzinga S., Verstappen F.W.A., and Bouwmeester H.J. 2007. Artemisinin and sesquiterpene precursors in dead and green leaves of *Artemisia annua* L. crops. *Planta Medica* 73(10): 1133–1139.

Lydon J., Teasdale J.R., and Chen P.K. 1997. Allelopathic activity of annual wormwood (*Artemisia annua*) and the role of artemisinin. *Weed Science* 45: 807–811.

Maggi M.E., Mangeaud A., Carpinella M.C., Ferrayoli C.G., Valladares G.R., and Palacios S.M. 2005. Laboratory evaluation of *Artemisia annua* L. extract and artemisinin activity against *Epilachna paenulata* and *Spodoptera eridania*. *Journal of Chemical Ecology* 31(7): 1527–1536.

Marchese J.A., Ferreira J.F.S., Rehder V.L.G., and Rodrigues O. 2010. Water deficit effect on the accumulation of biomass and artemisinin in annual wormwood (*Artemisia annua* L., Asteraceae). *Brazilian Journal of Plant Physiology* 22(1): 1–9.

Peplow M. 2016. Synthetic malaria drug meets market resistance. *Nature* 530(7591): 389–390.

Peterson R.L. and Farquhar M.L. 1996. Root hairs: Specialized tubular cells extending root surfaces. *Botanical Review* 62(1): 1–40.

Schmidt S.K. and Ley R.E. 1999. Microbial competition and soil structure limit the expression of allelochemicals in nature. In Inderjit, K.M.M. Dakshini, and C.L. Foy (eds) *Principles and Practices in Plant Ecology*. CRC Press, Boca Raton, FL, 339–351.

Simonnet X., Quennoz M., and Carlen C. 2011. Apollon, a new *Artemisia annua* variety with high artemisinin content. *Planta Medica* 77(12): 1369.

Stiles L.H., Leather G.R., and Chen P.K. 1994. Effects of two sesquiterpene lactones isolated from *Artemisia annua* on physiology of *Lemna minor*. *Journal of Chemical Ecology* 20(4): 969–978.

Strobel B.W., Jensen P.H., Rasmussen L.H., and Hansen H.C.B. 2005. Thujone in soil under *Thuja plicata*. *Scandinavian Journal of Forest Research* 20(1): 7–11.

Tellez, M.R., Canel, C., Rimando, A.M., Duke, S.O. 1999. Differential accumulation of isoprenoids in glanded and glandless *Artemisia annua* L. *Phytochemistry* 52(6): 1035–1040.

Wallaart T.E., Pras N., Beekman A.C., and Quax W.J. 2000. Seasonal variation of artemisinin and its biosynthetic precursors in plants of *Artemisia annua* of different geographical origin: Proof for the existence of chemotypes. *Planta Medica* 66(1): 57–62.

WHO. 2015. World malaria report 2015 fact sheet. World Health Organization. Geneva, Switzerland.

Zhang S.S., Zhu W.J., Wang B., Tang J.J., and Chen X. 2011. Secondary metabolites from the invasive *Solidago canadensis* L. accumulation in soil and contribution to inhibition of soil pathogen *Pythium ultimum*. *Applied Soil Ecology* 48(3): 280–286.

CHAPTER 8

Therapeutics of *Artemisia annua*
Current Trends

Shilpi Paul

CONTENTS

8.1 *ARTEMISIA ANNUA*: THE PLANT

Based on traditional medicine usage for thousands of years, a number of modern drugs have also been isolated from natural sources like plants. One such plants is *Artemisia*. Extensive documentation over the centuries about the plant *Artemisia*, with the first record dating from about 1100 BC (Wu Shi Er Bing Fang, containing 52 prescriptions), followed by the works such as the Shennong Herbal (~100 BC: 365 drugs) and the Tang herbal (659 AD: 850 drugs) against the treatment of different diseases, existed in Chinese literature (Cragg et al., 2001), and which are also referred to in the Chinese *materia medica*.

Several *Artemisia* species have been cited by early herbalists including Theophrastus in the third century BC (Einarson and Link, 1976), Pliny (Bostock and Riley, 1855–1857), and Dioscorides (Gunther, 1959) in the first century BC. Wormwood (probably the species *Artemisia judaica*) has also been mentioned in the Bible (Rev 8:10,11). In 340 AD, Ge Hong prescribed aerial part of *Artemisia* for the treatment of fever in the "Chinese Hand Book of Prescriptions for Emergency Treatments" and, in 1527, Li Shi Zhen, a Chinese herbalist/pharmacologist mentioned the use of *huang hua hao* (or yellow flower, later identified as *Artemisia annua*) for the treatment of children's fever and qinghao (*Artemisia apiacea*) as a treatment for the disease now known as malaria (Klayman, 1993).

The earliest mention of this herb dates back to 168 BC in the ancient recipes found in the tomb of the Mawangdui Han dynasty (Klayman, 1985). In 1967, Chinese researchers evaluated the effectiveness of several traditional herbal remedies against malaria and found hot water or ethanol extract of *A. annua* having no antimalarial effect (Klayman et al., 1984). In 1971, a low temperature extraction of *A. annua* with diethyl ether produced a complex antimalarial activity on both infected mice and monkeys. The main active principle, named artemisinin, was isolated this way and its structure defined in 1972 in China (Anon, 1979).

A. annua is native to Asia, most probably China (McVaugh and Andrson, 1984). In China, *A. annua* L. is commonly known as "Qinghao." In English, it is known as annual wormwood, sweet annie, or sweet wormwood, and it is a vigorous annual weed (Hall and Clements, 1923). It generally occurs naturally as a part of steppe vegetation in the northern part of the Chahar and Suiyuan provinces (40°N, 109°E) in China at 1000–1500 m above sea level (Wang, 1961). Presently, the species is widely distributed throughout the temperate region (Bailey and Bailey, 1976; Simon et al., 1984). The plant now grows wild in countries such as Argentina, Bulgaria, France, Hungary, Romania (where it is cultivated for its essential oil), Italy, Spain, the United States, and the former Yugoslavia (Bailey and Bailey, 1976; Gray, 1984; Klayman, 1989, 1993). Some high-yielding varieties, namely CIM-Arogya, Jeewanraksha, and Suraksha (Khanuja et al., 2005; Kumar et al., 2002; Paul et al., 2010), were developed by Indian researchers for large-scale production.

The plant is usually single stemmed, reaching about 2 m in height with alternate branches and alternate deeply dissected aromatic leaves ranging from 2.5 to 5.0 cm in length. Tiny greenish-yellow nodding capitula only 2 or 3 mm in diameter are enclosed by numerous imbricate bracts. Flowers in capitula are displayed in a loose panicle, containing numerous central bisexual florets and marginal pistillate florets, the latter extruding their stigmas prior to those of the central florets. The capitulum receptacle is glabrous, not chaffy and triangular in shape. Both florets and receptacle bear abundant 10-celled biseriate glandular trichomes, which are the source of arte-misinin and highly aromatic volatile essential oils (Ferreira et al., 1995) as shown in Figure 8.1.

Non-glandular, "T" shaped, trichomes are also present along with 10-celled bise-riate glandular trichome on leaves, stem, and inflorescence. The morphology and origin of the trichomes have been described in detail for leaves (Duke and Paul, 1993) and capitula (Ferreira and Janick, 1995) using light and/or scanning electron microscopy. It is a short day plant, very responsive to photoperiodic stimulus. The critical photoperiod is reported to be about 13.5 daylight hours, but there is a possi-bility of photoperiod × temperature interaction. The plants flower in early September with the production of mature seeds in October (Janick, 1995). Nuclear DNA content and other karyological characters in the population of *A. annua* were studied by Torrell and Valles (2001). In their study, the DNA per haploid genome was 1.75pg, the total karyotypic length was 19.58 µm, and the total haploid chromosome set length was 9.74 µm. They also reported that the annual species *A. annua*, showed the lowest amount of DNA as compared to the other species. It has approximately 35% lower C value than those of the perennial species.

Figure 8.1   *A. annua* in the field.

8.2 THERAPEUTIC COMPOUNDS OF *A. ANNUA*

The widespread occurrence of volatile and non-volatile monoterpenes and sesquiterpenes in the Asteraceae has led to the use of these compounds as possible taxonomic markers in the classification of this family as well as in the treatment of the genus *Artemisia*. Artemisinin and other cadinanes are unique to *A. annua* (Liersch et al., 1986). Similarly high concentrations of Artemisia ketone and its derivatives in the volatile oil of *A. annua* suggest their possible systematic significance since these compounds have not been found outside the tribes of the Asteraceae.

Artemisinin has been detected in aerial parts of the plants, that is, small green stem, buds, flowers, and seeds (Acton et al., 1985; Zhao and Zeng, 1985; Liersch et al., 1986). The leaves and inflorescence have 10-celled biseriate trichomes possessing artemisinin and essential oil. The effect of morphogenetic variation on oil and artemisinin production was studied by Gupta et al. (2002).

Sesquiterpene lactone (Artemisinin) was isolated for the first time in 1972 and its structure was established by combined spectral, chemical, and x-ray analysis (Liu et al., 1979; Zhongshan et al., 1985; Lepan et al., 1988). Its simple and quick analytical method for estimation was developed by Gupta et al. (1996).

More than 60 constituents have been reported in the distilled oil. The major oil constituents are Artemisia ketone, 68.5% (Charles et al., 1991); 1-8 cineole, 31.5% (Libbey and Sturtz, 1989); Camphor, 27.5% (Charles et al., 1991); germacrene D, 18.9% (Woerdenbag et al., 1994); camphene hydrate, 12% (Charles et al., 1991); α-pinene, 16% (Charles et al., 1991); β-pinene, 8.6% (Woerdenbag et al., 1994); myrcene, 7.5% (Woerdenbag et al., 1994); and artemisia alcohol, 7.5% (Woerdenbag et al., 1993).

8.2.1 Artemisinin and Its Derivatives

Chinese chemists isolated the substance responsible for its medicinal property from the leafy portions of *A. annua* L. in 1972. They named the crystalline compound "Qinghaosu." In the west, it is called artemisinin (Klayman, 1985). Its structure lacks a nitrogen-containing heterocyclic ring system. The peroxide moiety of artemisinin appears to be indispensable for chemotherapeutic activity. Muller et al. (2000) reported that 92% of the malarial patients showed disappearance of parasitemia within four days after the treatment of *A. annua* tea. Artemisinin is poorly soluble in water, hence its water-soluble derivatives are used for therapy.

When artemisinin is treated with borohydride, to give dihydroartemisinin (DHA), a lactol is formed in which the integrity of the peroxide group is retained and the schizonticidal activity is enhanced two-fold. In the pharmacokinetics study, DHA showed 12.5 $T^{1/2}$ (half-life), 7.40 Tmax time of Cmax and 593 Cmax (peak plasma concentration) in dose of 200 mg with artemether (Benakis et al., 1993). Maggs et al. (1997) reported that the principal biliary metabolite ($21 \pm 9.3\%$ of DHA dose) was the biologically inactive DHA glucuronide. The other metabolites are the product of reductive cleavage and rearrangement of the endoperoxide bridge, a process known to generate reactive radical intermediates and abolish antimalarial activity. Dimer of DHA showed the cytotoxicity to committed progenitor cells of the granulocyte monocyte lineage (CFU-GM) cells (Beekman et al., 1998). It was determined in plasma through gas chromatography–mass spectrometry by Mohamad et al. (1999). It was chemically synthesized from artemisinin by Jain et al. (2001). It may be dissolved in groundnut oil with mild heating at 80°C–90°C and cooled to room temperature. DHA, the active metabolite of the compound is known to alter the ribosomal organization and the endoplasmic reticulum as well as causing the dilation of nuclear envelope and disintegration of food vacuoles (Kakkilaya, 2002).

Artemether is formed by the conversion of DHA into methyl ester in a two-step procedure, and is stable at room temperature (Valecha and Tripathi, 1997). It was first synthesized by Li Yin in 1978 (Jansen and Zhimin, 1997). Mohmad et al. (1999) determined artemether in plasma through gas chromatography–mass spectrometry. It is available as an injection of 80 mg, which was soluble in oil for intramuscular use. The dose of this compound was 3.2 mg/kg intramuscularly as a loading dose followed by 1.6 mg/kg orally, daily for a maximum of seven days (Kakkilaya, 2002). β-Artemether is a novel antimalarial drug, which is more active than the parent compound, artemisinin (WHO, 1985). β-Artemether is produced in Belgium under the trade name is Artenam®. It is also marketed under the trade name Paluther® and is manufactured in India by Ipca laboratories under the trade name Larither®.

Artesunate is an ester derivative of artemisinin (Barradill and Filton, 1995; Valecha and Tripathi, 1997) it is more soluble in water. Zhao et al. (1986) reported that the activity of the cytochrome oxidase, which is located in the plasma, nuclear, and the food vacuole limiting membrane as well as in the mitochondria of the trophozoite of *Plasmodium berghei,* was inhibited completely by sodium artesunate at 1mM in the *in vitro* test and at 100 mg/kg *in vivo*. This enzyme appears to be a target for the antimalarial mechanism of action of artesunate and artemisinin. The drug is available as a powder; it should be first dissolved in 1 mL of 5% sodium bicarbonate and shaken for 2–3 mins. After complete dissolution, it is diluted with 5 mL of 5% dextrose or saline for intravenous use, or 2 mL for intramuscular use. An intravenous dose should be injected slowly at a rate of 3–4 mL/min. It is also available as a tablet, each tablet containing 50 mg of the drug. Adults and children over 6 months, 5 mg/kg orally on the first day followed by 2.5 mg/kg on the second and third days, in combination with mefloquine (15 mg/kg) in a single dose on the second day. In a few areas, a higher dose (25 mg/kg) of mefloquine may be required for a cure to be obtained. Oral artesunate is not recommended during pregnancy (Kakkilaya, 2002). Borrmann et al. (2003) suggested that artesunate alone is not sufficient to radically cure *P. falciparum* malaria, but it improves the efficacy of treatment with the addition of amodiaquine.

Artelinate/artelinic acid is prepared by two-step batches of 10 g artemisinin; first by using a reduction of artemisinin, which is converted into DHA, and then DHA is converted into a mixture of α- and β- artelinic acid (Shrimali et al., 1998).

Artether is prepared in two steps using artemisinin batches of 10 g. The first step in the preparation of arteether involving the reduction of artemisinin with sodium borohydride into DHA and subsequent esterification of DHA by Lewis acid catalyzed reaction, affording an epimeric (80:20 mixtures of β- and α- isomers) ether of DHA. Both the epimers are separated by column chromatography and crystallized to yield crystalline β arteether and oily α arteether (Jain et al., 2000). An arteether capsule containing 40 mg of the drug is now available. It is also available as an injection of 150 mg in 2 mL. The dosage is 3 mg/kg once a day for 3 days as deep intramuscular injection. On the basis of pharmacokinetic experiments on the derivatives of artemisinin, Kager et al. (1994) concluded that arteether may have the longest half-life of all the artemisinin derivatives studied so far. The study of Dutta et al. (1989) strongly advocated for the future clinical evaluation of arteether for the control of multidrug-resistant and high-risk *P. falciparum* cases in areas especially of chloroquine-, mefloquine-, and quinine-resistant malaria. The objectives of the WHOCHEMAL's programme encouraged the development of arteether as a single dose parenteral treatment for severe and complicated *P. falciparum* infection. The choice to develop β-arteether rather than artemether was based on the fact that arteether would be more lipophilic and its metabolic breakdown gives ethanol and not methanol, which would avoid the problems of methanol toxicity that can arise from the metabolic formation of formaldehyde and formic acid (Ritchi, 1985; Brossi et al., 1988). India has made a significant contribution in this worldwide effort, through the development of a highly efficacious variant of arteether. It developed alpha–beta (α–β) arteether (30:70 mixture of enantiomers) as a fast-acting blood schizontocide especially for the control of drug-resistant and cerebral malaria (Dutta et al., 1994). For the clinical trial, as per requirement, the Central Institute of Medicinal and Aromatic Plants (CIMAP) has prepared purified arteether and separated it into α- and β-enantiomers. Both the isomers of arteether are equally effective. α-arteether has advantage over β-arteether particularly with regard to higher solubility in lipids. Conversion of β-isomer to α-isomer is a tedious operation but Jain et al. (2000) have achieved it with anhydrous $FeCl_3$ under specific reaction conditions. Pharmaceutical grade of α–β-arteether (30:70) was formulated and supplied to the Central Drug Research Institute, Lucknow, India (a Council of Scientific and Industrial Research lab, established by the Government of India) for clinical trial. This drug is now being manufactured and marketed by Themis Medicare Pvt. Ltd., Mumbai, India, under the brand name E-MAL (Eradicate Malaria) since April 1997. The drug is being extensively used under National Antimalarial Programme (NAMP) since 1999.

Genovese et al. (2000) studied the behavior and toxicity of the artemisinin antimalarials arteether, artesunate, and artelinate in rats. They observed that in a histochemical examination of artesunate, artelinate, arteether, and sodium carbonate at equimolar doses, the brain cells of rats did not show damage with artesunate, artelinate, and sodium carbonate, but significant damage of brain cells was reported with arteether.

8.2.2 Essential Oil

The second important compound that is synthesized in the glandular trichome is essential oil. The compounds present in essential oil are limonene (1-methyl-4-[1-methylethenyl]-cyclohexene), 1,8-cineole (eucalyptol), limonene, myrcene, α-pinene, β-pinene, sabinene, and α-terpineol (Raguso et al., 2006). Some of these have shown activity against *Plasmodium* species. For example, 10% limonene inhibits isoprenoid biosynthesis in *Plasmodium* (Rodrigues et al., 2004) and development of the ring and trophozoite stages (Moura et al., 2001). Similarly, eucalyptol affects the trophozoite stage of *Plasmodium* (Su et al., 2008). It arrests protein isoprenylation in *P. falciparum*, stopping parasite development within 48 h (Moura et al., 2001). van Zyl et al. (2006) reported IC50 of 533 µmol/L against *in vitro Plasmodium* in these trials. Limonene and its metabolites remain in the plasma for at least 48 h (Miller et al., 2010), which is important for elimination of gametocytes and malaria transmission.

The volatile monoterpene α-pinene (4,6,6-trimethylbicyclo[3.1.1] hept-3-ene) is present in the plant at levels of up to 0.05% of dry weight (Bhakuni et al., 2001); it has an IC50 of 1.2 µmol/L, in the range of quinine at 0.29 µmol/L (van Zyl et al., 2006).

The other important compound of essential oil is 1,8-cineole (which has an anticoughing property). It may comprise up to 30% (0.24%–0.42% [V/DW]) of the essential oil in *A. annua* (Charles et al., 1990) and is a strong inhibitor of the pro-inflammatory cytokines tumor necrosis factor (TNF)-α and interleukin (IL)-6 and IL-8 (Juergens et al., 2004). It is effective against both chloroquine-resistant and chloroquine-sensitive *Plasmodium* strains, which are affected at the early trophozoite stage (Su et al., 2008). Similarly, eucalyptol (1,3,3-trimethyl-2-oxabicyclo [2,2,2] octane) is another volatile component which rapidly enters the blood when delivered either as an inhalant or orally (Kovar et al., 1987; Stimpfl et al., 1995).

Artemisia ketone (3,3,6-trimethyl-1,5-heptadien-4-one), is one of the major constituents of some cultivars of *A. annua*. Other ketones like curcumin (Akhtar et al., 2012) showed inhibitor activity against of β-hematin synthesis, hence it affects the hemozoin formation. Hemoglobin is important for *Plasmodium* survival and multiplication of merozoites inside the red blood cell. In the presence of this compound the parasite subsequently oxidizes Fe^{2+} in heme to Fe^{3+} forming hematin, a nontoxic insoluble polymeric crystal called β-hematin (also known as hemozoin), which also inhibits cell-mediated immunity against the parasite. Water extracts of *A. annua* also inhibit hemozoin synthesis (Akkawi et al., 2014).

The larger part of the essential oil is mono-, di-, and sesquiterpenes. Monoterpenes enhanced the antimalarial effect of artesunate and even reversed the observed resistance of *Plasmodium berghei* against artesunate (Liu et al., 2000). But no antilarval activity has been reported from such monoterpenes. Camphor also affects thymocyte viability and aids in developing malaria immunity through production of T cells (Roberts et al., 1977).

Some other sesquiterpenes, like nerolidol (3,7,11-trimethyl-1,6,10- dodecatrien-3-ol), showed IC50 of 0.99 µmol/L and were able to arrest development of the intraerythrocytic stages of the *Plasmodium* (van Zyl et al., 2006). In a report that Indians of the Amazon basin in Brazil treated malaria using the vapors of the leaves of *Viola surinamensis*; where nerolidol was identified as the active constituent leading to 100% growth inhibition at the schizont stage (Lopes et al., 1999). A plant variety from Ethiopia showed high levels of nerolidol (Muzemil et al., 2008) and found more in the stem than in leaves (Li et al., 2011).

The phytoconstituents of *A. annua* are given in Table 8.1 (Bhakuni et al., 2001). Other compounds found in the essential oil of Indian varieties of *Artemisia* are camphene hydrate (Charles et al., 1991), caryophyllene oxide, β-caryophyllene, chrysanthenone, α-copaene, β-cubebene, β-farnesene, fenchol, D-germacrene,

Table 8.1   Phytoconstituents of *A. annua*

S.no	Compound(s)	Plant Part	Reference(s)
Sesquiterpenes			
1	Abscisic acid	AP	Shukla et al. (1991)
2	Abscisic acid methyl ester	AP	Shukla et al. (1991)
3	Annuic acid	AP	Mishra et al. (1993b)
4	Artemisinic acid	EP	Mishra et al. (1993b)
		AP	Ahmad et al. (1994)
		SC	Gupta et al. (1996)
		HRC	Banerjee et al. (1997)
5	Quinghaosu 1	AP	Mishra et al. (1993)
		AP	Ahmad et al. (1994)
6	Arteannuin B	HRC	Banerjee et al. (1997)
		SC	Gupta et al. (1996)
7	Arteannuin C	AP	Mishra et al. (1986)
8	Artemisinic acid, epoxy	AP	Mishra et al. (1986)
9	Artemisinin	AP	Mishra et al. (1986)
		LF	Singh et al. (1986)
		AP	Shukla et al. (1992)
		SC	Gupta et al. (1996)
		LE	Gupta et al. (1996)
		EP	Jain et al. (1996)
		EP	Ram et al. (1997)
		HRC	Banerjee et al. (1997)
10	Artemisinin deoxy	EP	Khan et al. (1991)
11	Cadin-4-en-7-ol,3-iso-butyryl	AP	Ahmad et al. (1994)
12	Cadin-4 (15) -11-dien-9-one (38)	AP	Ahmad et al. (1994)
13	Cadin-4(7)-11-dien-12-al (39)	AP	Ahmad et al. (1994)
Triterpenes			
1	Amyrenone, alpha	AP	Ahmad et al. (1994)
2	Amyrin, alpha	AP	Ahmad et al. (1994)
3	Amyrin, beta	AP	Ahmad et al. (1994)
4	Beurenol	AP	Ahmad et al. (1994)
5	Oleanolic acid	AP	Ahmad et al. (1994)
6	Taraxasterone	AP	Ahmad et al. (1994)
7	Taraxerol acetate	RT	Agarwal et al. (1996)
Coumarin			
1	Scopolin	AP	Agarwal et al. (1996)
Flavonoid			
1	Chrysosplenetin	EP	Bhardwaj et al. (1985)
2	Chrysosplenol	AP	Mishra et al. (1986)
3	Chrysosplenol, 3-methoxy	AP	Mishra et al. (1986)
4	Flavone, 3-3'-5-trihydroxy-4-6-7-trimethoxy	AP	Mishra et al. (1986)

(Continued)

Table 8.1 (Continued) **Phytoconstituents of *A. annua***

S.no	Compound(s)	Plant Part	Reference(s)
Miscellaneous			
1	Hentriacontayl triacontanoate	AP	Bhakuni et al. (1990)
2	Nonacosane-1-ol	LF+ST	Bhakuni et al. (1990)
3	Octacosan-1-ol	EP	Khan et al. (1991)
4	Triacosan-8-on-23-ol,2-methyl	LF+ST	Bhakuni et al. (1990)
5	Triacontane-2-29, dimethyl	LF+ST	Bhakuni et al. (1990)
6	Tetratriacontane, N (Alkaloids)	AP	Khan et al. (1991)
7	Purine, 7-8-dihydro:6-(3'-methyi-butyl-amino)-2-hydroxy	AP	Shukla et al. (1991)
8	Zeatin	AP	Shukla et al. (1992)
9	Zeatin, dihydro:riboside	AP	Shukla et al. (1992)
10	Annphenone	AP	Singh et al. (1986)
11	Phthalate, bis-(hydroxy-2-methyl-propyl)	AP	Khan et al. (1991)
12	Hentriacontan-10ol-triacontanoate	EP	Mishra et al. (1986)
13	Fenchone	AP	Khan et al. (1991)
14	Mycerene beta hydroperoxide	EP	Mishra et al. (1986)
15	Mycerene beta hydroperoxide	AP	Ahmad and Mishra (1994)
16	Tricyclene	AP	Mishra et al. (1986)
17	Sitosterol, beta	RT	Agarwal et al. (1996)
	Stigmasterol	AP	Mishra et al. (1986)
		EP	Khan et al. (1991)
		AP	Ahmad et al. (1994)
		RT	Agarwal et al. (1996)

Source: Bhakuni, R.S., et al., *Current Science*, 80, 35–48, 2001.
AP: apical part; ST: stem; LF: leaf; RT: root; SC: suspension culture; EP: entire plant; HRC: hairy root culture.

hepta-3-trans-5diene-2-one, 6-methyl, α-humulene, limonene, longipinene, myrtenal, myrtenol, pinocarvone, sabinene, cis hydrate, α-thujone, terpinen-4-ol, α-terpinene, γ-terpinene, and α-thujene (Ahmad and Mishra, 1994; Bhakuni et al., 2001). The varieties of *A. annua* growing in different countries varied in their amounts of phytoconstituents, as displayed in Table 8.2.

8.2.3 Flavonoids

Other than artemisinin and essential oils, flavonoids are also important compounds of *A. annua*. More than 40 flavonoids (Ferreira et al., 2010) are reported to be found in *A.annua*; among them, artemetin, casticin, chrysosplenetin, chrysoplenol-D, cirsilineol, eupatorin, kaempferol, luteolin, myricetin, quercetin, and rutin showed weak therapeutic efficacy against *P. falciparum* malaria (Liu et al., 1992;

Table 8.2   Selected Compounds Isolated from the Essential Oil of *A. annua* from Different Countries

S. No	Compound(s)	Country(ies)	Percentage	Compound Type	Reference(s)
1	Artemisia alcohol	USA (CA)	5.2	–	Libbey and Sturtz (1989)
		China (cult)	7.5	–	Woerdenbana et al. (1992)
		China (cult)	7.5	–	Woerdenbag et al. (1993)
		Vietnam	0.1–0.6	–	Woerdenbag et al. (1994)
		India (cult)	0.155	–	Ahmad et al. (1994)
2	Artemisia ketone	England	61.0	Monoterpene	Banthorpe et al. (1971)
		Bulgaria	–	–	Toleva et al. (1974)
		Not stated	38.0	–	Banthorpe et al. (1977)
		USSR	–	–	Dembitskii et al. (1983)
		Not stated	–	–	Dembitskii et al. (1983)
		China ++	–	–	Liu et al. (1988)
		China (cult)	63.9	–	Libbey and Sturtz (1989)
		Vietnam	–	–	Nguyen et al. (1990)
		France +++	52.50	–	Chalchat et al. (1991)
		USA (IN)	68.5	–	Charles et al. (1991)
		China (cult)	63.9	–	Woerdenbag et al. (1993)
		USA (CA)	35.7	–	Brown (1994)
		Vietnam	0.1–4.4	–	Woerdenbag et al. (1994)
		India (cult)	58.84	–	Ahmad et al. (1994)
3	Camphor	Vietnam	21.80	–	Nguyen et al. (1990)
		France +++	27.5	–	Chalchat et al. (1991)
		USA (IN)	3.30	–	Charles et al. (1991)
		China (cult)	21.80	–	Woerdenbana et al. (1992)
		Vietnam (cult)	3.30	–	Woerdenbana et al. (1992)
		Vietnam (cult)	10.90	–	Woerdenbag et al. (1993)
		China (cult)	9.1–22.0	–	Woerdenbag et al. (1993)
		India (cult)	15.75	–	Woerdenbag et al. (1993)
		China	–	–	Hu et al. (1993)
		Vietnam	–	–	Woerdenbag et al. (1994)
4	Cineol 1-8	Vietnam (cult)	3.1	–	Allen et al. (1977)
		France +++	11.66	–	Dembitskii et al. (1983)
		Vietnam (cult)	3.1	–	Liu et al. (1988)
		Not stated	–	–	Libbey et al. (1989)
		USSR	–	–	Nguyen et al. (1990)
		USA (IN)	22.8	–	Chalchat et al. (1991)
		China ++	–	–	Charles et al. (1991)
		USA (CA)	31.5	–	Woerdenbag et al. (1993)
		India (cult)	–	–	Ahmad et al. (1994)
		Vietnam	–	–	Fulzele et al. (1995)
5	Pinene, alpha	India (cult)	0.39	–	Dembitskii et al. (1983)
		USA (CA)	11.2	–	Dembitskii et al. (1983)
		USA (IN)	16.0	–	Libbey et al. (1989)

(Continued)

Table 8.2 (Continued) Selected Compounds Isolated from the Essential Oil of *A. Annua* from Different Countries

S. No	Compound(s)	Country(ies)	Percentage	Compound Type	Reference(s)
		Vietnam	0.1–1.4		Charles et al. (1991)
		USSR	–		Woerdenbana et al. (1992)
		Not stated	–		Woerdenbag et al. (1994)
6	Pinene, beta	USA (CA)	1.8		Zhong et al. (1983)
		Not stated	–		Dembitskii et al. (1983)
		India (cult)	1.93		Dembitskii et al. (1983)
		USSR	–		Liu et al. (1988)
		China ++	–		Libbey et al. (1989)
		China	–		Ahmad et al. (1994)
		Vietnam	0.1–0.5		Ahmad et al. (1994)

Note: + = Leaf essential oil; ++= Inflorescence essential oil; +++= Aerial part essential oil.
Source: Bhakuni, R.S., et al., *Current Science,* 80, 35–48, 2001.

Elford et al., 1987; Lehane et al., 2008; Ganesh et al., 2012). Some of these flavonoids were shown to improve the IC50 of artemisinin against *P. falciparum in vitro* (Liu et al., 1992). Elford et al. (1998) reported that casticin (5-hydroxy-2-[3-hydroxy-4-methoxyphenyl]-3,6,7-trimethoxychromen-4-one) showed synergism with artemisinin, while it did not show synergistic activity with chloroquine. Combining casticin with artemisinin inhibited parasite-mediated transport systems that control influx of myo-inositol and L-glutamine in malaria-infected erythrocytes. The synergistic actions between flavonoids and artemisinin suggested that flavonoids are important for efficacious use of *A. annua*, consumed either as whole dried leaves or as tea.

Many flavonoids have antiplasmodial effects and inhibit *P. falciparum* growth in liver cells (Lehane et al., 2008). Flavonoids are known to persist in the body for more than five days; Ogwang et al. (2011, 2012) showed prophylactic effects from *A. annua* tea infusion using a once a week dose. Many dietary flavonoids inhibit *Plasmodium* growth *in vitro*, but amounts in diets are reportedly insufficient to offer protection against malaria (Lehane et al., 2008). Plants such as *A. annua* with high concentrations of flavonoids (e.g., up to 0.6%) may work in concert with artemisinin to prevent malaria when consumed regularly.

The flavonoid compound flavone luteolin (2-[3,4-dihydroxyphenyl]-5,7-dihydroxy-4-chromenone) is one of the most important compounds (0.0023% DW) in *Artemisia* (Bhakuni et al., 2001) and has been used against ailments such as cough, diarrhea, dysentery, diabetes, cancer, and malaria. It has an IC50 value around 11 μmol/L (Lehane et al., 2008) and is one of the more active antiplasmodial flavonoids found in *A. annua*, (Ganesh et al., 2012). It affects enzymes, like the NADPH-dependent *b*-ketoacyl-ACP reductase (FabG), which are potential antimalarial targets. Among 30 flavonoids studied, luteolin and quercetin had the lowest IC50 values, and also showed *in vitro* activity in the sub-micromolar range against multiple strains of *P. falciparum* (Tasdemir et al., 2006).

Similarly, isovitexin {5,7-dihydroxy-2-(4-hydroxyphenyl)-6-[(2S,3R,4R,5S,6R)-3,4,5-trihydroxy-6-(hydroxymethyl)oxan-2-yl]chromen-4-one} is another flavone, the 6-*C*-glucoside of apigenin, that was found in the *A. annua* tea infusion and has more than 100 mg/L comprising micromolar antiplasmodial activity (Carbonara et al., 2012; Suberu et al., 2013). Isovitexin inhibits lipid peroxidation and xanthine oxidase activity and protects cells from ROS damage with an overall LD50 >400 µmol/L (Lin et al., 2002) in malarial parasite.

8.2.4 Phenolic Acids

Some of the important phenolic acids in this category are rosmarinic (2"R")-2-[[(2"E")-3-(3,4-dihydroxyphenyl)-1-oxo-2-propenyl]oxy]-3-(3,4-dihydroxyphenyl) propanoic acid and chlorogenic (1*S*,3*R*,4*R*,5*R*)-3-{[(2*Z*)-3-(3,4-dihydroxyphenyl) prop-2-enoyl]oxy}-1,4,5-trihydroxy cyclohexane-carboxylic acid) acids. These compounds have strong antioxidants found in a wide variety of *A. annua* cultivars (Melillo de Magalhães et al., 2012). This study showed that these acids significantly inhibited the activity of CYP3A4 (one of the hepatic P450s responsible for metabolism of artemisinin to deoxyartemisinin, an inactive form of the drug) (Svensson et al., 1999). These and other phenolic acids are also present in the *A. annua* tea infusion (Suberu et al., 2013). In the tea infusion, it was observed that both phenolic acids have an IC50 of about 65 µmol/L and significantly reduced the secretion of cytokines IL-6 and IL-8, and thus had enhanced antimalarial activity while reducing inflammation (Melillo de Magalhães et al., 2012).

8.3 PHARMACOKINETICS OF ARTEMISININ

Due to short half-life of artemisinin (<60 min), the derivatives of artemisinins were used for therapy. DHA has a half-life of around 60 min. This is a significant part of why recrudescence is so high with artemisinins: the compounds rapidly kill the parasites. Within a few days of finishing the treatment, the small numbers of residual parasites may reproduce and create recrudescence of the disease. Use of longer-acting drugs combined with artemisinins is essential to long-term efficacy in patients with malaria.

The pharmacokinetics of artemisinin of a tea made from the *A. annua* cultivar *Artemis* showed maximum amounts of artemisinin. It was extracted by covering 9 g of dried leaf with 1 L of boiling water after 10 min away from the heat it showed 94.5 mg of artemisinin with 76% extraction efficiency.

Artemisinin auto-induces its own catabolism, primarily in the gut wall, but also in the liver. CYP3A4, 2B6, and 2C19 are the main enzymes involved. Over 7–10 days of ongoing administration of artemisinin, blood levels fall dramatically in healthy volunteers and patients with uncomplicated malaria (Svensson et al., 1999).

8.3.1 Artemisinin as Oral Supplement

It is reported that when artemisinin and its derivatives are given orally or rectally, DHA showed higher bioavailability in humans than artemisinin (Zhao et al., 1993).

The Cmax, Tmax, and T half-life for orally delivered DHA were 0.13–0.71 mg/L, 1.33 h, and approximately 1.6 h, respectively; for pure artemisinin, it was 0.09 mg/L, 1.5 h, and 2.27 h, respectively. Similarly, Alin et al. (1996) investigated orally delivered artemisinin and artemisinin-mefloquine combination therapy for treatment of *P. falciparum* malaria. In his report, infected and uninfected patients had similar pharmacokinetic parameters. After a single dose, the bioavailability of artemisinin was not altered. Ilet et al. (2005) also investigated artemisinin pharmacokinetics in patients with *falciparum* malaria and reported a dose of 9.1 mg/kg. Results obtained by Alin et al. (1996) are in conformity with these findings.

8.3.2 Artemisinin in the Form of Tea and Dried Leaf

Some studies were also carried out on supplementation of artemisinin in the form of tea infusion. In the report, artemisinin in the form of tea infusion contains 94.5 mg of artemisinin (Ashton et al., 1998). The short half-life of artemisinin in the tea infusion may account for higher recrudescence as compared to pure artemisinin. The tea infusion suggests more than two doses. In a report *A. annua* tea also showed potent antiplasmodial activity against 40 field isolates of *P. falciparum* collected in Pikine, Senegal (mean IC50 0.095 µg/mL) (Gueye et al., 2013). The pharmacokinetic studies of artemisinin and its derivatives are given in Table 8.3. Along with the artemisinin other antimalarial drugs are also in the market as shown in Table 8.4.

8.4 THERAPIES USING ARTEMISININ AND ITS DERIVATIVES AS THERAPY

8.4.1 Malaria

Artemisinin is one of the major constituents responsible for the antimalarial properties of *A. annua*, although other plants of this genus that do not contain artemisinin or have very low content, like *A. dracunculus, A. roxburghiana, A siversiana* (Manna et al., 2010; Paul et al unpublished data), *A. absinthium, A. herba-alba, A. apiacea, A. ludoviciana, A. abrotanum,* and particularly *A. afra*, have excellent antimalarial properties.

8.4.2 Leishmaniasis

Cutaneous leishmaniasis (CL) is endemic in about 98 countries worldwide. In a report, leaves and seeds of *A. annua* showed effects in *in vitro* and *in vivo* trials in male and female golden hamsters after three months of treatment. Animals which were treated after typical skin lesions showed 100% response (Restrepo et al., 2011; Robledo et al., 2013). These results suggest that tea from *A. annua* is highly effective against Leishmania parasites and could be a promising candidate for oral treatment of CL. An Indian *in vitro* study on the leaves and seeds of *A. annua* confirmed this efficacy against amastigotes and promastigotes of *Leishmania donovani*. No cytotoxic effect was noticed (Sen et al., 2007).

Table 8.3   **Pharmacokinetics Studies of Artemisinin**

S. no.	Drug type	Dose	Assay	Cmax (µg/L)	Tmax (h)	T $^{1/2}$ β (h)	Reference(s)
1	**Artemisinin**						
	Oral administration	500 mg	HPLC-ED	391	1.80	2.59	Duc et al. (1994)
	12 (HM)	500 mg	HPLC-ED	364	2.90	2.70	Titulaer et al. (1990)
	11 (MP)	400 mg	HPLC-UV	260	1	1.90	Zhao et al. (1987)
	10 (HM)	400 mg IM	HPLC-UV	209	3.40	3.83	Hassan et al. (1996)
	1 (HM)	10 mg/kg	HPLC-UV	1100	=4	2.3	Zhao et al. (1993)
	19 (MP)	500 mg day 1	HPLC-UV	588	2.4	2.2	Zhou et al. (1988)
	11 (MP)	500 mg day 6	RIA	116	3.1	4.8	
	Rectal administration	15 mg/kg	HPLC-ED	40	4.6	4.1	
	5 (HM)	10 mg/kg	HPLC-ED	170	11.3	4.0	
	9 (HM)	10 mg/kg		180	9.6		
	6 (MP)						
2	**Dihydroartemisinin**						
	3 (HM)	1.1 mg/kg PO	RIA	130	1.33	1.63	Zhao et al. (1993)
	3 (HM)	2.2 mg/kg PO	RIA	710	1.33	1.57	Bangchangk et al. (1994)
	5 (HM)	8 mg/kg PR	RIA	100	4.70	4.82	Benakis et al. (1993), and
	6 (HM)	200 mg artemether	HPLC-UV	379	6	10.6	Batty et al. (1996)
	6 (MP)	200 mg artemether	HPLC-UV	593	7.40	12.5	
	6 (Severe MP)	2 mg/kg artesunate IM	HPLC-ED	390		1.59	
	6 (Severe MP)	2 mg/kg artesunate IV	HPLC-ED	2020		1.59	
	6 (MP)	120 mg artesunate IV	HPLC-UV	/200		0.57	

(continued)

Table 8.3 (Continued)

S. no.	Drug type	Dose	Assay	Cmax (μg/L)	Tmax (h)	$T^{1/2}\,\beta$ (h)	Reference(s)
3	**Artesunate**						
	6 (Severe MP)	2 mg/kg IM	HPLC-ED	510		0.49	Benakis et al. (1993)
	6 (Severe MP)	2 mg/kg IV	HPLC-ED	2640+1800		0.49	Batty et al. (1996)
	6 (MP)	120 mg IV	HPLC-UV			3.5 MIN	
4	**Artemether**						
	6 (HM)	200 mg PO	HPLC-UV	118	3	3.1	Bangchang et al. (1994)
	6 (MP)	200 mg PO	HPLC-ED	231	3	4.2	Zhou et al. (1988)
	6 (HM)	6 mg/kg	HPLC-ED	145	5.2	7.7	
	6 (HM)	10 mg/kg		224	6.3	11.1	
5	**Arteether**						
	1 (HM)	3.6 mg/kg IM	HPLC-ED			23	Kager et al. (1994)
	1 (HM)	3.6 mg/kg and 1.6 mg/kg IM at 24,48,72,96 h	HPLC-ED			69.30	
6	Artelinic acid	10 mg/kg IV	HPLC-EC	12706+1010	11.3	1.35	Li et al. (1998)
		10 mg/kg IM	HPLC-EC	8032+1011		2.13	

Abbreviation: Cmax=peak plasma drug concentration, HM=healthy male volunteers, HPLC(ED, UV)=high-performance liquid chromatography (Electrochemical detection, ultraviolet detection), IV intra-venous, MP=patient with falciparum malaria, PO=oral, PR=rectal, RIA=radioimmunoassay, $T^{1/2}\,\beta$=elimination half-life, T max=time to Cmax.

Table 8.4 Antimalarial Drugs Exploited since 1930

Old Drug(s)	New Drug(s)
Cinchona alkaloids, pamaquine	Artemisinin
Mepacrine, chloroquine	Artesunate, arteether, artemether
Proguanil, amodiaquine	Pyronaridine
Pyrimethamine, primaquine	Mefloquine, halofantrine
Pyrimethamine-sulfa combination	Atovaquone-proguanil
	Artemether-lumefantrine
	Tafenoquine
	Sulfadoxine

Source: Silachamroon, U. et al., *Korean J. Parasitol.*, 40(1): 1–7, 2002.

8.4.3 Anthelmintic and Antitrematode Effects

A. annua is commonly named "wormwood" because its anthelmintic properties have been known for ages. It is mentioned in the Bible (Revelation 8:10). In Senegal, it has been noticed that *A. annua* is efficacious in the treatment of schistosomiasis (bilharziasis), which is the second-most significant disease in tropical Africa. Clinical trials are underway to confirm this finding. On the other hand, trematode infections negatively affect human and livestock health, and threaten global food safety. Only triclabendazole and praziquantel drugs are approved as human anthelmintics for trematodiasis. Crude extracts of *A. annua, Asimina triloba,* and *A. absinthium* were tested against adult *Schistosoma mansoni, Fasciola hepatica,* and *Echinostoma caproni in vitro.* The ethanolic extracts were more active (Ferreira et al., 2011).

8.4.4 Chagas' Disease (*Trypanosoma cruzi*)

It has been demonstrated that low doses of artemisinin are able to inhibit the development of *T. cruzi* and *Trypanosoma brucei rhodesiense* by inhibiting the ATPase activity. Nevertheless, further studies are needed to better assess the influence of artemisinin on membrane pumps in relation with calcium for these parasites (Mishina et al., 2007, Sinha et al., 2007). The University of Cumana in Venezuela in 2013, performed an evaluation of the impact of *A. annua* on epimastigotes of *T. cruzi.* After a treatment of seven days, the density of parasites significantly decreased. Two types of infusions were used: one with leaves from plants grown in Venezuela, the other with leaves from Luxembourg. A dose-dependent effect on proliferation was noticed. The infusion from Luxembourg had a stronger effect, even though the seeds from both locations were of the same origin (Cariaco et al., 2013).

8.4.5 Immunity and HIV

Several trials were carried out in Congo, Uganda, and India, and their preliminary results indicated that *A. annua* raises the CD4+ level. Low CD4+ values are indicative of a depressed immune system and often concomitant with HIV (Lubbe et al., 2012).

8.4.6 Cancer

Ten years ago, several research teams claimed that artemisinin was an excellent cure against different forms of cancer. Although, their claims were based exclusively on *in vitro* trials. No *in vivo* trial with artemisinin or its derivatives has ever been run against cancer. These observations need confirmation by clinical *in vivo* trials (Bachmeier et al., 2011).

8.4.7 Dengue and Other Viruses

The species *A. annua* possesses a wide variety of antiviral effects mainly against herpes simplex virus, coronavirus, and dengue virus. In 2014, a female patient with dengue fever in Vanuatu and her relatives were cured after consumption of *A. annua* infusion (where the herb originated in Luxembourg) (P. Lutgen, personal communication). Some *in vitro* data had already been published (Zandi et al., 2011). These findings need confirmation by clinical trials in accordance with World Health Organization (WHO) protocols.

8.4.8 Antimicrobial

Different extracts and metabolites isolated from *A. annua* have been tested *in vitro* against several Gram-positive and Gram-negative bacteria and fungi (Table 8.1), demonstrating activity in some microorganisms mainly *Bacillus subtilis*, *Salmonella enteritidis*, and *Candida albicans*.

8.4.9 Other Traditional Veterinary, Agricultural, or Insecticidal Uses of *A. annua*

As the purpose of this review is to cover mainly the applications of *A. annua* in human health, we will just mention the best-known other uses. Although the plant contains many nutrients that will boost their health and immune system, mammals often avoid it because of its bitterness. *A. annua* has a bactericidal effect in contaminated water (Lutgen et al., 2008) and reduces the frequency of gastrointestinal diseases. This plant is known as repellents for insects in agriculture or in houses when used in fumigation. They also have insecticidal properties (Tripathi et al., 2000; Karahroodi et al., 2009; Zamani et al., 2011). Essential oils of *A. annua* have fungicidal properties (Soylu et al., 2005). Unfortunately, the *Artemisia* plants also have allelopathic properties, which may negatively impact on other plants growing in their vicinity.

8.5 DOSING

A. annua and artemisinins are not recommended for use as monotherapy in order to prevent resistance developing, because recrudescence is so high in monotherapy. Presently, WHO recommends using only the semisynthetic derivatives of artemisinin known as DHA (converted from artemisinins by humans), artemether, and artesunate for uncomplicated malaria. Though these drug combinations have been documented to cure acute malaria patients effectively, it is questionable whether continuing a synthetic pharmaceutical approach will be sustainable ecologically or economically. The historical record suggests a high rate of resistance to semisynthetic quinine derivatives around the world. Quinine is an alkaloid from various species of the Amazonian tree *Cinchona* spp.

Mixing dried *A. annua* leaf into millet porridge and cooking them together has been shown to be a palatable and stable delivery form for the herb to children in Kenya. Pharmacokinetic and clinical efficacy profiles of this delivery form have not been published. Further research is warranted given the sustainability of this approach.

WHO Guidelines for synthetic artemisinins combination therapy:

- 1-DHA 4 mg/kg bw for 3 d + piperaquine, 18 mg/kg bw for 3 d, but no registered/ quality form available
- 2-Artemether (20 mg) + lumefantrine (120 mg), 1–4 caps bid for 3 d (with fat; worldwide)
- 3-Artesunate 4 mg/kg bw + amodiaquine 10 mg base/kg bw qd for 3 d (worldwide)
- 4-Artesunate 4 mg/kg bw qd for 3 d + mefloquine 25 mg base/kg bw qd for 2–3 d (Asia only)
- 5-Artesunate 4 mg/kg bw qd for 3 d + sulfadoxine-pyrimethamine 25/1.25 mg base/kg bw single dose

8.6 RESISTANCE

In Cambodia, resistance to DHA already appears widespread. It is not clear if resistance was developed from whole plant extracts or not. This is unlikely as they contain multiple compounds and not a single agent, including the aforementioned methoxylated flavonoids that appear to prevent development of resistance (Noedl et al., 2010).

8.7 MODE OF ACTION

A number of studies have been conducted to discover the mode of action of drugs on parasites, as well as in humans. Biochemical action of artemisinin depends on two sequential steps. The first step, activation, comprises the iron-mediated cleavage of the endoperoxide bridge to generate unstable, organic free radicals and/or other electrophilic species. The second step, alkylation involves the formation of a covalent bond between the drug and the malarial protein as shown in Figure 8.2 (Meshnick et al., 1994).

Iron has the property to catalyze the decomposition of both hydrogen peroxide and organic peroxides into free radicals. This activity is found in both free iron

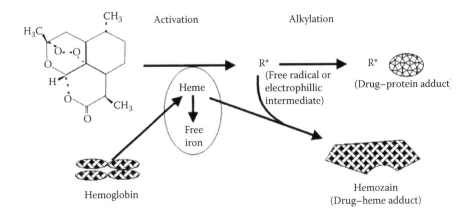

Figure 8.2 Proposed mechanism of action of artemisinin. (After Meshnick, S.R., *Trans. Roy. Soc. Trop. Med. Hyg.,* 88(1), 31–32, 1994.)

and heme-bound iron. Posner and Oh (1992) elucidated the iron-mediated decomposition mechanism of artemisinin. The observation that iron chelators, which bind free iron, are antagonistic suggested that free iron might be important for activation (Kamchonwongpaisan et al., 1992; Meshnick et al., 1993). The free heme pool in the red blood cells has been hypothesized to be important in this activation step (Ferreira et al., 1997).

After the drug is converted to a reactive free radical, it can then form covalent bonds with proteins. The reaction of both (C14) artemisinin and (H3) DHA with protein has been studied in two model systems, human serum and isolated red cell membranes. In the human serum albumin studies using radiolabeled artemisinin it was shown to be covalently bound to the free amino groups (Yang et al., 1993). This radiolabeled drug was also taken up by the isolated red cell membranes, where it formed covalent bonds with various proteins. When artemisinin was incubated with intact erythrocytes, there was no uptake or protein alkylation (Meshnick, 1994).

Very high concentrations of artemisinin (0.1–1 mM) cause the oxidation of red blood cell membrane protein and a decrease in the red cell deformability, especially in the presence of exogenous heme (Scott et al., 1989; Meshnick et al., 1993). This suggests that the drug may affect the plasma membrane function in the infected erythrocytes. However, no such effect has been seen at the therapeutic drug concentrations. The endoperoxide represents a new class of antimalarial agents, which are entirely unrelated to any of the other currently available drugs. Insight into their mechanisms of action can aid in the design of new and more effective drugs (Meshnick, 1994). At high doses, artemisinin can be neurotoxic, but toxicity has not been found in clinical studies (Meshnick, 2002). Not only artemisinin of *A.annua*, but also its essential oil and ethanol extract affect the central nervous system. Essential oil has a high acute toxicity and is a possible cholinergic agent. The ethanolic extract shows a possible central activity as a dopaminergic and cholinergic agent, and does not present a significant acute toxicity (Perazzo et al., 2003).

8.8 CONCLUSION

A. annua is the only source of artemisinin. The traditional and low-cost method for cultivating *A. annua* in developing countries and controlled agriculture practices would probably fulfill the demand for artemisinin. In parallel, great care must be taken during field harvest and post-harvest storage, so as not to affect the quality of the product. WHO has established good agricultural practices specifically to grow *A. annua* for purposes of artemisinin extraction (WHO, 2006), for general medicinal plants (WHO, 2003), and to minimize contamination of herbal medicines.

ACKNOWLEDGMENTS

The author is thankful to CIMAP, GBPIHED, and SERB, New Delhi for their constant support and encouragement.

REFERENCES

Acton N., Klayman D.L., Rollman I.J. 1985. Reductive electrochemical HPLC assay for artemisinin 'quinghaosu'. *Planta Med.* 51:445–446.

Agarwal P.K., Bishnoi V. 1996. Sesquiterpenoids from *Artemisia annua* [13]NMR shielding behaviour. *J. Sci. Indust. Res.* 55:17–26.

Ahmad A., Mishra L.N. 1994. Terpenoid from *Artemisia annua* and constituents of its essential oil. *Phytochemistry* 37:183–186.

Akhtar, F., Rizvi, M.M., Kar, S.K. 2012. Oral delivery of curcumin bound to chitosan nanoparticles cured *Plasmodium yoelii* infected mice. *Biotechnol. Adv.* 30:310–320.

Akkawi M., Jaber S., Abu-Remeleh Q., Ogwang P.E., Lutgen P. 2014. Investigations of *Artemisia annua* and *Artemisia sieberi* water extracts inhibitory effects on β-hematin formation. *Medicin. Arom. Plants* 3(1). doi: 10.4172/2167-0412.1000150.

Alin M.H., Ashton M., Kihamia C.M., Mtey G.J., Björkman A. 1996. Clinical efficacy and pharmacokinetics of artemisinin monotherapy and in combination with mefloquine in patients with falciparum malaria. *Br. J. Clin. Pharmacol.* 141:587–592.

Allen K.V., Banathrope D.V., Charlwood B.V., Voller C.M. 1977. Biosynthesis of artemisia ketone in higher plants. *Phytochemistry* 16:79–83.

Anon 1979. Quinghaosu antimalarial coordinating research group. Antimalarial studies on Quinghaosu. *Chin. Med. J.* 92:811–816.

Ashton M., Gordi T., Trinh N.H., Nguyen V.H., Nguyen D.S., Nguyen T.N., Dinh X.H., Johansson M., Le D.C. 1998. Artemisinin pharmacokinetics in healthy adults after 250, 500 and 1000 mg single oral doses. *Biopharm. Drug Dispos.* 19:245–250.

Bachmeier B., Fichtner I., Killian P., Kronski E., Pfeffer U., Efferth T. 2011. Development of resistance towards artesunate in MDA-MB-231 human breast cancer cells. *PLoS One.* 6(5):e20550.

Bailey L.H., Bailey E.Z. 1976. *Hortus Third.* Macmillan, New York.

Banerjee S., Zehra M., Gupta M.M., Kumar S. 1997. Regeneration of plants from *Agrobactrium rhizogenes* transformed hairy roots. *Planta Med.* 63:467–469.

Bangchangk N.A., Karbwang J., Thomas C.G. 1994. Pharmacokinetics of artemether after oral administration to healthy Thaimales and patients with acute, uncomplicated falciparum malaria. *Ba. J. Clin. Pharmacol.* 37:249–253.

Banthorpe D.V., Baxendale D., Gatford C., William S.R. 1971. *Planta Med.* 20:147. In: Bhakuni RS, Jain DC, Sharma RP and Kumar S. 2001. Secondary metabolites of *Artemisia annua* and their biological activity. *Current Science* 80:35–48.

Banthorpe D.V., Charlwood B.V., Greaves G.M., Voller C.M. 1977. *Phytochemistry.* 16:1387. In: Bhakuni R.S., Jain D.C., Sharma R.P. and Kumar S. 2001. Secondary metabolites of *Artemisia annua* and their biological activity. *Curr. Sci.* 80:35–48.

Barradell L.B., Fitton A. 1995. Artesunate: A review of its pharmacology and therapeutic efficacy in the treatment of malarial. *Drugs.* 50(4):714–741.

Batty K.T., Davis T.M.E., Thu L.T.A. 1996. Selective high performance liquid chromatography determination of artesunate and α- and β-dihydroartemisinin in patients with falciparum malaria. *J. Chromatogr. B. Biomed. Appl.* 677:345–350.

Beekman A.C., Pieter K.W., Herman J.W., Uden W.V., Pars N., Antonius W.T.K., Farouk S.E., Ahmed M.G., Hakan V.W. 1998. Artemisinin derivatived sesquiterpene lactones as potential antitumour compounds: Cytotoxic action against bone marrow and tumour cells. *Planta Med.* 64:615–619.

Benakis A., Paris M., Plessas C. 1993. Pharmacokinetics of sodium artesunate after iii and iv administration (abstract). *Am. J. Trop. Med. Hyg.* 293 Suppl:293.

Bhakuni R.S., Jain D.C., Sharma R.P., Kumar S. 2001. Secondary metabolites of *Artemisia annua* and their biological activity. *Curr. Sci.* 80:35–48.

Bhakuni R.S., Jain D.C., Shukla Y.N., Thakur R.S. 1990. Lipid constituents from *Artemisia annua. J. Indian Chem. Soc.* 67:1004–1005.

Bhardwaj D.K., Jain R.K., Jain S.C., Manchanda C.K. 1985. Constitution of *Artemisia annua* flavones. *Proc. Indian Nat. Sci. Acad.* A51:741–745.

Borrmann S., Adegnika A.A., Missinou M.A., Binder R.K., Issifou S., Schindler A., Matsiegui P.B. et al., 2003. Short course artesunate treatment of uncomplicated *Plasmodium falciparum* malaria in gabon. *Antimicrob. Agents Chemother.* 47(3):901–904.

Bostock J., Riley H.T. (Transl.). 1855–1857. *The Natural History of Pliny* (6 Vols.). H.G. Bohn, London (vol 5, Book XXV, Chapter 36, p. 106–107)

Brossi A., Venugopalan B., Grepe L.D., Yeh H.J.C., Flippen-Anderson J.L., Buchs P., LUO X.D., Milhous W., Peters W. 1988. Arteether, a new antimalarial drug synthesis and antimamalrial properties. *J. Med. Chem.* 31:645–650.

Brown G.D. 1994. Secondary metabolism in tissue cultures of *Artemisia annua. J. Nat. Prod.* 57:975–977.

Carbonara T., Pascale R., Argentieri M.P., Papadia P., Fanizzi F.P., Villanova L., Avato P. 2012. Phytochemical analysis of herbal tea from *Artemisia annua* L. *J. Pharm. Biomed. Anal.* 62:79–86.

Cariaco S., Yusmaris J. 2013. Caracterización de *Artemisia annua* cultivada bajo condiciones ambientales típicas del noreste Venezolano y evaluación de su acción sobre epimastigotes de *Trypanosoma cruzi*. [Cumaná-Venezuela]: Universidad de Oriente; 2013.

Chalchat J.C., Garry R.P., Michet A., Gorunovic M., Stosic D. 1991. A contribution to chemotaxonomy of *Artemisia annua* L. Asteraceae. *Acta. Pharm. Jugosl.* 41:233–236.

Charles D.J., Cebert E., Simon J.E. 1991. Characterisation of essential oil of *Artemisia annua* L. *J. Ess. Oil Res.* 3:33–39.

Charles D.J., Simon J.E., Wood K.V., Heinstein P. 1990. Germplasm variation in artemisinin content of *Artemisia annua* using an alternative method of artemisinin analysis from crude plant extracts. *J Nat. Prod.* 53:157–160.

Cragg G.M., David, J. 2001. Natural product drug discovery in the next milinium. *Pharm. Biol.* 39:8–17.

Dembitskii A.D., Krotova G.I., Kuchkhidze N.M., and Yakobashvili N.Z. 1983. *Maslo-Zhir. Promst.* 3: 31. In: Bhakuni R.S., Jain D.C., Sharma R.P. and Kumar S. 2001. Secondary metabolites of *Artemisia annua* and their biological activity. *Curr. Sci.* 80:35–48.

Duc D.D., de Vries P.J., Le Nguyen X.K., Nguyen B., Kager P.A., van Boxtel C.J. 1994. The pharmacokinetic of a single dose of artemisinin in healthy Vietnamese subjects. *Am. J. Trop. Med. Hyg.* 51:785–790.

Duke S.O., Paul R.N. 1993. Development and fine structure of glandular trichomes of *Artemisia annua* L. *Int J. Plant Sci.* 154:107–118.

Dutta G.P., Asthana O.P., and Gupta K.C. 1994. New potential antimalarials: Arteether α/β, a fast acting blood schizontocide. In *Tropical Diseases: Molecular Biology and Control Strategy.* Eds S. Kumar, Sen A.K., Dutta G.P., and Sharma R.N. Publication and Information Directorate (CSIR), New Delhi, pp. 301–317.

Dutta G.P., Bajpai R., and Vishwakarma R.A. 1989. Antimalarial efficacy of Arteether against multiple drug resistant strains of *Plasmodium yoelii nigeriensis. Pharma. Res.* 21(4):415–419.

Einarson B., Link G.K.K. (Transl) 1976. *Theophrastus: De Causis Plantarum* (3 vol) Harvard University Press. Cambridge, MA.

Elford B.C., Roberts M.F., Phillipson J.D., Wilson R.J. 1987. Potentiation of the antimalarial activity of qinghaosu by methoxylated flavones. *Trans. R. Soc. Trop. Med. Hyg.* 81: 434–436.

Ferreira J., Peaden P., and Keiser J. 2011. In vitro trematocidal effects of crude alcoholic extracts of *Artemisia annua, A. absinthium, Asimina triloba,* and *Fumaria officinalis*: Trematocidal plant alcoholic extracts. *Parasitol. Res.* 109(6):1585–92.

Ferreira J.F., Luthria D.L., Sasaki T., and Heyerick A. 2010. Flavonoids from *Artemisia annua* L. as antioxidants and their potential synergism with artemisinin against malaria and cancer. *Molecules* 15:3135–3170.

Ferreira J.F.S., Janick J. 1995. Production and detection of artemisinin from *Artemisia annua. Acta. Hort.* 390:41–49.

Ferreira J.F.S., Simon J.E., and Janick J. 1995. Developmental studies of *Artemisia annua:* Flowering and artemisinin production under greenhouse and field conditions. *Planta Med.* 61:167–170.

Ferreira J.F.S., Simon J.E., and Janick J. 1997. *Artemisisa annua*: Botany, horticulture, pharmacology. *Horticulture Rev.* 19:319–371.

Fulzele D.P., Heble M.R., and Rao P.S. 1995. Production of terpenoids from *Artemisia annua* L plantlet cultures in bioreactors. *J. Biotchnol.* 40:139–143.

Ganesh D., Fuehrer H.P., Starzengrüber P., Swoboda P., Khan W.A., Reismarnn J.A., Mueller M.S., Chiba P., and Noedl H. 2012. Antiplasmodial activity of flavonol quercetin and its analogues in *Plasmodium falciparum:* Evidence from clinical isolates in Bangladesh and standardized parasite clones. *Parasitol Res.*; 110:2289–2295.

Genovese R.F., Newman D.B., and Brewer T.G. 2000. Behavioural and neural toxicity of the artemisinin antimalarial, arteether, but not artesunate and artelinate: In rats. *Pharmacol. Biochem. Behav.* 67:37–44.

Gray A. 1984. *Synoptical Flora of North America,* 1 Part II. Smithsonian Institution, Washington, DC University Press, Wiley, Cambridge, UK.

Gueye P.E.O., Diallo M., Deme A.B., Badiane A., Dior D.M., Ahouidi A., Abdoul A.N., and Dieng T., et al. 2013. Tea *Artemisia annua* inhibits *Plasmodium falciparum* isolates collected in Pikine, Senegal. *Af. J. Biochem. Res.* 7:107–113.

Gunther R.T. 1959. In: *The Greek Herbal of Dioscorides* (illustrated by Byzantine A.D. 512, Translated by John Goodyear A.D. 1655 edited and first printed A.D 1933). Hafner, New York, pp. 259–263; 357–360.

Gupta M.M., Jain D.C., Mathur A.K., Singh A.K., and Verma R.K. 1996. Isolation of a high artemisinic acid containing plants of *Artemisia annua*. *Planta Med.* 62:280–282.

Gupta S.K., Singh P., Bajpai P., Ram G., Singh D., Gupty M.M., Jain D.C., Khanuja,. S.P.S., and Kumar S. 2002. Morphogrenetic variation for artemisinin and volatile oil in *Artemisia annua*. *Ind. Crops Prod.* 16:217–224.

Hall H.M., and Clements F.E. 1923. The phylogenetic method in taxonomy. *The North American species of Artemisia, Chrysanthamnus and Atriplex*. Carnegie Institution, Washington, DC.

Hassan A., Ashton M., and Kihamia C.M. 1996. Multiple dose pharmacokinetics of oral artemisinin and comparison of its efficacy with that of oral artesunate in falciparum malaria patients. *Trans. R. Soc. Trop. Med. Hyg.* 90:61–65.

Hu S., Xu Z., Pan J., Hou Y. 1993. *J. Res. Educ. Indian Med.* 3: 9. In: Joshi, S.P., Ranjekar, P.K. and Gupta, V.S. 1999. Molecular markers in plant genome analysis. *Curr. Sci.* 77(2):230–240.

Ilet K.F., Batty K.T. 2005. Artemisinin and its derivatives. In: Yu V.L., Edwards G., McKinnon P.S., Peloquin C.A., Morse G., editors. *Antimicrobial Therapy and Vaccines*. ESun Technologies: Pittsburgh. 981–1002.

Jain D.C., Bhakuni R.S., Gupta M.M., Sharma R.P., Kahol A.P., Dutta G.P., and Kumar S. 2000. Domestication of *Artemisia annua* plant and development of new antimalarial drug Arteether in India. *J. Sci. Ind. Res.* 59:1–11.

Jain D.C., Bhakuni R.S., Sharma R.P., Kumar S., and Dutta G.P. 2001. Formulation of dihydroartemisinin for the control of wide spectrum of malaria. United States Patent No 6,214,864

Jain D.C., Mathur A.K., Gupta M.M., Singh A.K., Verma R.K., Gupta A.P., and Kumar S. 1996. Isolation of high artemisinin yeilding clones of *Artemisia annua*. *Phytochemistry* 43:993–1001.

Janick J. 1995. Annual wormwood. Center for New Crop and Plant Products. Purdue University.

Jansen F.H., Zhimin Y. 1997. Who discovered artemisinin: In the meeting with Youyang Government at Wulingshan Pharma Factory, Chongqing. Fhj@dafra.be.

Juergens U.R., Engelen T., Racké K., Stöber M., Gillissen A., and Vetter H. 2004. Inhibitory activity of 1,8-cineol (eucalyptol) on cytokine production in cultured human lymphocytes and monocytes. *Pulm. Pharmacol. Ther.* 17:281–287.

Kager P.A., Schultz M.J., Zijlstra E.E., van den Berg B., van Boxtel C.J. 1994. Arteether administration in human: preliminary studies of pharmacokinitic safety and tolerance. *Trans. R. Soc. Trop. Med. Hyg.* 88:S1/31/S1/32.

Kakkilaya B.S. 2002–2004. malariasite.com

Kamchowongpaisan S., Vanitcharoen N., and Yutha Y. 1992. The mechanism of action artemisinin (qinghaosu) In: *Lipid Soluble Antioxidants: Biochemistry and Clinical Application* Ong, A.S.H. and Packer L. (eds) Basel. Birkhauser Verlag pp. 363–372.

Karahroodi Z.R., Moharramipour S., and Rahbarpour A. 2009. Investigated repellency effect of some essential oils of 17 native medicinal plants on adults *Plodia interpunctella*. *Am. Eurasian J. Sustain Agric.* 2009; 3(2):181–4.

Khan M., Jain D.C., Bhakuni R.S., Zaim M., and Thakur R.S. 1991. *Plant. Sci.* 75: 161. In: Bhakuni R.S., Jain D.C., Sharma R.P. and Kumar S. 2001. Secondary metabolites of *Artemisia annua* and their biological activity. *Curr. Sci.* 80:35–48.

Khanuja S.P.S., Paul S., Shasany A.K., Gupta A.K., Darokar M.P., Gupta M.M., Verma R.K., et al. Genetically tagged improved variety 'Cim-Arogya' of *Artemisia annua* for high artemisinin yield. *J. Med. Arom. Plant Sci.* 27:520–524.

Klayman D.L. 1985. Quinghaosu (artemisinin): An antimalarial drug from China. *Science.* 228:1049–1055.

Klayman D.L. 1989. Weeding out malaria. *Nat. Hist.* 18–26.

Klayman D.L. 1993. *Artemisia annua:* From weed to respectable ant malarial plant. 242–255. In: A.D. Kinghorn and M.F. Baladrin (eds), *Human Medicinal Agents from Plants.* American Chemical Society, Washington, DC.

Klayman D.J., Lin A.J., Acton N., Scovill J.P., Hoch J.M., Milhous W.K., and Theorides A.D. 1984. Isolation of artemisinin (qinghaosu) from *Artemisia annua* growing in United States. *J. Nat. Prod.* 47:715–717.

Kovar K.A., Gropper B., Friess D., and Ammon H.P. 1987. Blood levels of 1,8-cineole and locomotor activity of mice after inhalation and oral administration of rosemary oil. *Planta Med.* 53:315–318.

Kumar S., Gupta S. K., Gupta M. M., Verma R. K., Jain D. C., Shasany A.K., Darokar M. P., and Khanuja S. P. S. 2002. Patent 6 393,763. Method for maximization of artemisinin production by the plant *Artemisia annua.*

Lehane A.M., Saliba K.J. 2008. Common dietary flavonoids inhibit the growth of the intraerythrocytic malaria parasite. *BMC Res Notes.* 1:26.

Lepan I., Golic L., and Jupeij M. 1988. Crystal and molecular structure of quinghaosu: A redetermination. *Acta Pharm. Jugosl.* 38:71–77.

Li Q.G., Peggins J.O., Fleckenstein L.L., Masonic K., Heiffer M.H., and Brewer T.G. 1998. The pharmacokinetics and bioavailability of dihydroartemisinin, arteether, artmether, artesunic acid and artelinic acid in rats. *J. Pharm. Pharmacol.* 50:173–182.

Li Y., Hu H.B., Zheng X.D., Zhu J.H., and Liu L.P. 2011. Composition and antimicrobial activity of essential oil from the aerial part of *Artemisia annua.* *J Med. Plants Res.* 5:3629–3633.

Libbey L.M., Sturtz G. 1989. Unusual essential oil grown in Oregon, II: *Artemisia annua* L. *J Essent. Oil Res.* 1:201.

Liersch R., Soicke H., Stehr C., and Tullner H.U. 1986. Formation of artemisinin in *Artemisia annua* during one vegetation period. *Planta Med.* 52:387–390.

Lin C.M., Chen C.T., Lee H.H., and Lin J.K. 2002. Prevention of cellular ROS damage by isovitexin and related flavonoids. *Planta Med.* 68:365–367.

Liu A.R., Yu Z.Y., Lu L.L., Sui Z.Y. 2000. The synergistic action of guanghuoxiang volatile oil and sodium artesunate against *Plasmodium berghei* and reversal of SA-resistant *Plasmodium berghei Chinese Journal of Parasitology & Parasitic Diseases.* 18:76–78.

Liu J.M., Ni M.Y., and Fan J.F. 1979. Structure and reaction of arteannuin. *Acta. Chem. Sci.* 37:129–143.

Liu K.C., Yang S.L., Roberts M.F., Elford B.C., and Phillipson J.D. 1992. Antimalarial activity of *Artemisia annua* flavonoids from whole plants and cell cultures. *Plant Cell Rep.* 11:637–640.

Liu Q., Yang Z., Sa G., and Wang X. 1988. Prelimnary analysis on chemical constituents of essential oil from inflorescence of *Artemisia annua* L. *Acta. Bot.* 30:223–225.

Lopes N.P., Kato M.J., Andrade E.H., Maia J.G., Yoshida M., Planchart A.R., and Katzin A.M. 1999. Antimalarial use of volatile oil from leaves of *Virola surinamensis* (Rol.) Warb. by Waiãpi Amazon Indians. *J. Ethnopharmacol.* 67:313–319.

Lubbe A., Seibert I., Klimkait T., van der Kooy, F. 2012. Ethnopharmacology in overdrive: The remarkable anti-HIV activity of *Artemisia annua*. *J. Ethnopharmacol.* 141(3):854–9.

Lutgen P., Michels B., and Auditors L.S., Fir, B. Volleker, I. 2008. Bactericidal properties of *Artemisia annua* tea and dosimetry of artemisinin in water by fluorescence under UV light. Proceedings of the International Conference, "*Maladies tropicales, aspects humanitaires et économiques*," Luxembourg, June 3–4 2008. 73–78.

Maggs J.L., Stephen M., Laurence P.B., Paul M.O.N., and Park B.K. 1997. The rat biliary metabolites of dihydroartemisinin, an antimalarial endoperoxide. *Am. Soc. Pharmacol. Exp. Ther.* 25(10):1–8.

Mannan A., Ahmed I., Arshad W., Asim M.F., Qureshi R.A., Hussain I., and Mirza B. 2010. Survey of artemisinin production by diverse *Artemisia* species in northern Pakistan. *Malaria.* 9:310–318.

McVaugh R. and Andrson W.R. 1984. *Flora Novo- Galiciana: A Descriptive Account of Vascular Plants of Western Mexico*, 12 (Compositae). University of Michigan Press, Ann Arbor.

Melillo de Magalhães, P., Dupont I., Hendrickx A., Joly A., Raas T., Dessy S., Sergent T., Schneider Y.J. 2012. Anti-inflammatory effect and modulation of cytochrome P450 activities by *Artemisia annua* tea infusions in human intestinal Caco-2 cells. *Food Chem.* 134:864–871.

Meshnick S.R. 1994. The mode of action of antimalarial endoperoxides. *Trans. Roy. Soc. Trop. Med. Hyg.* 88(1):31–32.

Meshnick S.R. 2002. Artemisinin: Mechanism of action, resistance and toxicity. *Int. J. Parasitol.* 32:1655–1660.

Meshnick S.R., Yang Y.Z., Lima V., Kuypers F., Kamchonwongpaisan S., and Yuthavong Y. 1993. Iron-dependent free radical generation and the antimalarial artemisinin (qinghaosu). *Antimicrob. Agent Chemother.* 37:1108–1114.

Miller J.A., Hakim I.A., Chew W., Thompson P., Thomson C.A., and Chow H.H. 2010. Adipose tissue accumulation of d-limonene with the consumption of a lemonade preparation rich in d-limonene content. *Nutr. Cancer* 62:783–788.

R.K., Meade J.C. 2007. Artemisinins inhibit Trypanosome cruzi and Trypanosoma crucei rhodesiense in vitro growth. *Antimicrob. Agents Chemother.* 51:1852–1854.

Mishra L.N. 1986. Arteannuin C. a sesquiterpene lactone from *Artemisia annua*. *Phytochemistry* 25:2892–2893.

Mishra L.N., Ahmad A., Thakur R.S., and Jakupovic J. 1993a. Bisnor-cadinanes from *Artemisia annua* and definitive ³CNMR assignments or β-arteether. *Phytochemistry.* 33:1461–1464.

Mishra L.N., Ahmad A., Thakur R.S., Lotter H., and Wagner H. 1993b. Crystal structure of artemisinic acid: A possible biogenetic precuosor of antimalarial artemisinin from *Artemisia annua*. *J. Nat. Prod.* 56:215.

Mohamed S.S., Khalid S.A., Ward S.A., Wan T.S.M., Tang H.P.O., Zheng M., Haynes R.R., Edwards G. 1999. Simultaneous determination of artemether and its major metabolite dihydroartemisinin in plasma by gas chromatography- mass spectrometry- selected ion monitoring. *J. Chromatogr.* 251–260.

Moura I.C., Wunderlich G., Uhrig M.L., Couto A.S., Peres V.J., Katzin A.M., and Kimura E.A. 2001. Limonene arrests parasite development and inhibits isoprenylation of proteins in *Plasmodium falciparum*. *Antimicrob. Agents Chemother.* 45:2553–2558.

Mueller M.S., Karhagomba I.B., Hirt H.M., and Wemakor E. 2000. The potential of *Artemisia annua* L. as a locally produced remedy for malaria in the tropics: agricultural, chemical and clinical aspects. *J. Ethnopharmacol.* 73:487–493.

Muzemil A. 2008. Determination of artemisinin and essential oil contents of *Artemisia annua* L. grown in Ethiopia and in vivo antimalarial activity of its crude extracts against *Plasmodium berghei* in mice. Ethiopia: MS Thesis in Medicinal Chemistry, Addis Ababa University.

Nguyen X.D., Leelercq P.A., Dinh H.K., and Nuuyen M.T. 1990. *Tap. Chi. Duoc. Hoc.* 2: 11. In: Bhakuni R.S., Jain D.C., Sharma R.P. and Kumar S. 2001. Secondary metabolites of *Artemisia annua* and their biological activity. *Curr. Sci.* 80:35–48.

Noedl H., Se Y., Sriwichai S., Schaecher K., Teja-Isavadharm P., Smith B., Rutvisuttinunt W., et al. 2010. Artemisinin resistance in cambodia: A clinical trial designed to address an emerging problem in Southeast Asia. *Clin. Inf. Dis.* 51(11). e82–e89.

Ogwang P.E., Ogwal-Okeng J., Kasasa S., Ejobi F., Kabasa D., Obua C. 2011. Use of *Artemisia annua* L infusion for malaria prevention: Mode of action and benefits in a Ugandan community. *Br. J. Pharm. Res.* 1:124–132.

Ogwang P.E., Ogwal J.O., Kasasa S., Olila D., Ejobi F., Kabasa D., and Obua C. 2012. *Artemisia annua* L. infusion consumed once a week reduces risk of multiple episodes of malaria: A randomised trial in a Ugandan community. *Trop. J. Pharm. Res.* 13:445–453.

Paul S., Khanuja S.P.S., Shasany A.K., Gupta M.M., Darokar M.P., Saikia D., and Gupta A.K. 2010. Enhancement of artemisinin content through four cycles of recurrent selection with relation to heritability, correlation and molecular marker in *Artemisia annua* L. *Planta Med.* 76:1468–1472.

Perazzo F.F., Carvalho J.C., Carvalho J.E., and Rehder V.L.G. 2003. Central properties of the essential oil and the crude ethanol extract from arial part of *Artemisia annua* L. *Euphytica.* 497–502.

Poshner O. 1992. A rgiospecifically oxygen-18 labelled 1,2,4-trioxane: A simple chemical model system to probe the mechanism for the antimalarial activity of artemisinin (qinghaosu). *J. Am. Chem. Soc.* 114:8328–8329.

Raguso R.A., Schlumpberger B.O., Kaczorowski R.L., and Holtsford T.P. 2006. Phylogenetic fragrance patterns in Nicotiana sections Alatae and Suaveolentes. *Phytochemistry.* 67:1931–1942.

Ram M., Gupta M.M., Dwivedi S., and Kumar S. 1997. *Planta Med.* 63: 372. In: Bhakuni, R.S., Jain, D.C., Sharma, R.P. and Kumar S. 2001. Secondary metabolites of *Artemisia annua* and their biological activity. *Curr. Sci.* 80:35–48.

Restrepo A.M., Muñoz D.L., Upegui Y.A., Lutgen P., Vélez I.D., and Robledo S.M. 2011. Utilidad de *Artemisia annua* como tratamiento en leishmaniasis cutánea, estudios in vivo e in vitro. *Biomédica.* 31(Sup 3, II Tomo):412–418.

Ritchi J.M. 1985. The aliphatic alcohols. In Gilman A.G., Goodman L.S., Rall T.W., Murrad G. (eds) *Pharmacological Basis of Therapeutics* 7th edn, New York, Macmillan, pp. 372–386.

Roberts D.W., Rank R.G., Weidanz W.P., and Finerty J.F. 1977. Prevention of recrudescent malaria in nude mice by thymic grafting or by treatment with hyperimmune serum. *Inf. Immun.* 16:821–826.

Robledo S.M., Lutgen P., Restrepo A.M., Muñoz D.L., Upegui Y.A., and Velez I.D. 2013. Therapeutic response of *Artemisia annua* tea in the treatment of cutaneous leishmaniasis studies in vivo. Fifth World Congress on Leishmaniasis. Porto Galinhas, Pernambuco, Brazil. p. 469.

Rodrigues G.H., Kimura E.A., Peres V.J., Couto A.S., Aquino D.F.A., and Katzin A.M. 2004. Terpenes arrest parasite development and inhibit biosynthesis of isoprenoids in *Plasmodium falciparum. Antimicrob. Agents Chemother.* 48:2502–2509.

Scott M.D., Meshnick S.R., William R.A., Chill D., Pan H.Z., Lubin B., and Kuypers F. 1989. Quinghaosu-mediated oxidation in normal and abnormal erythrocytes. *J. Lab. Clin. Med.* 114:401–406.

Sen R., Bandyopadhyay S., Dutta A., Mandal G., Ganguly S., Saha P, et al. 2007. Artemisinin triggers induction of cell-cycle arrest and apoptosis in *Leishmania donovani* promastigotes. *J. Med. Microbiol.* 56(9):1213–8

Shrimali M., Bhattacharya A.K., Jain D.C., Bhakuni R.S., and Sharma R.P. 1998. Sodium artelinate: A potential antimalarial. *Indian J. Chem.* 37B:1161–1163.

Shukla A., Abad F.A.H., Shukla Y.N., Sharma S. 1992. *Plant Growth Regu.* 11: 165. In: Bhakuni, R.S., Jain, D.C., Sharma, R.P. and Kumar, S. 2001. Secondary metabolites of *Artemisia annua* and their biological activity. *Curr. Sci.* 80:35–48.

Shukla A., Farooqui A.H.A., Shukla Y.N. 1991. Growth inhibitor from *Artemisia annua. Indian Drugs* 28:376–377.

Silachamroon U., Krudsood S., Phophak N., and Looareesuwan S. 2002. Management of malaria in Thailand. *Korean J. Parasitol.* 40(1):1–7.

Simon J.E., Chadwick A.F., and Craker L.E. 1984. *Herbs: An Indexed Bibliography: The Scientific Literature on Selected Herbs, and Aromatic and Medicinal Plants.* Archon Books, Hamden, CT.

Singh A., Kaul V.K., Mahagan V.P., Singh A., Mishra L.N., Thakur R.S., and Husain A. 1986. Introduction of *Artemisia annua* and isolation of artemisinin: A promising antimalarial drug. *Indian. J. Pharma. Sci.* 48:137–138.

Sinha Y.V., Krishna S., Haynes R.K., and Meade J.C. 2007. Artemisinins inhibit *Trypanosoma cruzi* and *Trypanosoma brucei* rhodesiense in vitro growth. *Antimicrob. Agents Chemother.* 51(5):1852–4.

Soylu E., Yigitbas H., Tok F., Soylu S., Kurt S., Baysal Ö, et al. 2005. Chemical composition and antifungal activity of the essential oil of *Artemisia annua* L. against foliar and soilborne fungal pathogens. *J Plant Dis. Prot.* 112(3):229–39.

Stimpfl T., Nasel B., Nasel C., Binder R., Vycudilik W., and Buch Bauer, G. 1995. Concentration of 1,8-cineol in human blood during prolonged inhalation. *Chem. Senses.* 20:349–350.

Su V., King D., Woodrow I., McFadden G., and Gleadow R. 2008. *Plasmodium falciparum* growth is arrested by monoterpenes from eucalyptus oil. *Flavour. Frag J.* 23:315–318.

Suberu J.O., Gorka A.P., Jacobs L., Roepe P.D., Sullivan N., Barker G.C., and Lapkin A.A. 2013. Anti-plasmodial polyvalent interactions in *Artemisia annua* L. aqueous extract: Possible synergistic and resistance mechanisms. *PLoS One.* 8:e80790.

Svensson U.S., Ashton M. 1999. Identification of the human cytochrome P450 enzymes involved in the in vitro metabolism of artemisinin. *Br. J. Clin. Pharmacol.* 48:528–535.

Tasdemir D., Lack G., Brun R., Rüedi P., Scapozza, L., and Perozzo R. 2006. Inhibition of *Plasmodium falciparum* fatty acid biosynthesis: evaluation of FabG, FabZ, and FabI as drug targets for flavonoids. *J. Med. Chem.* 49:3345–3353.

Titulaer H.A.C., Zuidema J., Kager P.A., Wetsteyn J.C.F.M., Lugt C.B., Merkus F.W.H.M. 1990. The pharmacokinetics of artemisinin after oral, intramuscular and rectal administration to volunteers. *J. Pharm. Pharmacol.* 42:810–813.

Toleva P. D., Ognyanov I. V., Karova E. A., and Georgiev E.V., 1974. *Int Congress Essential Oils* (PAP) 6th Allured Publ Corp., vol. 117, p. 7.

Torrell M., Valles J. 2001. New or rare chromosome counts in *Artemisia* L. (Asteraceae, Anthemideae) and related genera from Armenia and Iran. *Bot. J. Linnean Soc.* 135:51–60.

Tripathi A., Prajapati V., Aggarwal K., Khanuja S., and Kumar S. 2000. Repellence and toxicity of oil from *Artemisia annua* to certain storedproduct beetles. *J. Econ. Entomol.* 93(1):43–7.

Valecha N., Tripathi K, D. 1997. Artemisinin: Current status in malaria India. *J. Pharmacol.* 29:71–75.

van Zyl R.L., Seatlholo S.T., van Vuuren S.F., and Viljoen A.M. 2006. The biological activities of 20 nature identical essential oil constituents. *J. Essent. Oil Res.* 18:129–133. Special Edition.

Wang C.W. 1961. *The Forest of China with a Survey of Grassland and Desert Vegetation* P. 171–187. In Havered Uni Marria Moors Cabot Foundation No 5 Harvard University Cambridge MA.

WHO (World Health Organization). 1985. Tropical disease research: Global Partnership Seventh Programme Report 2: malaria. Geneva: WHO.

WHO (World Health Organization). 2003. WHO guidelines on good agricultural and collection practices (GACP) for medicinal plants. Geneva: WHO. Available from: http://whqlibdoc.who.rnt/publications/2003/9241546271.pdf.

WHO (World Health Organization). 2006. WHO monograph on good agricultural and collection practices (GACP) for *Artemisia annua* L. Available from: http://www.who.int/malaria/publications/atoz/9241594438/en/.

Woerdenbag H.J., Bos R., Salomons M.C., Hendriks H., Pras N., and Malingre T. M. 1992. *Plant. Med. Suppl.* 58: A682. In: Bhakuni R.S., Jain D.C., Sharma R.P., and Kumar S. 2001. Secondary metabolites of *Artemisia annua* and their biological activity. *Curr. Sci.* 80:35–48.

Woerdenbag H.J., Pars N., Chan N.G., Bang B.T., Bos R., Van Vden W., Van Y., Bol N.V., Batterman S., and Lugt C.B. 1994. Artemisinin related sesquiterpene and essential oil in *Artemisia annua*. During a vegetation period in Vietnam. *Planta Med.* 60:272–275.

Yang Y.Z., Asawamahasakdaw and Meshnick S.R. 1993. Alkylation of human albumen by the ant malarial artemisinin. *Biochem. Pharmacol.* 46:336–339.

Zamani S., Sendi J.J., and Ghadamyari M. 2011. Effect of *Artemisia annua* L. (Asterales: Asteraceae) essential oil on mortality, development, reproduction and energy reserves of plodia interpunctella. *J. Biofertil Biopestici.* 2(105).

Zandi K., Teoh B., Sam S., Wong P., Mustafa M., and Abubakar S. 2011. Antiviral activity of four types of bioflavonoid against dengue virus type-2. *Virol J.* 560–4.

Zhao K.C., Song Z.Y. 1993. Pharmacokinetics of dihydroqinghaosu in human volunteers and comparison with qinghaosu. *Acta Pharmaceutical Sinica.* 28:342–346.

Zhao S. 1987. High performance liquid chromatographic determination of artemisinin (qinghaosu) in human plasma and saliva. *Analyst.* 12:661–664.

Zhao S.S. and Zeng M.Y. 1985. Spektrometrische Hochdruck-Flüssigkeits-Chromato graphische (HPLC) Untersuchungen zur Analytik von Qinghaosu. *Planta Med.* 51:233–237.

Zhao Y., Hanton W.K., and Lee K.H. 1986. Antimalarial agents, 21,2 artesunate, an inhibitor of cytochrome oxidase activity in *Plasmodium berghei. J. Nat. Prod.* 49:139–142.

Zhong Y., Cui S., 1983. *Chung Yao T'ung Pao,* 8: 31. In: Bhakuni R.S., Jain D.C., Sharma R.P. and Kumar S. 2001. Secondary metabolites of *Artemisia annua* and their biological activity. *Curr. Sci.* 80:35–48.

Zhongshan W., Naksnima T.T., Kopecky K.R., and Molina J. 1985. Qinghaosu: H and C^{13} nuclear magnetic resonance spectral assignments and luminiscence. *Can. J. Chem.* 63:3070–3074.

Zhou Z.M., Huang Y.X., Xie G.H., Sun Y., Wang Y., Fu L., Jian H., Guo Z., and Li G. 1988. HPLC with polarograph detection of artemisinin and its derivatives and application: The method to the pharmacokineticc study of artemether. *J. Liq. Chromatogr.* 11:1117–1137.

CHAPTER 9

Various Applications of *Artemisia annua* L. (Qinghao)

**Himanshu Misra, Mauji Ram, Ashish Bharillya, Darshana Mehta,
Bhupendra Kumar Mehta, and Dharam Chand Jain**

CONTENTS

9.1 INTRODUCTION

Artemisia annua L. (Asteraceae), also known as *Qinghao* (Chinese), annual or sweet wormwood, or sweet Annie, is an annual herb native to Asia, predominantly China. The natural product artemisinin was isolated from the sweet wormwood plant *A. annua* in 1972 during a program started by the Chinese army in the 1960s. *A. annua* is endemic to the northern parts of Chahar and Suiyuan provinces of China, where it is known as "*quinghao,*" and has been used to treat chills and fever for more than 2000 years. The remarkable story of the discovery of artemisinin and the detection of its antimalarial activity by Chinese scientists represents one of the great medical breakthroughs of the latter half of the 20th century. Through a collaborative effort, collectively referred to as "Project 523," the Chinese prepared dihydroartemisinin (DHA), artemether, and artesunate in the 1970s. These derivatives, with others including artemisone, arteether, and artelinic acid, are generally known as "artemisinins" (Figure 9.1) and are making a crucial contribution to the management of malaria, one of our most serious infections (Haynes, 2006; Krishna et al., 2008; White, 2008a). Nowadays, artemisinin and its derivatives have become essential components of antimalarial treatment, and the artemisinin-based combinatory therapies (ACTs) are recommended by the WHO to treat especially multidrug-resistant forms of malaria. Other studies also point out that, despite their huge potential to reduce or block malaria transmission, ACTs cannot be relied on to prevent it altogether. New approaches have been proposed, including co-administration of ACTs with other antimalarial drugs such as primaquine and ferroquine (Mustfa et al., 2011; Chaudhari et al., 2013; Jelinek, 2013; Ogutu, 2013; Pareek et al., 2013; Price, 2013; Wells et al., 2013).

These features have prompted various scientists around the world to evaluate the potential of artemisinin and its derivatives to control other human diseases. In addition to its antimalarial properties, many researchers have confirmed that artemisinin and its related compounds have many other important medicinal properties, including antitumor, antimicrobial, antiallergic, and immune function regulation properties, as well as combating both neurodegenerative disorders and trauma-induced neuronal injuries (Yao et al., 2012; Sullivan, 2013).

Artemisinin derivatives are active against *Schistosoma mansoni* and *Schistosoma japonicum in vitro* and in experimental animals (Sano et al., 1993; Xiao and Catto, 1989). This is of mechanistic interest since schistosomes, like malaria parasites, degrade hemoglobin and produce hemozoin (Homewood et al., 1972). Activity has also been demonstrated against *Leishmania major* (Yang and Liew, 1993), *Toxoplasma gondii* (Chang and Pechere, 1988; Holfels et al., 1994; Ke et al., 1990; Yang et al., 1990), and *Pneumocystis carinii* both *in vitro* (Merali and Meshinick, 1991) and *in vivo* (Chen and Maibach, 1994). Artemisinin derivatives have immunosuppressive activity (Chen et al., 1994; Tawfik et al., 1990) and potential anticancer activity (Lai and Singh, 1995; Woerdenbag et al., 1993). The concentrations or doses of artemisinin derivatives that are necessary for these alternate activities *in vitro*

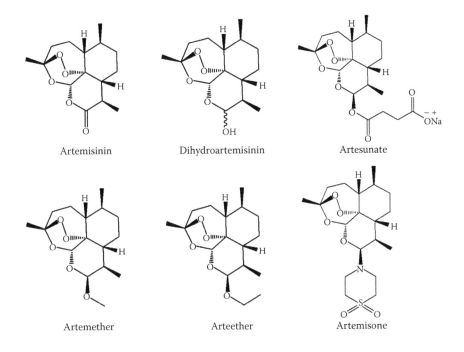

Figure 9.1   Artemisinin and related analogues 'artemisinins.'

and *in vivo* are substantially higher than those required for antimalarial activity. The cytotoxic effect of artemisinin is specific to cancer cells because most cancer cells express a high concentration of transferrin receptors on the cell surface and have a higher iron ion influx than normal cells via the transferrin mechanism. Artemisinin tagged to transferrin via a carbohydrate chain has also been shown to have high potency and specificity against cancer cells. The conjugation enables targeted delivery of artemisinin into cancer cells (Nakase et al., 2008).

Artemisinin contains an endoperoxide group that can be activated by intracellular iron to generate toxic radical species. Cancer cells overexpress transferrin receptors (TfR) for iron uptake, while most normal cells express nearly undetectable levels of TfR. Artemisinin-tagged transferrin (ART-Tf) conjugate kills the prostate carcinoma cell line by the mitochondrial pathway of apoptosis (Nakase et al., 2009). Artemisinin shows unique anticancer activity by an iron-dependent mechanism. Artemisinin was covalently conjugated to a transferrin receptor-targeting peptide, HAIYPRH, that binds to a cavity on the surface of transferrin receptor. This enables artemisinin to be co-internalized with receptor-bound transferrin. The iron released from transferrin can activate artemisinin to generate toxic radical species to kill cells. The artemisinin-peptide conjugates showed potent anticancer activity against Molt-4 leukemia cells with a significantly improved cancer/normal cells selectivity (Oh et al., 2009).

9.1.1 Phytoconstituents of *A. annua*

A. annua has become the subject of intensive phytochemical evaluation following the discovery of the antimalarial drug artemisinin (Wang et al., 2011). Phytochemical analysis has identified various compounds including steroids, coumarins, phenolics, flavonoids, purines, triterpenoids, lipids, aliphatic compounds, monoterpenoids (Emadi, 2013; Cafferata et al., 2010; Ferreira et al., 2010), essential oils, alkaloids, and glycosides (Zanjani et al., 2012). Major terpene derivatives such as Artemisia ketone (Tellez et al., 1999), artemisinic alcohol, arteannuin B, and myrcene hydroperoxide have also been identified. A few compounds are also present in essential oil (Verdian-rizi et al., 2008; Brown, 2010). Essential oils contain both non-volatile and volatile constituents. The volatile components of the essential oil are camphene, 1-camphor, iso-Artemisia ketone, β-camphene, β-caryophyllene, β-pinene, Artemisia ketone, 1,8-cineole, camphene hydrate, cuminal (WHO Monographs, 2006; Willcox, 2009; Das, 2012), 1,8-cineole camphor, germacrene D, camphene hydrate, alpha-pinene, beta-caryophyllene, myrcene, and artemisia alcohol (Liao et al., 2006; Ferreira and Janick, 2009). The non-volatile component of the essential oil contains sesquiterpenoids (Brown, 2010), flavonoids, coumarins, β-galactosidase, β-glucosidase, β-sitosterol, and stigmasterol (Willcox, 2009; Cafferata et al., 2010; Das, 2012). It also contains erythritol (50.30%), camphor (7.25%), pinocarveol (4.13%), and diethoxyethane (2.18%) (Haghighian et al., 2008). Scopoletin belongs to the group of coumarins that have been found in *A. annua* extracts (Tzeng et al., 2007). Scopoletin (coumarin), scopolin (coumarin glycoside), domesticoside (phloroacetophenone), chrysosplenol-D (flavonoid), and norannuic acid (bisnor-cadinane) are vital phytoconstituents (Emadi, 2013; Cafferata et al., 2010). The first time artemisinin (sesquiterpene lactone) was isolated from *A. annua* was in 1972 (Geldre et al., 2000; Ogwang et al., 2012). Artemisinin is a rare sesquiterpene lactone endoperoxide of the cadinane series (Laughlin, 2002). Artemisinin and essential oil levels in the leaves of *A. annua* ranged from 0.01% to 1.4% and 0.04% to 1.9%, respectively (Damtew et al., 2011). Mouton and Kooy (2014) prepared tea infusions of *A. annua* leaves using deionized water which were subjected to liquid chromatography–mass spectrometry (LC–MS) analysis, followed by semi-preparative fractionation and nuclear magnetic resonance (NMR) analysis to confirm the identities of the major compounds. Eleven major compounds were identified, including chlorogenic acids, feruloylquinic acids, flavonols, and coumarins, of which two compounds, *cis-* and *trans-*melilotoside, are new compounds from *Artemisia* spp. The melilotosides are known to be active against diarrhea-causing pathogens, and therefore might explain the traditional use of *A. annua* to treat diarrhea.

9.1.2 Presence of Artemisinin in the *Artemisia* Genus

The leaves of *A. annua* are the only natural source of artemisinin and other vital secondary metabolites that can be further used for the production of derivatives of pharmacological importance (Laughlin, 2002; Willcox et al., 2004; Brown, 2010; Cafferata et al., 2010). Although presence of artemisinin is supposed to be in the

Artemisia genus, there are reports describing presence or production of artemisinin in species other than *A. annua* (Misra et al., 2014). Presence of artemisinin has also been reported in *A. cina, A. sieberi, A. absinthium, A. dubia, A. indica, A. apiacea, A. lancea, A. abrotanum, A. pallens, A. moorcroftiana, A. vestita, A. vulgaris, A. sieversiana, A. roxburghiana* var. *roxburghiana, A. roxburghiana* var. *gratae, A. parviflora, A. dracunculus* var. *dracunculus, A. dracunculus* var. *persica, A. aff. Tangutica, A. bushriences, A. japonica,* and *A. scoparia* (Misra et al., 2014 and references therein; Singh and Sarin, 2010). El-Naggar et al. (2013) reported extremely high artemisinin concentration (4.85–4.90% w/w) in their study performed in the Egyptian desert, which for the first time is being seen as a new promising cultivating area for the production of pharmaceutical artemisinin. Additionally, our study revealed that artemisinin was present, in variable amounts, in aerial parts (decreasing order of artemisinin: leaves > side branches > main stem) and could not be detected in roots (Misra et al., 2014). The leaves from the same plant may have different artemisinin contents according to their localization along the stem: upper leaves contain significantly more artemisinin than middle and lower ones (Diemer and Griffee, 2013).

9.1.3 Awards and Recognition

In the 1970s, research to develop antimalarial drugs led to the discovery of artemisinin, a compound which is extracted from *A. annua*, by Chinese scientist Youyou Tu, for which she shared the 2015 Nobel Prize in Physiology or Medicine. The prize was divided, one half jointly to William C. Campbell and Satoshi Ōmura "for their discoveries concerning a novel therapy against infections caused by roundworm parasites," and the other half to Youyou Tu "for her discoveries concerning a novel therapy against Malaria" (The Nobel Foundation, 2015).

In 2011, Youyou Tu was awarded the Lasker DeBakey Clinical Research Award for the discovery of artemisinin and its use in the treatment of malaria, a medical advance that has saved millions of lives across the globe, especially in the developing world. The future benefits of many seminal discoveries in basic biomedical sciences are not always obvious in the short run, but for a handful of others, the impact on human health is immediately clear. Such is the case for Youyou Tu and her colleagues' discovery of artemisinin (also known as Qinghaosu) for the treatment of malaria. Artemisinin has been the frontline treatment since the late 1990s and has saved countless lives, especially among the world's poorest children (Miller and Su, 2011).

9.2 THERAPEUTIC POTENTIALS OF *A. ANNUA*

9.2.1 Antimicrobial Activity

Tang et al. (2000) isolated two cadinane derivatives, arteannuin B and artemisinin, oleanolic acid, β-sitosterol, and stigmasterol, and the four flavones artemetin, bonanzin, eupalitin, and crysosplenetin during phytochemical investigation of antifungal fractions derived from chloroform extract of *A. annua*. All isolates were tested *in vitro* for antifungal activity. Arteannuin B, a main sesquiterpenoid in *A. annua*, showed antifungal activity against one human (*Candida albicans*, minimum inhibitory concentration [MIC]: 100 µg/mL) and four plant pathogenic fungi (*Gaeumannomyces graminis var. tritici, Rhizoctonia cerealis, Gerlachia nivalis*, and *Verticillium dahliae*, MICs: 150, 100, 150, and 100 µg/mL, respectively) whereas others showed no antifungal activity. The MIC value of ketoconazole to *C. albicans* was 1.0 µg/mL, and those of triadimefon to *G. graminis* var. *tritici* and *R. cerealis*, 150 and 100 µg/mL.

Listeria monocytogenes, Bacillus cereus, and *Staphylococcus aureus* were the most sensitive bacteria with MICs of 0.14, 0.8, and 0.62 µL/mL, respectively (Tripathi et al., 2000). The essential oil of *A. annua* also presented toxic repellent and development inhibitory activities against two economically important stored product insects, *Tribolium castaneum* and *Callosobruchus maculatus*. Adult beetles were repelled significantly by the oil of *A. annua* at 1% concentration and in the filter paper arena test (Tripathi et al., 2000).

A. annua, well recognized for its production of the antimalarial drug artemisinin, is seldom attacked by any of phytopathogenic fungi, which could be partially associated with the presence of endophytes. The investigation of Liu et al. (2001) aimed at disclosing whether the endophytes inside *A. annua* produce antifungal substances. A total of 39 endophytes were isolated and fermented, and the fermented broth was evaluated *in vitro* for antifungal activity against crop-threatening fungi *Gaeumannomyces graminis var. tritici, Rhizoctonia cerealis, Helminthosporium sativum, Fusarium graminearum, Gerlachia nivalis*, and *Phytophthora capsici*. These plants pathogens still cause wheat take-all, sharp eyespot, common rot, scab, snow mold, and pepper phytophthora blight, respectively. Out of 39 endophytes investigated, 21 can produce *in vitro* substances that are inhibitory to all or a few of the tested phytopathogens whereas the rest yielded nothing active. Moreover, the most active broth of endophyte IV403 was extracted with ethyl acetate and *n*-butanol, and comparisons of the antifungal activity of the extracts indicated that the major active metabolites were ethyl acetate extractable (Liu et al., 2001).

The essential oil of *A. annua* aerial parts, consisting of camphor (44%), germacrene D (16%), trans-pinocarveol (11%), β-selinene (9%), β-caryophyllene (9%), and Artemisia ketone (3%), was screened for its antimicrobial activity against *Escherichia coli, S. aureus, Enterococcus hirae, C. albicans, and Saccharomyces cerevisiae*. The essential oil remarkably inhibited the growth of tested Gram (+) bacteria *E. hirae* and both tested fungi (*C. albicans and S. cerevisiae*). This oil has shown an antioxidant activity equivalent to 18% of the reference compound (α-tocopherol) (Juteau et al., 2002).

Rasooli et al. (2003) studied the kinetics of microbial destruction by the *A. annua* oil. The oil was found to have antimicrobial (particularly antibacterial) effect. GC and GC/MS analysis of the oil revealed 17 components. The major components were Artemisia ketone (24.2%), alpha-pinene (12.1%), 1,8-cineole (9.8%), camphore (8.4%), alpha-selinene (7.5%), and borneol (6.0%). The high monoterpenes hydrocarbons seem to contribute for strong antimicrobial activity of *A. annua*.

Gupta et al. (2009) investigated the crude extracts obtained from the aerial parts of *A. annua* Linn. (Asteraceae) for their antibacterial activity by using agar well diffusion assays against five Gram-positive bacteria (*S. aureus, Bacillus sub-tilis, Bacillus pumilus, B. cereus,* and *M. luteus*) and three Gram-negative bacteria (*E. coli, Salmonella typhi,* and *Pseudomonas aeruginosa*). Of the various extracts, the methanol extract showed the strongest activity against most bacteria used in this study. The most sensitive organism was *M. luteus.* The results showed that *S. aureus* required ~0.25 mg/mL of the methanol extract for inhibition. The high-performance thin-layer chromatography (HPTLC) fingerprint of the methanol extract after derivatization with anisaldehyde sulfuric acid reagent showed a maximum number of separated components. Thin-layer chromatography (TLC) bioautography of the methanol extract showed that the area of inhibition around compounds differentiated at Rf=0.32, Rf=0.42, Rf=0.46, Rf=0.77, and Rf=0.87 against *S. aureus.* This is the first report of the antibacterial activity of *A. annua* against food-borne bacteria (Gupta et al., 2009).

The antimicrobial activities of ethanol, methanol, and hexane extracts from *A. absinthium, A. annua,* and *A. vulgaris* were studied. Plant extracts were tested against five Gram-positive bacteria, two Gram-negative bacteria, and one fungal strain. The results indicated that *A. annua* alcoholic extracts are more effective against tested microorganisms. However, all plants extracts have moderate or no activity against Gram-negative bacteria. The results confirm the use of *Artemisia* species extracts in traditional medicine as treatment for microbial infections (Poiată et al., 2009).

Essential oil of *A. annua,* cultivated in Bosnia, had Artemisia ketone (30.7%) and camphor (15.8%) as per gas chromatography–mass spectrometry (GC–MS) analysis. This oil was tested for radical-scavenging ability using the stable 2,2-diphe-nyl-1-picrylhydrazyl (DPPH) radical, the 2,2'-azino-bis-(3-ethylbenzothiazoline)-6-sulphonic acid (ABTS) radical, for reducing power ability with a test based on the reduction of ferric cations, for reducing ability of hydroxyl radical in oxygen radical absorbance capacity (ORAC) assay, and for metal chelating ability using the fer-rozine assay. In all tests, essential oil did not show a prominent antioxidant activ-ity, but still comparable with thymol, an already known antioxidant. The screening of essential oil against Gram-positive, Gram-negative bacteria and fungi, using the agar diffusion method. All tested microorganisms were inhibited by essential oil. Moreover, authors are claiming that this is the first report of antimicrobial activity of essential oil of *A. annua* against *Haemophilus influenza, Enterococcus faeca-lis, Streptococcus pneumoniae, Micrococcus luteus,* and *Candida krusei* microbial strains. Additionally, the antioxidant, antibacterial, and antifungal activity of essen-tial oil of *A. annua* from Bosnia is presented, for the first time, in this study (Ćaver et al., 2012).

Désirée et al. (2013) carried out experiments to assess the *in vitro* antimicro-bial potential of the essential oil extracted by hydrodistillation from the variety of *A. annua* grown in West Cameroon. This evaluation was conducted by testing the microbial growth inhibition through agar diffusion, minimal inhibitory and minimal lethal concentrations. Tested microorganisms included bacteria isolates belonging to the following categories: *S. aureus, E. coli, Salmonella enteritidis, Shigella flexneri, Proteus mirabilis, P. aeruginosa, Klebsiella pneumoniae,* and *Vibrio cholerae*. This activity was also tested on a dimorphic fungal species, *C. albicans*. Data analysis revealed that the essential oil possessed an intrinsic antimicrobial activity that was potentiated by the solvent (DMSO). Inhibition zone diameters varied from 6 (*P. aeruginosa* and *S. flexneri*) to 45 mm (*V. cholerae*). It was also observed that *V. cholerae* was susceptible to the lowest concentration of the essential oil used (0.3 mg/mL), while *P. aeruginosa* was shown to tolerate the highest (80 mg/mL). Also, the minimal inhibitory and lethal concentrations were equal (MLC/MIC = 1), implying the absolute lethal property of the oil. This lethal potential on fungi, Gram-negative and Gram-positive bacteria makes of this plant an appropriate candidate for new conventional antimicrobial drug production and infectious disease prevention. Well exploited, it might be used to control the current epidemics of *V. cholerae*-associated cholera in Cameroon. Additional studies should also be conducted to lay down reliable basis for comprehensive test interpretations that consider correlations between these *in vitro* test results and the ones that would be obtained with conventional antimicrobials (Désirée et al., 2013).

Massiha et al. (2013) performed study to determine the antibacterial effects of aqueous, chloroform, methanol and ethanol extracts of *A. annua* against eight bacterial species. Antimicrobial activity, minimum inhibitory concentration, and minimum bactericidal activity of the essential oil and extract were performed by agar disc diffusion and microdilution broth methods. The obtained results showed antibacterial activity of the organic and chloroformic extracts of *A. annua* against the tested microorganisms (*S. aureus, E. coli, S. aureus, B. cereus, Bacillus* spp., *E. faecalis,* UPEC, and *P. aeruginosa*). Presence of tannins, saponins, alkaloids, amino acids, phenolic compounds, quinines, and terpenoids were identified in the composition of the obtained extract using mass gas-chromatograph. The best result for the minimum inhibitory concentration and minimum bactericidal concentration was reported for the 32 mg/mL of chloroform extract. The results indicate the fact that the extracts and essential oils of the plants can be useful as medicinal or preservatives composition.

Recently, Bilia et al. (2014) reviewed numerous antimicrobial properties of essential oil from *A. annua*.

9.2.2 Antipyretic, Analgesic (Antinociceptive), Anti-Inflammatory Activity, and Antiallergic (Antibronchial Asthma) Activity

The antipyretic, heat-resistant, anti-inflammatory, analgesic, and bacteriostasis effects of water extracts ethyl acetate and *n*-butyl alcohol extracts of *A. annua* are reported by Huang et al. (1993). Animal experiments have demonstrated that *qinghao* acid is one of the actively bacteriostatic constituents. Scopoletin is one of the anti-inflammatory constituents of *A. annua* (Bilia et al., 2014). Artemisinin, DHA, and arteether have been found to exhibit marked suppression of humeral responses in mice at high dose level. These agents did not alter the delayed-type hypersensitivity response to sheep erythrocytes, and were not found to possess any anti-inflammatory activity when tested on carrageenan-induced edema (Tavlik et al., 1990).

A. annua is a source of both essential oil (1.4%–4.0%), depending on chemotype, and other substances such as sesquiterpene lactones, flavonoids, polyalkynes, and coumarins. The essential oil composition has been studied thoroughly and about 60 components have been identified; camphor, *Artemisia* ketone, germacrene D, and 1,8-cineole are usually the main components. The ethanolic extract of aerial parts of *A. annua* (200 mg/kg) was found to be effective against chemical as well as thermal stimuli (Abid et al., 2013). So, the results agree with the traditional use of plants as an analgesic agent.

Allergic asthma is a chronic airway disorder characterized by airway inflammation, mucus hypersecretion, and airway hyperresponsiveness (AHR) (Galli et al., 2008). Artesunate is a clinically effective drug for both uncomplicated and severe malaria (Krishna et al., 2008; Rosenthal, 2008; Woodrow et al., 2005). Some recent findings (Cheng et al., 2011) support a novel therapeutic use of artesunate in the treatment of asthma. Cheng et al. (2011), for the first time, investigated the effects of artesunate on various aspects of ovalbumin (OVA)-induced T helper type 2 (TH2)-mediated allergic airway inflammation in an *in vivo* mouse asthma model to explore the anti-inflammatory mechanism of action of artesunate. Results revealed that artesunate, a semisynthetic derivative of artemisinin, has effectively reduced OVA-induced inflammatory cell recruitment into BAL fluid, IL-4, IL-5, IL-13, eotaxin production, pulmonary eosinophilia, mucus hypersecretion, and AHR in a mouse asthma model potentially via inhibition of the phosphoinositide 3-kinase (PI3K)/Akt signaling pathway.

Recently, Favero et al. (2014) used enriched sesquiterpene lactone fraction (Lac-FR), containing artemisinin (1.72%) and deoxyartemisinin (0.31%), for evaluation of the pharmacological potential of Lac-FR on different nociceptive and inflammatory experimental animal models based on *in vivo* chemical-induced behavioral assays. Lac-FR was administered by intraperitoneal injection producing a relevant reduction in the reaction time of the animals in both phases of the formalin test, significantly reduced the sensitivity to mechanical allodynia stimulus, reduced the paw edema caused by carrageenan injection, and promoted high antinociceptive activity in the tail flick model suggesting relationship with the opioid system. For the first time, results presented herein provided consistent data to support the potential use of these lactones for pain relief as revealed by chemical-induced nociception assays in mice.

9.2.3 Sedative Activity

To establish the scientific basis of reported ethnomedicinal use of *A. annua* as a sedative agent, Emadi et al. (2011) gathered plants of *A. annua* from Gilan Province in Iran. Methanol extract of aerial parts was partitioned into chloroform, petroleum ether, and ethyl acetate. All the three fractions were administered intraperitoneally (i.p.) in male mice with different concentrations (50, 100, and 200 mg/kg), and for evaluation of sedative activity immobility time was determined. In an effort to clarify the mechanism of action, flumazenil (3 mg/kg, i.p.) as a benzodiazepine (BZD) receptor antagonist was injected 15 min before chloroform fraction (200 mg/kg, i.p.). Compared with the control group (saline-treated mice), the chloroform fraction significantly increased immobility time in a dose-dependent manner. Flumazenil decreased immobility time induced by chloroform fraction significantly; therefore, the present study suggests that *A. annua* growing in Iran has sedative effects, which are probably mediated via BZD receptors pathways (Emadi et al., 2011).

9.2.4 Anti-Arthritis Activity

Experimental studies have revealed that water-soluble artemisinin derivative SM905 (obtained from *A. annua*) suppresses the inflammatory and Th17 responses which cause the improvement in collagen-induced arthritis. These studies have been carried out on collagen-induced arthritis (CIA) by type II bovine collagen model (CII) in DBA/1 mice through oral administration of artemisinin derivatives SM905. Incidence of disease and severity were observed regularly. Gene expression and T helper (Th) 17/Th1/Th2 type cytokine production levels have also been examined. Observations of this study revealed that the SM905 compound plays a key role as it delayed the onset of disease, hence reducing the incidence of arthritis. Furthermore, it also reduces the overexpression of a variety of pro-inflammatory cytokines and chemokines (Das, 2012; Sadiq et al., 2014).

9.2.5 Antihypertensive Activity

Antihypertensive capacity of aqueous extracts prepared from aerial parts of *A. annua*, collected from Iran have been assessed by using *in vivo* models of male albino Wistar rats that were administered, subchronically, with doses of 100 mg/kg/0.5 mL saline daily and 200 mg/kg/0.5 mL saline daily for 4 weeks. Results revealed that the aqueous extract of *A. annua* significantly inhibited the phenylephrine-induced contraction and simultaneously stimulated the endothelium-dependent relaxation of rat aortic rings in Krebs solution. Furthermore, the effect of *A. annua* aqueous extract on the vascular reactivity of aortic rings in rats treated with 200 mg/kg was greater than that of rats treated with 100 mg/kg/0.5 mL saline daily. The beneficial effect of long-term and/or subchronic *A. annua* aqueous extract treatment on contraction and endothelium-dependent relaxation responses may be specific for aortas of rats. Additionally, data of rat's weights before and after experimental periods showed an increase in the weight of rats treated with aqueous extract in comparison to normal saline rats ($p < 0.001$–$p < 0.0005$). On the other hand, the serum glucose of extract-treated rats was decreased significantly compared to normal rats ($p < 0.01$–$p < 0.005$). Finally, the authors concluded that aqueous extract of *A. annua* could attenuate the α_1-adrenoceptor agonist-induced contraction of rat aorta and produce a direct or indirect vasorelaxant effect. Since it can reduce the risk factors of some cardiovascular disorders including hypertension, it is highly recommended that *A. annua* aqueous extract be administered as a complementary therapeutic regimen for patients with cardiovascular abnormalities (Mojarad et al., 2005).

9.2.6 Neuroprotective Activity

The discovery of new compounds that can stimulate regeneration and differentiation of neurons could be of great importance for developing therapeutics against both neurodegenerative disorders, such as Alzheimer's disease and Parkinson's disease, and trauma-induced neuronal injuries, such as cerebral ischemia and spinal cord injury. Growth of neurite processes from the cell body is a critical step in neuronal development, regeneration, differentiation, and response to injury. The discovery of compounds that can stimulate neurite formation would be important for developing new therapeutics against these diseases (Martínez et al., 2014).

Artemisinins (artemisinin and its derivatives) containing an endoperoxide bridge, such as artemisinin and DHA, induced growth of neurite processes at concentrations that were slightly cytotoxic; artemisinin had the most potent effect among them (Sarina et al., 2013). Deoxyartemisinin, which lacks the endoperoxide bridge (Figure 9.2), was ineffective. Artemisinin upregulated phosphorylation of extracellular signal-regulated kinase (ERK) and p38 mitogen-activated protein kinase (MAPK), critical signaling molecules in neuronal differentiation (Martínez et al., 2014).

The activation of NF-κB and NLRP3 inflammasome is implicated in neuroinflammation, which is closely linked to Alzheimer's disease. Shi et al. (2013) investigated the effect of artemisinin on Alzheimer's disease in five-month-old APPswe/Ps1E9 transgenic mice. These results revealed that artemisinin treatment decreased neuritic plaque burden, did not alter β-amyloid transport across the blood–brain barrier, inhibited β-secretase activity, and inhibited NF-κB activity and NLRP3 inflammasome activation in transgenic mice. These studies suggest that targeting NF-κB activity and NLRP3 inflammasome activation offers a valuable intervention for Alzheimer's disease (Martínez et al., 2014).

9.2.7 Plant Growth Regulatory Activity (Herbicidal Activity)

Researchers have found plant growth-inhibitory activity in artemisinin, with potential as a herbicide (Duke et al., 1987; Chen and Leathers, 1990). Artemisinin reduced growth of the roots in lettuce and several weed species by about 50% at 33 µM. Bagchi et al. (1997) have also reported plant growth regulatory activity in artemisinin and its one

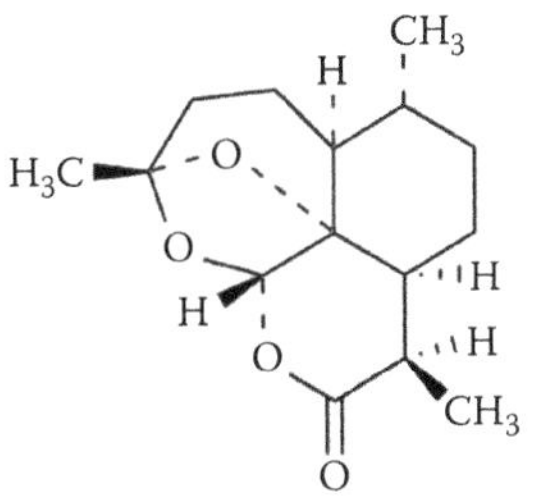

Figure 9.2   Structure of deoxyartemisinin.

semisynthetic derivative. The compounds bis(1-hydroxy-2-methylpropyl) phthalate, abscisic acid, and abscisic acid methyl ester isolated from *A. annua* were also found to possess plant growth regulatory activity (Shukla et al., 1992). These results indicated that artemisinin or artemisinin-derived compounds can be used in agriculture as herbicides.

Bagchi et al. (1998) experimented to find out more effective herbicidal compounds than artemisinin; the known natural precursors of artemisinin in the *A. annua* plant, artemisinic acid and arteannuin B and the semisynthetic derivative of artemisinin, arteether, were tested for their effect on seed germination and seedling growth of several mono- and dicotyledonous plants. The plants against which tests were performed include *A. annua*, *Lactuca sativa* (Asteraceae), *Raphanus sativus* (Brassicaceae), *Portulaca oleracea* (Portulacaceae), *Amaranthus blitum* (Amaranthaceae), *Secale cereale*, and *Hordeum vulgare* (Poaceae). Arteether was found to be highly effective among the tested compounds. It produced 50% inhibition in root growth at 1 ppm in *S. cereale*, *H. vulgare*, and shoot growth in *L. sativa*. In *P. oleracea*, arteether caused inhibitions in both shoot and root growth at 1 ppm. Arteether also retarded seedling growth in *A. blitum* and *A. annua* at 10 ppm. However, at this concentration, it produced only shoot growth retardation in *S. cereale* and *H. vulgare*. Artemisinin showed equivalent or lower activity than arteether in most of the test plants. At lower concentrations, all the tested *A. annua* compounds promoted seedling growth in many examined species. The growth promoting activity of artemisinic acid and arteannuin B was more pronounced than artemisinin and arteether.

Bharati et al. (2012) examined the phytotoxic effect of artemisinin on the chloroplast photoelectron transport activity both under *in vitro* and *in vivo* experimental conditions. The results from our *in vitro* and *in vivo* studies suggested the presence of two different modes of action of the compound on photosynthetic electron transport. Artemisinin is primarily an energy transfer inhibitor of isolated thylakoid membranes. However, the *in vivo* studies indicate that the compound impairs the thylakoid electron flow as an inhibitor of secondary quinine (QB) of Photosystem II (PS II). The inhibition was determined to be caused by a yet uncharacterized artemisinin metabolite rather than its direct interaction with thylakoid membrane. Thus, artemisinin may act as a natural prophytotoxin.

9.2.8 Antiulcerogenic Activity

The gastric ulcer is a disease that attacks many in the population. Lifestyle factors such as diet, stress, alcoholic drinks, and drug abuse are known to worsen ulcer conditions. Therefore, there is an urgent need for new, efficient drugs for gastric ulcer prophylaxis (Dias, 1997). Foglio et al. (2002) demonstrated that the Artemisinin extraction by-product exhibited intense antiulcerogenic activity in ulcer models induced by indometacin and ethanol, comparable to the standard drug carbenoxolone. According to the authors, artemisinin did not provide cytoprotection under the experimental models tested. Only the dihydro-epideoxyarteannuin B and deoxyartemisinin decreased the ulcerative lesion index produced by ethanol and indomethacin in rats. Thus, these compounds have proved to be promising drugs in ulcer control.

Dias et al. (2001) examined antiulcer activity of crude ethanol extract (CEE) and prepurified fractions of *A. annua*. For this, dried, ground aerial parts were extracted with ethanol, and this extract was treated to lead acetate solution, filtered and extracted with chloroform, dried over magnesium sulfate, and evaporated to dryness to give the sesquiterpene lactone fraction (SLF). SLF was purified over silica gel 60 based column chromatography using a hexane/ethyl acetate gradient into three fractions: nonpolar fraction (NP), medium-polar fraction (MPF), and polar fraction (PF). These were submitted to gas chromatography and biological assays. The CEE (500 mg/kg) inhibited the ulcerative lesion index (ULI) by 53.8%, when administered orally to an indomethacin-induced ulcer model, whereas SLF (i.e., purified CEE) inhibited ULI by 86.1% on the same experimental model. The same dose of SLF, when administered subcutaneously to an ethanol-induced ulcer model, inhibited ULI by 59.8%. The purified fractions NP, MPF, and PF, when administered orally (500 mg/kg) to an indomethacin-induced ulcer model, inhibited ULI by 88.3%, 57.7%, and 31.1%, respectively, compared with cimetidine (100 mg/kg; ULI inhibition 88.3%). GC-MS analysis of the NP, MPF, and PF revealed that the sesquiterpene lactones were concentrated in the NP and MPF; therefore, dose response effects of MPF were studied at 10, 25, 50, 125, 250, and 500 mg kg^{-1}, in ethanol-induced ulcer model, inhibited 34.4%, 27.9%, 88.7%, 93.1%, 98.4%, and 98.0%, respectively, which was compared with Carbenoxolone (200 mg/kg; ULI inhibition 85.2%). The participation of nitric oxide was evaluated on an ethanol-induced ulcer model, which had a previous administration of L-NAME (a NO-synthase inhibitor). Under these conditions MPF maintained the antiulcerogenic activity, suggesting that nitric oxide could not be involved in the antiulcerogenic activity. When the animal group were treated with N-ethylmaleimide (an alkylator of sulfhydryl groups) using the same experimental model, the MPF maintained its antiulcerogenic activity, suggesting that the pharmacological mechanism is not related to non-protein sulfhydryl compounds. On the ethanol-induced ulcer with previous indomethacin treatment, the MPF lost its antiulcerogenic activity indicating that the active compounds of *A. annua* increase the prostaglandin levels in the gastric mucosa. This hypothesis was reinforced by an increase of adherent mucus production by gastric mucosa, produced by the MPF on the hypothermic restraint stress-induced ulcer model (Dias et al., 2001).

9.2.9 Contraceptive Activity

Ethanol (98%) extract of *A. annua* leaves (containing 1.098% artemisinin) was prepared to ascertain the contraceptive claim of *A. annua*. The impacts of ethanolic extract on selected female fertility indices were investigated using healthy, sexually mature female Wistar rats (180–220 g) allotted into four experimental study groups of six rats each. The control group received normal saline, while the *A. annua*-treated groups received 100, 200, and 300 mg/kg of *A. annua* ethanolic extract for 2 weeks, followed by mating with proven fertile males (1:1). The rats were allowed to carry the pregnancy to term. At birth and weaning periods, selected reproductive outcome and fertility indices were determined. The results showed that *A. annua* ethanolic extract significantly reduced litter size, reproductive outcome, and fertility indices compared with the control ($p < 0.05$). These results imply that *A. annua* could serve as a prospective contraceptive agent in addition to its antimalarial activity (Abolaji et al., 2014).

9.2.10 Antiparasitic Activity

Oriakpono et al. (2012) prepared the metholic extracts of *A. annua*. Preliminary phytochemical screening of methanol extract revealed the presence of terpenoids, essential oils, flavonoids, and coumarins. Oriakpono et al. (2012) evaluated for the antiparasitic properties of methanolic extracts of *A. annua*. For this, 750 10-day-old juvenile *Sarotherodon melanotheron* fish samples were obtained and exposed to conditions that enhanced parasite proliferation for 1 week, after which fish showing signs of parasite burden were isolated and parasite load was determined by counting under a stereo-microscope. They were then separated into five groups based on samples with the same number of parasites and exposed to the extract. Results of bioactivity of extracts revealed a time-dependent dislodgement of parasites from point of attachment, coagulation, and death of parasite cells exposed to the extracts. Thus, the secondary metabolites of *A. annua* and its essential oils have antiparasitic potentials against monogenetic trematodes as found in the case of monogenean parasites of *S. melanotheron*.

Khosravi et al. (2011) studied effects of methanol extract of *A. annua* against lesser mulberry pyralid, *Glyphodes pyloalis* Walker (Lepidoptera: Pyralidae), which is a monophagous and dangerous pest that has recently been observed in Guilan province, northern Iran. His group investigated effects of methanolic extract on toxicity, and biological and physiological characteristics of this pest under controlled conditions ($24 \pm 1°C$, $75 \pm 5\%$ RH, and 16:8 L:D photoperiod). The effects of acute toxicity and sublethal doses on physiological characteristics were performed by topical application. The LC_{50} and LC_{20} values on fourth instar larvae were calculated as 0.33 g and 0.22 g leaf equivalent/mL, respectively. The larval duration of fifth instar larvae in LC_{50} treatment was prolonged (5.8 ± 0.52 days) compared with the control group (4.26 ± 0.29 days). However, larval duration was reduced in LC_{20} treatment. The female adult longevity in LC_{50} dose was the least (4.53 ± 0.3 days), while longevity among controls was the highest (9.2 ± 0.29 days). The mean fecundity of adults after larval treatments with LC_{50} was recorded as 105.6 ± 16.84 eggs/female, while the control was 392.74 ± 22.52 eggs/female. The percent hatchability was reduced in all treatments compared with the control. The effect was also studied of extract in 0.107, 0.053, 0.026, and 0.013 g leaf equivalent/mL on biochemical characteristics of this pest. The activity of α-amylase and protease 48 h post-treatment was significantly reduced compared with the control. Similarly, lipase, esterase, and glutathione S-transferase activity were significantly affected by *A. annua* methanolic extract.

Chagas et al. (2011) investigated the *in vitro* action of *A. annua* ethanol extract and its four fractionated extracts (lyophilized water extract, water extract basified with $NaHCO_3$ at 0.1%, ethanol [96°GL] followed by concentration in a rotary evaporator, and dichloromethane followed by concentration in a rotary evaporator) on tick (parasite) *Rhipicephalus (Boophilus) microplus*. The extracts were not effective on the larvae at concentrations tested (3.1–50 mg/mL). The crude concentrated extracts showed greater efficacy on engorged females (EC_{50} of 130.6 mg/mL and EC_{90} of 302.9 mg/mL) than derived extracts.

Mites, particularly phytophagous pest mites, are among the most important harmful organisms in fruit trees, forest, vegetables, flowers, crops, and so on. Because of small individual size, short generation cycle, and high mutation rate, mites were thought of as a kind of pest, which was more difficult to be controlled. Now, mites resistance to many of the chemical pesticides are reported due to frequent and irrational drug use. Developing green, safe, and environmental pesticides is an effective way for solving the lack of excellent acaricides (Sendi and Khosravi, 2014).

Artemisinin is a potent antimalarial even against chloroquine and quinine-resistant *Plasmodium falciparum* and other malaria-causing parasites. Its activity is based on an unusual mode of action, leading to the alkylation of malarial-specific proteins (Fesen et al., 1994).

Liu et al. (1992) performed a study in which cell suspension cultures developed from *A. annua* plant exhibited antimalarial activity, against *P. falciparum in vitro* both in the *n*-hexane extract of the cell culture medium and in the chloroform extract of the cells. Trace amounts of artemisinin may be responsible for antimalarial activity of the *n*-hexane fraction, but only methoxylated flavonoids artemetin, chrysosplenetin, chrysosplenol-D, and cirsilineol can account for the activity of the chloroform extract. These purified flavonoids were found to have IC_{50} values at $2.4–6.5 \times 10^{-5}$ M against *P. falciparum in vitro* compared with an IC_{50} value of about 3×10^{-8} M for purified artemisinin. At concentrations of 5×10^{-6} M these flavonoids were not active against *P. falciparum* but did have a marked and selective potentiating effect on the antiplasmodial activity of artemisinin.

François et al. (1993) conducted an *in vitro* comparative examination for potential antiplasmodial activity against *P. falciparum* (NF 54/64, clone A1A9) between sesquiterpene lactones (including artemisinin) and other compounds in organic extracts of *A. annua*. Hexane and methanol extracts of *A. annua* leaves were successively made and used at final concentrations of 0.5 mg FW/mL. In case of *A. annua* callus tissue, a 10 mL methanol extract (50 mg FW/mL) was further partitioned over 20 mL water/methanol and chloroform phases. The methanol extract was used at a final concentration of 0.5 mg FW/mL, and the fractions were diluted 200 times. The presence of artemisinin and/or its derivatives in the extracts and fractions was determined by using an immuno-enzymatic assay (ELISA). The results revealed that the organic leaf extracts, that is, *n*-hexane and methanol extracts (in which artemisinin and/or natural derivatives were present) showed very pronounced antiplasmodial activity (100% growth inhibition of *P. falciparum*), while the methanolic extract of callus and its two fractions, water/methanol and chloroform (containing undetectable amounts of artemisinin and /or natural derivatives), revealed lesser activity, that is, 25.96%, 22.33%, and 24.55% growth inhibition of *P. falciparum*, respectively. These results have indicated that their lesser activity was caused by unknown constituent(s) other than artemisinin, and thus, artemisinin is strongly contributing to the antiplasmodial activity of the leaves extract (*n*-hexane and methanol).

Mueller et al. (2000) conducted experiments on the *A. annua* cv. *Artemis* leaves (grown in tropical Africa), having 0.63%–0.70% artemisinin on a dry weight basis. Approximately 40% of this artemisinin could be extracted by simple tea preparation. Five malaria patients who were treated with *A. annua* tea showed a rapid disappearance of parasitemia within 2–4 days. An additional trial with 48 malaria patients showed a disappearance of parasitemia in 44 patients (92%) within 4 days. Both trials showed marked improvement of symptoms; therefore, *A. annua* tea justifies the positive effect in the treatment of malaria in endemic areas, as suggested by the Chinese Pharmacopoeia.

During an ongoing investigation into *A. annua* for the treatment of malaria, Mouton et al. (2013) decided to study the possibility that synergism might enhance the efficacy of artemisinin. The main objective of the study was to test tea infusions and nonpolar extracts prepared from different *A. annua* varieties against *P. falciparum in vitro* in order to determine if synergism will increase the effectiveness of artemisinin in the samples as compared to pure artemisinin. The group found that the IC_{50} of artemisinin in the tea and nonpolar extracts was not significantly different to the IC_{50} of pure artemisinin. The study showed that the year and country of harvest or storage conditions did not have any influence on the activity and that it narrowly followed the concentration of artemisinin in all the extracts. So, based on these *in vitro* results, artemisinin seems to be the only active antiplasmodial compound in *A. annua*.

Artemisinin showed antitrypanosomal activity with an IC_{50} value of 35.91 μg/mL and with a selectivity index of 2.44. Deoxyqinghaosu was only present in *A. annua* and absent in the other three *Artemisia* species. Deoxyqinghaosu was the principal volatile component (20.44%) of the dichloromethane extract of *A. annua*. the dichloromethane extract from aerial part of *A. abyssinica* should be considered for further study for the treatment of trypanosomiasis (Nibret and Wink, 2010).

Avian coccidiosis is one of the most economically important diseases of the poultry industry, caused by apicomplexan parasites belonging to the genus *Eimeria*. There are seven species in this genus that affect chickens, with *E. tenella* being one of the most pathogenic (McDouglad and Reid, 1991). Infection with *E. tenella* is followed by caecal lesions (petechiae, thickening, ecchymoses, accumulation of blood, and caseous necrotic material in the caecum), accompanied with bloody diarrhea (Iacob and Duma, 2009). Intensive poultry production systems depend on chemoprophylaxis with anticoccidial drugs to combat infection. Anticoccidial drugs have been used for over 60 years, and their extensive use has led to the development of drug-resistant *Eimeria* spp. strains (Chapman, 1997; Harfoush et al., 2010; Jenkins et al., 2010). Drug-resistant strains are responsible for subclinical coccidiosis and economic losses due to poor weight gain and high food consumption. It was estimated that the 2003–2004 economic losses in India were 68.08% related to reduced body weight gain and 22.7% related to increased feed conversion ratio (FCR) (Bera et al., 2010). Regarding the anticoccidial effects of *A. annua* in chickens, past studies indicated that both artemisinin and *A. annua* can be effective against *Eimeria* spp. (Allen et al., 1998; Arab et al., 2006; Naidoo et al., 2008; Youn and Noh, 2001).

Drăgan et al. (2014) conducted a floor-pen study to evaluate the anticoccidial effect of *A. annua* and *Foeniculum vulgare* on *E. tenella* infection. Five experimental groups were established: negative control (untreated, unchallenged); positive control (untreated, challenged); a group medicated with 125 ppm lasalocid and challenged; a group medicated with *A. annua* leaf powder at 1.5% in feed and challenged; and a group treated with the mixed oils of *A. annua* and *F. vulgare* in equal parts, 7.5% in water and challenged. The effects of *A. annua* and oil extract of *A. annua* and *F. vulgare* on *E. tenella* infection were assessed by clinical signs, mortality, fecal oocyst output, feces, lesion score, weight gain, and feed conversion. Clinical signs were noticed only in 3 chickens from the lasalocid group, 6 from the *A. annua* group, and 9 from the *A. annua* and *F. vulgare* group, but were present in 19 infected chickens from the positive control group. Bloody diarrhea was registered in only 2 chickens from the *A. annua* group, but in 17 chickens from the positive control group. Mortality also occurred in the positive control group (7/20). Chickens treated with *A. annua* had a significant reduction in fecal oocysts (95.6%; P=0.027) and in lesion score (56.3%; P=0.005) when compared to the positive control. At the end of the experiment, chickens treated with *A. annua* leaf powder had the highest body weight gain (68.2 g/day), after the negative control group, and the best feed conversion (1.85) among all experimental groups. Our results suggest that *A. annua* leaf powder (Aa-p), at 1.5% of the daily diet post-infection, can be a valuable alternative for synthetic coccidiostats, such as lasalocid.

Schistosomiasis, a dreadful disease caused by parasitic trematode worms in humans as well as in animals is widespread in the world especially developing countries. It is considered second only to malaria as a major target disease of WHO (1993). Various species of fresh water snails act as intermediate host of schistosomiasis and the, snails belonging to family Planorbidae are known to act as intermediate host of both human and animal schistosomiasis. The fresh water snail *Indoplanorbis exustus* (Mollusca: Gastropoda) is widely distributed in India and acts as intermediate host of *Schistosoma nasale*, *Schistosoma spindale*, and *Schistosoma indicum* (Agrawal et al., 1991), the causative agents of animal schistosomiasis and other trematodal diseases. One of the possible approaches to eradicate or control this problem is to interrupt the life cycle of the parasitic trematodes by eliminating the snails, which are essential to their life cycle. Past studies were initiated with the use of certain chemical and synthetic compounds against these vectors (Preet, 2010 and references therein). At present, niclosamide (Bayluscide®, Bayer, Leverkusen, Germany) is the only registered molluscicide recommended by WHO for snail-control programmes (WHO, 1993). Since the compound is toxic to fish and other compounds also contaminate the aquatic environment, the use of natural products seems to be promising, as they have a shorter environmental half-life. The realization that plants could provide a cheap and readily available means of snail control has led to the screening of many plant species (Preet, 2010 and references therein). Preet (2010) carried out scientific investigations on the molluscicidal and cercaricidal potential of *A. annua* leaves, under laboratory conditions, to prevent the transmission of trematodal diseases via the snail host, *Indoplanorbis exustus.* Dried leaves were sequentially extracted (through soxhlet) with petroleum ether (60°–80°), carbon tetra chloride, and methanol for 48 h to get the crude extracts. The LC_{50} values of different crude extracts were tested for their molluscicidal activity against immature (Group I), young mature (Group II), and adult (Group III) stages of *I. exustus* using Probit analysis. Snail size and different extracts of *A. annua* strongly influenced the pattern of mortality and it was observed to be toxic to all size/age classes of snails. Of the three extracts tested, carbon tetra chloride was most toxic against all the age classes (I, II, III) with the LC_{50} of 8.14, 8.65, and 5.93 ppm, respectively. All the extracts also exhibited cercaricidal properties against two species of distome cercariae and the values were fairly below the molluscicidal range thus complementing a target specific approach toward control of trematodal infections.

Zhang et al. (2008) aimed to s determine the best extraction techniques, most suitable solvent, optimal plant part, and the acaricidal activities of *A. annua.* For this, his group collected leaves, stems, and roots of different periods of *A. annua* in April, May, June, July, and September and prepared petroleum ether (30°C–60°C), ethanol, acetone, and water parallel and sequenced extracts, then the acaricidal activity determined, by side-capillary method, against *Tetranychus cinnabarinus* of all extracts. The results indicated that the acaricidal bioactivities elevated as the development of *A. annua* plant at the concentration of 5 mg/mL. The general tendency exhibited the sequence of July > June > May > April, but September decreased compared with July; however, the most effective extracts in 5 months were all acetone parallel extract of *A. annua* leaf, and the corrected mortalities treated after 48 h ranged from 74% to 100%. The median lethal concentration (LC_{50}) against *T. cinnabarinus* of acetone parallel extracts of *A. annua* leaves in September, July, June, May, and April were 0.5986, 0.4341, 0.8376, 0.9443, and 1.3817 mg/mL, respectively, treated after 48 h. The acetone parallel extract of *A. annua* leaf in July was the most toxic to *T. cinnabarinus* and corrected mortality was 100% after 48 h.

In another study, Zhang et al. (2012) evaluated acaricidal activity of *A. annua* plant parts (leaf, stem, and root) collected in June and July against *Petrobia harti.* The biological activity of the plants collected in July was better than those collected in June, and the activity of acetone leaf extracts in July was higher than other extracts. The LC_{50} value of the acetone leaf extract in July was 0.4715 mg/mL compared with 0.9083 mg/mL in June.

Histomonosis is a parasitic disease in gallinaceous birds, primarily affecting turkeys and chickens. It causes severe lesions in the cecum and the liver and can lead to high mortality rates, especially in turkeys (McDougald, 2005). Thøfner et al. (2012) have taken dry leaves and three extracts (hexane, dichloromethane, and methanol leaf extracts) from *A. annua,* as well as the main antimalarial constituent of this plant, artemisinin, and evaluated for the first time for their antihistomonal activities *in vitro* against six different clonal cultures of *Histomonas meleagridis.* Four of the tested materials displayed *in vitro* activity against all protozoal clones. However, neither artemisinin nor Ext-DCM that were tested *in vivo* could prevent experimental histomonosis in turkeys or chickens at the given concentrations, although the clonal culture used for this investigation was one of the *in vitro* tested clones. Thus, the results of this study clearly demonstrated the importance of defined *in vivo* experiments to assess and verify *in vitro* results.

Studies have been conducted by implying various parameters of assessment of antifeedant activity for crude extracts of *A. annua*. Deterrency, growth regulatory effect, and ovicidal potential strongly recommend it as a good antifeedant herb (Haghighian et al., 2008), an anthelmintic, and an anti-insecticidal agent (Khosravi et al., 2011; Vicidomini, 2011). Some studies have reported that crude extracts of *A. annua* contain Artemisinin and its derivatives, which act as natural pesticides (WHO Library Cataloguing-in-Publication Data, 2006; Huang et al., 2010; Weathers et al., 2011). Recently, Anshul et al. (2013) evaluated methanolic extract of powdered *A. annua* leaves and different compounds isolated from the extract for toxicity and inhibition and disruption of growth and development of the African pod borer, *Helicoverpa armigera* (Hübner) (Lepidoptera: Noctuidae). Methanol extract of *A. annua* and eight known constituent compounds (artemisinic acid, artemisinin, scopoletin, arteannuin B, deoxyartemisinin, artemetin, and isomeric flavonoids [casticin and chrysosplenetin]) were bio-assayed for larval mortality, abnormal development, and growth inhibition. Among *A. annua* constituent compounds, the isomeric flavonoids exhibited a strong reduction in mean larval weight (58.5%), and growth inhibition (50.0%) as compared to the control. The mean weight of treated larvae reached only 0.026 g compared to the 0.270 g in the control and at par with larvae treated with 2% neem seed kernel extract (0.035 g) and 0.02% w/w azadirachtin (0.059 g). Thus, extracts of *A. annua* and its isomeric flavonoids appear to have potential for developing novel biopesticides.

Tripathi et al. (2001) isolated 1,8-Cineole from *A. annua* and tested against *Tribolium castaneum* (Herbst) for contact toxicity, fumigant toxicity, and antifeedant activity. The adults of *T. castaneum* were more susceptible than larvae to both contact and fumigant toxicity of 1,8-Cineole, and LD_{50} and LC_{50} values of 108.4 µg/mg body weight of adult insect and 1.52 mg/L air were found, respectively. Compound 1,8-Cineole applied to filter paper at a concentration of 3.22–16.10 mg cm^2 significantly ($p < 0.05$) reduced the hatching of *T. castaneum* eggs and the subsequent survival rate of the larvae. Adult emergence was also reduced by 1,8-Cineole. Feeding deterrence of 81.9% was achieved in *T. castaneum* adults by using a concentration of 121.9 mg/g food, whereas larvae showed 68.8% at the same concentration.

Leishmania, a unicellular trypanosomatid protozoan parasite, is the causative organism of leishmaniasis, which comprises a wide disease spectrum ranging from localized, self-healing, cutaneous lesions to disfiguring mucocutaneous leishmaniasis, and the visceral form, which can be fatal if neglected (Murray et al., 2005). In the past decade, an unprecedented increase in unresponsiveness to antimonials, the first line of treatment in visceral leishmaniasis, has been observed, importantly in the Indian subcontinent, a major endemic area of visceral leishmaniasis (Croft et al., 2006). In the absence of an effective vaccine, chemotherapy remains the sole weapon in the arsenal against leishmaniasis (Murray et al., 2005). Current treatment modalities are limited and have the potential to develop resistance and possess unacceptable toxicity (Sundar and Chatterjee, 2006).

Shahbazfar et al. (2012) carried out study on antileishmanial effects of artemisinin and artemisinin–iron combination treatment, in different concentrations (0.15, 0.3, 0.6, and 1.2 µg/mL) on to the culture of MDCK (Madin-Darby canine kidney) cells with and without iron (86 µg/dL). All the changes were controlled and photographed every 12 h using an invert microscope. Biochemical parameters showed cellular reaction and injury in a concentration-dependent manner. Cell injury was more severe in the iron-added groups. Microscopic exams showed cell and nuclear swelling, granular degeneration, vacuole and vesicle formation, cellular detachment, piknosis, karyorrhexis, cellular necrosis, and inhibition of new mitosis. Shahbazfar et al. (2012) believe that Artemisinin, as an herbal-based drug, can be used for treatment of leishmaniasis in dogs, but that its use requires further study regarding the influence of the drug on the liver, as the main metabolizing organ, and the nervous system, as the organ most influenced by this drug family. Overall, the drug does have some side effects, and the belief that it is completely safe since it is an herbal medication is not true. Its use, thus, necessitates caution and supervision.

Artemisinin showed antileishmanial activity in both promastigotes and amastigotes, with IC_{50} values of 160 and 22 µM, respectively, and, importantly, was accompanied by a high safety index (>22-fold) (Sen et al., 2007). The leishmanicidal activity of artemisinin was mediated via apoptosis as evidenced by externalization of phosphatidylserine, loss of mitochondrial membrane potential, *in situ* labeling of DNA fragments by terminal deoxyribonucleotidyltransferase-mediated dUTP nick end labeling (TUNEL), and cell-cycle arrest at the sub-G_0/G_1 phase. Taken together, these data indicate that artemisinin has promising antileishmanial activity that is mediated by programmed cell death and, accordingly, merits consideration and further investigation as a therapeutic option for the treatment of leishmaniasis.

Visceral leishmaniasis, the second most dreaded parasitic disease after malaria, is currently endemic in 88 countries. Very recently, Indian researchers have demonstrated *in vitro* and *in vivo* leishmanicidal effects of essential oil, obtained from hydrodistillation of *A. annua* leaves, against *Leishmania donovani*. A GC-MS study of this essential oil revealed the presence of camphor (52.06%) and β-caryophyllene (10.95%) as the most abundant compounds. The essential oil exhibited significant leishmanicidal activity with 50% inhibitory concentration of 14.63 ± 1.49 µg/mL and 7.3 ± 1.85 µg/mL, respectively, against the promastigotes and intracellular amastigotes. The essential oil revealed no cytotoxic effects against mammalian macrophages even at 200 µg/mL. Additionally, intra-peritoneal administration of the essential oil (200 mg/kg. b.w.) to infected BALB/c mice reduced the parasite burden by almost 90% in the liver and spleen with significant reduction in weight and no hepato- or nephrotoxicity was observed. Such leishmanicidal agents from inexpensive natural sources may provide a new lead for the treatment of visceral leishmaniasis (Islamuddin et al., 2014).

Schistosomiasis continues to rank—following malaria—second in the world's parasitic diseases, in terms of the extent of endemic areas and the number of people infected. There is no vaccine available yet, and the current mainstay of control is chemotherapy with praziquantel used as the drug of choice. In view of the concern about the development of tolerance and/or resistance to praziquantel, there is need for research and development of novel drugs for the prevention and cure of schisto-somiasis (Utzinger et al., 2001). Interestingly, derivatives of artemisinin, which are already effectively used in the treatment of malaria, also exhibit antischistosomal properties. Significant advances have been made with artemether, the methyl ether derivative of artemisinin.

The antischistosomal activity of artemether was discovered by Chinese scientists two decades ago; the detailed laboratory studies carried out for the susceptibility of, and effect on, the different developmental stages of *Schistosoma japonicum*, *Schistosoma mansoni*, and *Schistosoma haematobium* to artemether (Le et al., 1982; Xiao et al., 2000a,b).

Plants in the genus *Artemisia* have traditionally been used as anthelmintics and whole plants and plant extracts have demonstrated activity against gastrointestinal nematodes in several studies. Artemisinin derivatives have also shown efficacy against some trematodes, including *Fasciola hepatica.*

In vitro, artemether (at 10–30 µg/mL) caused severe tegument lesions in adult *Fasciola gigantica* 24 h after treatment. Artemether's effect on tegument was comparable to triclabendazole, and both affected immune modulation, osmoregulation, and nutrient absorption of *F. gigantica* (Shalaby et al. 2009). In sheep infected with *F. hepatica*, a single dose of 40 mg/kg (i.v.) artesunate reduced egg count and worm burden by 69% and 77%, respectively, while artemether (i.m.) at 40 mg/kg reduced fecal egg count (eggs per gram, EPG) and worm burden by 97.6% and 91.9%, respectively (Keiser et al. 2010).

The study carried out by Cala et al. (2014) evaluated the anthelmintic activity of *A. annua* crude extracts *in vitro* and compared the most effective extract with artemisinin in sheep naturally infected with *Haemonchus contortus*. *A. annua* leaves extracted with water, aqueous 0.1% sodium bicarbonate, dichloromethane, and ethanol were evaluated *in vitro* by the egg hatch test (EHT) and with the bicarbonate extract only for the larval development test (LDT) using *H. contortus*. The *A. annua* water, sodium bicarbonate (SBE), ethanol, and dichloromethane extracts tested *in vitro* contained 0.3%, 0.6%, 4.4%, and 9.8% of artemisinin, respectively. The sodium bicarbonate extract resulted in the lowest LC99 in the EHT (1.27 µg/mL) and in a LC99 of 23.8 µg/mL in the LDT. Following *in vitro* results, the SBE (2 g/kg body weight [BW]) and artemisinin (100 mg/kg BW) were evaluated as a single oral dose in naturally infected Santa Inês sheep. *A. annua* extract dose (micrograms per milliliter) inhibited 99% of egg hatching (LC99) of gastrointestinal nematodes of sheep.

Zamani et al. (2011) have conducted experiment on the effects of different concentrations (15%, 11%, 8%, 5.5%, and 4%) of *A. annua* essential oil on 17-day-old larvae of Indian meal moth *Plodia interpunctella* (Hübner) in laboratory conditions. LC_{25}, LC_{50}, and LC_{75} were estimated at 5.96%, 8.4%, and 11.3%, respectively, after 24 h. Sublethal doses revealed that *A. annua* essential oil reduced adult emergence, longevity of male and female insects, and the fecundity and fertility of females. The energy reserves (i.e., protein, carbohydrate, and lipid contents) of treated larvae were significantly reduced in comparison with the controls. Insect mortality was independent of sex.

Malaria is one of the most dangerous diseases occurring due to mosquito bites. Malaria is a communicable disease, caused by sporozoan parasites, of the genus *Plasmodium* and transmitted to man by certain species of infected female anopheles mosquitoes. Control of mosquitoes is something of utmost importance in the present day with the rising number of mosquito-borne illnesses. A child dies of malaria every 12 s (WHO, 2003). "The world's most dangerous animal is the mosquito," according to a BBC World Service health program; malaria now infects approximately 110 million people annually, causing 2–3 million deaths, and with increasing drug resistance, the problem is worsening, while attempts to control mosquitoes with pesticides have proved ineffective (Namdeo et al., 2006; Handerson et al., 2009). Mosquitoes are carriers of many harmful diseases like West Nile Virus disease (WNV), malaria, dengue fever, chikungunya, Lyme disease, and so on (White, 2008b). The only ways to get away from these diseases are to protect from mosquito bites, inhibit mosquitoes' breeding in the initial stage, and maintain cleanliness in surroundings. There are many ways to avoid mosquito bites, like wearing long-sleeved shirts and long pants trucked into socks while working outdoors. While indoors, stay in air-conditioned or screened areas or use bed nets; windows should also be protected by nets (Ro et al., 2006). Another important method to remain away from mosquitoes is, use of repellents. Recently, Yimer and Sahu (2014) used the essential oil of *A. annua* as mosquito repellent. The extracted oil was used with eucalyptus oil, neem oil, and rose oil. The better performance was shown by a combination of *A. annua* oil and eucalyptus oil, with results showing up to 2 h without a single bite by female mosquitoes.

9.2.11 Antiviral Activity

The first report on artemisinin derivatives being used as anti-HIV agents was published by Jung and Schinazi (1994), who embarked on a study of artemisinin trioxane derivatives and concluded that anti-HIV activity was common in artemisinin trioxane derivatives and further evaluation was needed for producing a potential anti-HIV treatment. In African countries, a tea infusion prepared from *A. annua* has been used for the treatment of malaria only for the past 10–20 years. Additionally, several informal claims in Africa exist stating that the *A. annua* tea infusions are also able to inhibit HIV. Based on the previous, Lubbe et al. (2012) evaluated antiviral activity of *A. annua* tea infusions against HIV through scientific investigation for the first time. Two independent cellular systems (*A. annua* and *Artemisia afra*) have been used for toxicity studies. The *A. annua* tea infusion exhibits highly significant activity at a very low concentration (2.0 µg/mL), while the chemically closely related species *A. afra* (not containing artemisinin) showed a similar level of anti-HIV activity. But artemisinin was found inactive at a higher concentration (25 µg/mL). Similarly, no cellular cytotoxic effects were observed at a higher concentration of tea infusion. Therefore, this *in vitro* study revealed that artemisinin plays a limited role and may act synergistically against anti-HIV activity (Lubbe et al., 2012). Some other *in vitro* studies have claimed about inhibitory effects for the hepatitis B virus (WHO Monographs, 2006). Currently, artemisinin and its derivatives have become the subject of scientific studies to investigate their potential against a number of viruses (Ferreira and Janick, 2009) with the aim of advanced combination antiviral therapies (Weathers and Towler, 2012). Artemisinin has been tested against various viruses; Efferth et al. (2008) reported on the broad spectrum of antiviral activity of artemisinin and its semisynthetic derivative artesunate that includes the inhibition of certain viruses, such as human cytomegalovirus and other members of the *Herpesviridae* family (e.g., herpes simplex virus type I and Epstein–Barr virus), hepatitis B virus, and bovine viral diarrhea virus. Artesunate showed inhibition of HIV at levels of 600 nM, but no reports on the activity of artemisinin against HIV was given. Jung and Schinazi (1994) reported on the anti-HIV activity of artemisinin, with EC_{50} and IC_{50} greater than 100 µM. Benedikt et al. (2005) have taken a patent on artemisinin and some of its derivatives against various viruses. According to this patent, the activity of artemisinin was given to be insignificant against HIV-1 and HIV-2. Karamoddini et al. (2011) showed that methanolic extracts of *A. annua* collected from Iran during the month of September, showed good *in vitro* activity against the KOS strain of herpes simplex virus type 1 (HSV1) in HeLa cells. Extracts of *A. annua* showed high virus inhibitory activity against tobamoviruses on their test hosts reacting hypersensitively. The virus inhibitory agent(s) occurring in *A. annua* plants was isolated by conventional methods, characterized and identified as sitosterol and stigmaterol (Khan et al., 1991).

The ethanolic extract of *A. annua* has been evaluated for its immunosuppressive activity. Ethanolic extract significantly suppressed concanavalin A (Con A) and lipopolysaccharide (LPS)-stimulated splenocyte proliferation *in vitro*, and this activity increases with increased dose. Results showed that the ethanol extract of *A. annua* could suppress the cellular and humoral response (Das, 2012; Sadiq et al., 2014). Immunosuppressive potential has been linked to flavonoids present in leaves which are able to modulate the immune response (Ferreira et al., 2010; Sadiq et al., 2014).

Recently, several compounds derived from TCM remedies, in particular, artemisinin and its semisynthetic derivative artesunate, have demonstrated antiviral activities (Efferth et al., 2002). The antiviral effect against hepatitis B virus (HBV) of artemisinin, its derivative artesunate, and other compounds highly purified from Traditional Chinese Medicine (TCM) remedies were investigated. HBV production by permanently transfected HepG2 2.2.15 cells was determined by measuring the release of surface protein (HBsAg) and HBV-DNA after drug exposure (0.01–100 µM) for 21 days (Romero et al., 2005). They induced strong inhibition of viral production at concentrations at which host cell viability was not affected (artemisinin and artesunate). Moreover, artesunate in conjunction with lamivudine had synergic anti-HBV effects, which warrants further evaluation of artemisinin/artesunate as antiviral agents against HBV infection (Romero et al., 2005). Artemisinin and artesunate are found to be active against HBV (Wohlfarth and Efferth, 2009).

Obeid et al. (2013) reported that artemisinin (ART) is an inhibitor of *in vitro* hepatitis C virus (HCV) subgenomic replicon replication. ART also exerts antiviral activity in hepatoma cells infected with full-length infections HCV JFH-1. Obeid et al. also identified a number of ART analogues that are up to 10-fold more potent and selective as *in vitro* inhibitors of HCV replication than ART. The iron-donor hemin only marginally potentiates the anti-HCV activity of ART in HCV-infected cultures. Carbon-centered radicals have been shown to be critical for the antimalarial activity of ART. ART and its analogues may possibly exert their anti-HCV activity by the induction of reactive oxygen species (ROS). The combined anti-HCV activity of ART or its analogues with L-N-Acetylcysteine (L-NAC) (a molecule that inhibits ROS generation) was studied. L-NAC significantly reduced the *in vitro* anti-HCV activity of ART and its derivatives. Taken together, the *in vitro* anti-HCV activity of ART and its analogues can, at least in part, be explained by the induction of ROS; carbon-centered radicals may not be important in the anti-HCV effect of these molecules.

9.2.12 Cytotoxic Activity

Artemisinin, with its worldwide-accepted applications in malaria therapy, is one of the showcase success stories of phytomedicine of the past few decades. Artemisinin-type compounds are also active toward other protozoal or viral diseases, as well as cancer cells *in vitro* and *in vivo*. Nowadays, *A. annua* tea is used as a self-reliant treatment in developing countries. The unsupervised use of *A. annua* tea has been criticized as fostering the development of artemisinin resistance in malaria and cancer due to insufficient artemisinin amounts in the plant when compared with standardized tablets with isolated artemisinin or semisynthetic artemisinin derivatives. However, artemisinin is not the only bioactive compound in *A. annua*. In the present investigation, Efferth et al. (2011), we analyzed different *A. annua* extracts. Dichloromethane extracts were more cytotoxic (range of IC_{50}: 1.8–14.4 µg/mL) than methanol extracts toward *Trypanosoma b. brucei* (TC221 cells). The range of IC_{50} values for HeLa cancer cells was 54.1–275.5 µg/mL for dichloromethane extracts and 276.3–1540.8 µg/mL for methanol extracts. Cancer and trypanosomal cells did not reveal cross-resistance among other compounds of *A. annua*, namely the artemisinin-related artemisitene and arteanuine B, as well as the unrelated compounds scopoletin and 1,8-cineole. This indicates that cells resistant to one compound retained sensitivity to another one. These results were also supported by microarray-based mRNA expression profiling showing that molecular determinants of sensitivity and resistance were different between artemisinin and the other phytochemicals investigated.

A study by Singh and Lai (2001) revealed that artemisinin becomes cytotoxic in the presence of ferrous iron. Since iron influx is high in cancer cells, artemisinin and its analogues selectively kill cancer cells under conditions that increase intracellular iron concentrations. Singh and Lai reported that after incubation with holotransferrin, which increases the concentration of ferrous iron in cancer cells, DHA, an analogue of artemisinin, effectively killed a type of radiation-resistant human breast cancer cell *in vitro*. The same treatment had considerably less effect on normal human breast cells. Since it is relatively easy to increase the iron content inside cancer cells *in vivo*, administration of artemisinin-like drugs and intracellular iron-enhancing compounds may be a simple, effective, and economical treatment for cancer.

Since cancer cells uptake relatively large amount of iron compared to normal cells, they are more susceptible to the toxic effect of artemisinin. Artemisinin reacts with iron to form free radicals that kills cancerous cells. In another study, Lai et al. (2005) attached artemisinin and iron-carrying plasma glycoprotein transferrin. Transferrin is transported into cells through a receptor-mediated endocytosis and cancer cells express significantly more transferrin receptors on their cell surface. This would enhance the toxicity and selectivity of artemisinin toward cancer cells, so Lai et al. synthesized a compound in which transferrin conjugates with artelinic acid (artemisinin analogue) through the N-glycoside chains on the C-domain. The resulting conjugate-compound (tagged-compound) was characterized by spectroscopic techniques, chemiluminescence, and HPLC. This compound was tested on a human leukemia cell line (Molt-4) and normal human lymphocytes. Lai et al. found that the conjugate-compound was very potent and selective in killing cancer cells. Therefore, such artemisinin-conjugates may be very useful chemotherapeutic agents for cancer treatment.

Zheng et al. (1994) reported significant cytotoxic activity of artemisinin and quercetagetin-6,7,3′,4′-tetramethylether against P-388, A-549, Ht-29, MCF-7, and KB tumor cells. Deoxyartemisinin, artemisinic acid, arteannuin B, stigmasterol, friedelin, friedelin-3α-ol, and artemetin were ineffective in the previously mentioned system. Beekman et al. (1997, 1998) found stereochemistry-dependent cytotoxicity in artemisinin and its semisynthetic analogues. Since artemisinin is a novel molecule by its chemical structure and mode of action, it is thus a new lead compound, which can be exploited for further drug development.

Artemisinin and its analogues are naturally occurring antimalarials that have also shown potent anticancer activity. In primary cancer cultures and cell lines, they inhibited cancer proliferation, metastasis, and angiogenesis. In xenograft models, exposure to artemisinins substantially reduces tumor volume and progression. However, the rationale for the use of artemisinins in anticancer therapy must be addressed by a greater understanding of the underlying mechanisms involved in their cytotoxic effects. The primary targets for artemisinin and the chemical base for its preferential effects on heterologous tumor cells needs to be elucidated (Crespo-Ortiz and Wei, 2012).

Worku et al. (2013) evaluated the *in vitro* effect of alcoholic extracts of *A. annua* on proliferation/viability of 1321N1 astrocytoma, MCF-7 breast cancer, THP-1 leukemia, LNCaP, Du-145, PC-3 prostate cancer cells, and on *T. brucei* cells. The proliferation of tumor cells was evaluated by WST-1 assay, and the viability/behavior of *T. brucei* by cell counting and light microscopy. Ethanol extract significantly inhibited tumor growth at 3 µg/mL in LNCaP and THP-1 cells. Ethanol extract inhibited proliferation of *T. brucei* cells in a concentration-dependent manner. Microscopic analysis revealed that 95% of the *T. brucei* cells died when exposed to 33 µg/mL ethanol extract for 6 h. This demonstrates the antitumor efficacy of these extracts, as well as their ability to dampen viability and proliferation of *T. brucei*, suggesting a common mechanism of action on highly proliferative cells, most probably by targeting cell metabolism.

9.2.13 Radical Inhibitory/Scavenging (Antioxidant) Activity

A. annua is a good source of different nutritional constituents and antioxidants (Das, 2012). Studies indicate that crude organic extracts of aerial parts have high antioxidant capacity, which is most probably because the leaves contains a high content and variety of flavonoids, including the newly reported C-glycosyl flavonoid as a possible component of the antioxidants. Flavonoids and the essential oil content present in *A. annua* impart antioxidant properties. Therefore, *A. annua* has ranked at the top of the list among other medicinal plants, based on the highest potential as an antioxidant (Ferreira and Janick, 2009; Juteau et al., 2002). Major groups of hydroxylated and polymethoxylated flavonoids have been identified, which further include chrysosplenol-D, cirsilineol, eupatin, chrysosplenetin, cirsilineol, casticin, and artemetin (Ferreira et al., 2010). Research studies have identified five bioactive flavonoids, 5-hydroxy-3,7,4′-trimethoxyflavone, 5-hydroxy-6,7,3′,4′-tetramethoxyflavonol, blumeatin, 5,4′-dihydroxy-3,7,3′-trimethoxyflavone, and quercetin, respectively, and were further subjected to structural analysis (Yang et al., 2009). Artemisinin itself has not been used in mainstream clinical practice due to its poor bioavailability when compared to its analogues. Besides the antimalarial activity of *A. annua*, the herb was reported to have *in vitro* antioxidative activity. Kim et al. (2014) investigated the protective effect of aqueous ethanol extract of *Qinghao* (AA extract) against D-galactose–induced oxidative stress in C57BL/6J mice. Feeding an AA extract-containing diet lowered the serum levels of malondialdehyde and 8-OH-dG that are biomarkers for lipid peroxidation and DNA damage, respectively. Furthermore, AA extract–feeding enhanced the activity of NQO1, a typical antioxidant marker enzyme, in tissues such as the kidney, stomach, small intestine, and large intestine. In conclusion, AA extract was found to have antioxidative activity in a mouse model.

Chukwurah et al. (2014) evaluated the antioxidant capacity of leaves of *A. annua*, for which leaves were extracted with four solvents (absolute ethanol, absolute methanol, 70% ethanol, and 70% methanol), and the extracts obtained were studied by five complementary *in vitro* antioxidant test systems using ascorbic acid (vitamin C) and rutin as standard references. Results of the investigation suggested that extracts remarkably inhibited lipid peroxidation (79.81%–86.70%) and erythrocyte haemolysis (40.02%–49.91%). Their IC_{50} values for hydroxyl, nitric oxide, and hydrogen peroxide radical-scavenging activities ranged from 2.39 to 3.81 mg/mL (superior to standards), 107.24–144.29, and 28.53–53.20 µg/mL, respectively. The 70% alcohol extracts generally showed better antioxidant activity than absolute alcohol extracts.

9.2.14 Enzyme Inhibitory Effects

The flavonoid compounds fisetin and patuletin-3,7-dirhamnoside isolated from *A. annua* were found to exhibit the potential for inhibiting the non-peptide angiotensin-converting enzyme (Lin et al., 1994; Bhakuni et al., 2001).

The greatest Acetylcholinesterase (AChE)-inhibitory activity was exhibited by the essential oil of flowers of the plant collected from the post-flowering phase

($IC_{50} = 0.13 \pm 0.02$ mg/mL) compared with the essential oil of flowers collected in pre-flowering stage (1.25 ± 0.09 mg mL) and full-flowering stage (2.92 ± 0.16 mg/mL). The major constituents identified in the post-flowering oil were camphor (16.62%), caryophyllene (16.27%), β-caryophyllene oxide (15.84%), β-farnesene (9.05%), and (−)-spathulenol (7.21%), while the major constituents of the pre-flowering essential oil were β-myrcene (37.71%), 1,8-cineol (16.11%), and camphor (14.97%) and full-flowering oil contained predominantly caryophyllene (19.4%), germacrene D (18.1), camphor (15.84%), 1,8-cineol (10.6%), and (Z)-β-farnesene (9.43%). This study shows that the high concentration of 1,8-cineole and the low concentration of camphor in the oil may result in an increase in its anticholinesterase activity (Yu et al., 2011).

9.2.15 Miscellaneous Activities

The leaves of *A. annua* are known to have an anti-smoking-habit effect produced by cigarette-type stop-smoking aids made of *Artemisia* (100%) that do not contain nicotine. *Artemisia*'s stop-smoking cigarette has been claimed to greatly reduce withdrawal symptoms that cause cravings to smoke when a piece is taken 10–15 times a day for about 4–5 weeks (Donghwa Pharm Co. Ltd., 2014).

Okafor et al. (2012) reported the corrosion inhibitory action of ethanol (EEAA), acid (AEAA), and toluene (TEAA) extracts of *A. annua* and artemisinin (ART) on mild steel corrosion in sulfuric acid solutions, which was investigated by gravimetric and gasometric techniques. The extracts and artemisinin functioned as good inhibitors, and their inhibition efficiencies (%IE) followed the trend: EEAA > AEAA > TEAA > ART. The %IE increased with an increase in inhibitor concentration and decreased with an increase in temperature. The enhanced %IE values of the extracts were attributed to the synergistic effect of the components of the plant extracts with ART. The adsorption of the inhibitors was consistent with Langmuir isotherm. Physisorption is proposed as the mechanism of inhibition.

REFERENCES

Abid M., Anupam K., Khan N.A., Islam G., Gahlot K., and Ghosh A.K. 2013. Pharmacological evaluation of *Artemisia annua* for antinociceptive activity in rats. *J. Appl. Pharm. Sci.* 3(5): 61.64.

Abolaji A.O., Eteng M.U., Ebong P.E., Dar A., Farombi E.O., and Choudhary M.I. 2014. *Artemisia annua* as a possible contraceptive agent: A clue from mammalian rat model. *Nat. Prod. Res.* 28(24): 2342–2346.

Agrawal M.C., Banerjee P.S., Shah H. L. 1991. Five mammalian schistosome species in an endemic focus in India. *Trans. Roy. Soc. Trop. Med. Hyg.* 85(2): 231.

Allen P.C., Danforth H.D., and Augustine C. 1998. Dietary modulation of avian coccidiosis. *Int. J. Parasitol.* 28(7): 1131–1140.

Anshul N., Bhakuni R. S., Gaur R., and Singh D. 2013. Isomeric flavonoids of *Artemisia annua* (Asterales: Asteraceae) as insect growth inhibitors against *Helicoverpa armigera* (Lepidoptera: Noctuidae). *Florida Entomol.* 96(3): 897–903.

Arab H.A., Rahbari S., Rassouli A., Moslemi M.H., and Khosravirad F. 2006. Determination of artemisinin in *Artemisia sieberi* and anticoccidial effects of the plant extract in broiler chickens. *Trop. Anim. Health Prod.* 38(6): 497–503.

Bagchi G.D., Jain D.C., and Kumar S. 1997. Arteether: A potent plant growth inhibitor from *Artemisia annua. Phytochemistry* 45(6): 1131–1133.

Bagchi G.D., Jain D.C., and Kumar S. 1998. The phytotoxic effects of the artemisinin related compounds of *Artemisia annua. J. Med. Aromat. Plant Sci.* 20: 5–11.

Beekman A.C., Barentesen A.R.W., Woedenbag H.J., Uden U.V., Pras N., Konings A.W.T., El-Feraly F.S., et al. 1997. Stereochemistry-dependent cytotoxicity of some artemisinin derivatives. *J. Nat. Prod.* 60(4): 325–330.

Beekman A.C., Wierenga P.K., Woedenbag H.J., Uden W.V., Pras N., Konings A. W.T., El-Feraly F.S., et al. 1998. Artemisinin-derived sesquiterpene lactones as potential antitumour compounds: Cytotoxic action against bone marrow and tumour cells. *Plant Med.* 64(7): 615–619.

Benedikt S., Van Hemel J., and Vandenkerckhove J. 2005. Use of endoperoxides for the treatment of infections caused by flaviviridae, including hepatitis C, bovine viral diarrhea and classical swine fever virus, United States Patent 20050059647.

Bera A.K., Bhattacharya D., Pan D., Dhara A., Kumar S., and Das S.K. 2010. Evaluation of economic losses due to coccidiosis in poultry industry in India. *Agric. Econ. Res. Rev.* 23: 91–96.

Bhakuni R.S., Jain D.C., Sharma R.P., and Kumar S. 2001. Secondary metabolites of *Artemisia annua* and their biological activity. *Curr. Sci.* 80(1): 35–48.

Bharati A., Kar M., and Sabat S.C. 2012. Artemisinin inhibits chloroplast electron transport activity: Mode of action, *PLoS ONE* 7(6): e38942.

Bilia A.R., Santomauro F., Sacco C., Bergonzi M.C., and Donato R. 2014. Essential oil of *Artemisia annua* L.: An extraordinary component with numerous antimicrobial properties. *Evidence-Based Complementary Altern. Med.* Article I.D: 139819 (1–7). (http://dx.doi.org/10.1155/2014/159819)

Brown G.D. 2010. The biosynthesis of artemisinin (Qinghaosu) and the phytochemistry of *Artemisia annua* L. (Qinghao). *Molecules* 15(15): 7603–7698.

Cafferata L.F.R., Gatti W.O., and Mijailosky S. 2010. Secondary gaseous metabolites analyses of wild *Artemisia annua* L. *Mol. Med. Chem.* 21: 48–52.

Cala A.C., Ferreira J.F.S., Carolina A., Chagas S., Gonzalez J. M., Rodrigues R.A.F., Foglio M.A. et al. 2014. Anthelmintic activity of *Artemisia annua* L. extracts *in vitro* and the effect of an aqueous extract and artemisinin in sheep naturally infected with gastrointestinal nematodes. *Parasitol. Res.* 113(6): 2345–2353.

Ćaver S., Maksimović M., Vidic D., and Parić A. 2012. Chemical composition and antioxidant and antimicrobial activity of essential oil of *Artemisia annua* L. from Bosnia. *Ind. Crops Prod.* 37(1): 479–485.

Chagas M.C. de S, Georgetti C.S., Olivo de Carvalho C., de Sena Oliveira M.C., Rodrigues R.A., Foglio M.A., and de Magalhães P.M. 2011. *In vitro* activity of *Artemisia annua* L. (Asteraceae) extracts against *Rhipicephalus (Boophilus) microplus. Rev. Bras. Parasitol Vet. Jaboticabal.* 20(1): 31–35.

Chang H.R., and Pechere J.C. 1988. Arteether, a qinghaosu derivative, in toxoplasmosis. *Trans. Roy. Soc. Trop. Med. Hyg.* 82: 867.

Chapman H.D. 1997. Biochemical, genetic and applied aspects of drug resistance in *Eimeria* parasites of the fowl. *Avian Pathol.* 26(2): 221–244.

Chaudhari H., Mehta J.B., Chaudhari K., and Farrow J. 2013. Treatment of cerebral malaria and acute respiratory distress syndrome (ARDS) with parenteral artesunate. *Tenn. Med.* 106(5): 41–43.

Chen P.K., and Leather G.R. 1990. Plant growth regulatory activities of artemisinin and its related compounds. *J. of Chem. Ecol.* 16(6): 1867–1876.

Chen H., Maibach H.I. 1994. Topical application of artesunate on guinea pig allergic contact dermatitis. *Contact Dermatitis* 30(5): 280–282.

Chen Y.T., Ma L., Mei Q., Tang Y., and Liao X.G. 1994. An experimental trial of artemether in treatment of *Pneumocystis carinii* in immunosuppressed rats. *Chin. Med. J.* 107(9): 673–677.

Cheng C., Ho W.E., Goh F.Y., Guan S.P., Kong L.R., Lai W.Q., Leung B.P., and Wong W.S.F. 2011. Anti-malarial drug artesunate attenuates experimental allergic asthma via inhibition of the phosphoinositide 3-kinase/Akt pathway. *PLoS ONE* 6(6): e20932.

Liu K.C-S.C., Yang S.-L., Roberts M.F., Elford B.C., and Phillipson J.D. 1992. Antimalarial activity of *Artemisia annua* flavonoids from whole plants and cell cultures. *Plant Cell Rep.* 11(12): 637–640.

Chukwurah P.N., Brisibe E.A., Osuagwu A.N., and Okoko T. 2014. Protective capacity of *Artemisia annua* as a potent antioxidant remedy against free radical damage. *Asian Pac. J. Trop. Biomed.* 4 (Suppl. 1): S92–S98.

Crespo-Ortiz M.P., and Wei M.Q. 2012. Antitumor activity of artemisinin and its derivatives: From a well-known antimalarial agent to a potential anticancer drug. *J. Biomed. Biotechnol.* 2012: 18.

Croft S.L., Sundar S., and Fairlamb A.H. 2006. Drug resistance in leishmaniasis. *Clin. Microbiol. Rev.* 19(1): 111–126.

Damtew Z., Tesfaye B., and Bisrat D. 2011. Leaf, essential oil and artemisinin yield of artemisia (*Artemisia annua* L.) as influenced by harvesting age and plant population density. *World J. Agri. Sci.* 7(4): 404–412.

Das S. 2012. *Artemisia annua* (Qinghao): A pharmacological review. *Int J Pharm Sci Res* 3(12): 4573–4577.

Désirée C.K.R., René F.K.P., Jonas K., Bibiane D.T., Roger S.M., and Lazare K. 2013. Antibacterial and antifungal activity of the essential oil extracted by hydro-distillation from *Artemisia annua* grown in West-Cameroon. *Br. J. Pharmacol. Toxicol.* 4(3): 89–94.

Dias P. 1997. Atividade antiulcerogênica dos extratos brutos e das frações semi-purifi cadas de *Artemisia annua* L. Dissertação de mestrado, Universidade Estadual de Campinas. Campinas, São Paulo.

Dias P.C., Foglio M.A., Possenti A., Nogueira D.C.F., and De Carvalho J.E. 2001. Antiulcerogenic activity of crude ethanol extract and some fractions obtained from aerial parts of *Artemisia annua* L. *Phytother. Res.* 15(8): 670–675.

Diemer P. and Griffee P. 2013. *Artemisia annua*; the plant, production and processing and medicinal applications, http://www.mmv.org/sites/default/files/uploads/docs/artemis-inin/2007 event/12_Diemer-Griffee_Artemisia_annuapaper.pdf.

Donghwa Pharm Co. Ltd. 2014. Artemisia no smoking. Donghwa Pharm Co. Ltd. (Korea), http://www.drugs.com/otc/131218/artemisia-no-smoking.html.

Drăgan L., Györke A., Ferreira J.F.S., Pop I.A., Dunca I., Drăgan M., Mircean V., et al. 2014. Effects of *Artemisia annua* and *Foeniculum vulgare* on chickens highly infected with *Eimeria tenella* (Phylum Apicomplexa). *Acta Vet. Scand.* 56: 22.

Duke S.O., Vaughn K.C., Croom Jr. E.M., and Elsohly H.N. 1987. Artemisinin, a constituent of annual wormwood (*Artemisia annua*), is a selective phytotoxin. *Weed Sci.* 35: 499.

Efferth T., Herrmann F., Tahrani A., and Wink M. 2011. Cytotoxic activity of secondary metabolites derived from *Artemisia annua* L. towards cancer cells in comparison to its designated active constituent artemisinin. *Phytomedicine* 18(11): 959–969.

Efferth T., Marschall M., Wang X., Huong S.M., Hauber I., Olbrich A., Kronschnabl M., et al. 2002. Antiviral activity of artesunate towards wild-type, recombinant and ganciclovir-resistant human cytomegaloviruses. *J. Mol. Med.* 80: 233–242.

Efferth T., Romero M.R., Wolf D.G., Stamminger T., Marin J.J.G., and Marschall M. 2008. The antiviral activities of artemisinin and artesunate. *Clin. Infect. Dis.* 47(6): 804–811.

El-Naggar E.-M.B., Azazi M., Švajdlenka E., and Žemlička M. 2013. Artemisinin from minor to major ingredient in *Artemisia annua* cultivated in Egypt. *J. Appl. Pharm. Sci.* 3(8): 116–123.

Emadi. 2013. Phytochemistry of *Artemisia annua.* http://edd.behdasht.gov.ir/uploads/178_340. emadi.pdf.

Emadi, F., Yassa, N., Hadjiakhoondi, A., Beyer C., and Sharifzadeh, M. 2011. Sedative effects of Iranian *Artemisia annua* in mice: Possible benzodiazepine receptors involvement. *Pharm. Biol.* 49(8): 784–788.

Favero F.de F., Grando R., Nonato, F.R., Sousa, I.M., Queiroz, N.C., Longato, G.B., Zafred, R.R. et al. 2014. *Artemisia annua* L.: Evidence of sesquiterpene lactones' fraction anti-nociceptive activity. *B MC Complement. Altern. Med.* https://hort.purdue.edu/newcrop/CropFactSheets/artemisia.pdf 14: 266.

Ferreira J. and Janick J. 2009. Annual wormwood (*Artemisia annua* L.). New Crop FactSHEET. Purdue University. Available from: www.hort.purdue.edu/newcrop/CropFactSheets/artemisia.pdf.

Ferreira J.F.S., Luthria D.L., Sasaki T., and Heyerick A. 2010. Flavonoids from *Artemisia annua* L as antioxidants and their potential synergism with artemisinin against malaria and cancer. *Molecules* 15(5): 3135–3170.

Fesen M.R., Pommier Y., Leteurtre F., Heroguchi S., Yang J. and Kohn K.W. 1994. Inhibition of HIV-1 integrase by flavones, caffeic acid phenethyl ester (CAPE) and related compounds. *Biochem. Pharmacol.* 48(3): 595–608.

Foglio M.A., Dias P.C., Antônio M.A., Possenti A., Rodrigues R.A.F., Silva E.F., Rehder V.L.G., et al 2002. Antiulcerogenic activity of some sesquiterpene lactones enriched fraction isolated from *Artemisia annua. Planta Med.* 68(6): 515–518.

François G., Dochez C., Jaziri M., and Laurent A. 1993. Antiplasmodial activities of sesquiterpene lactones and other compounds in organic extracts of *Artemisia annua. Planta Med.* 59 (Suppl. issue): A677–A678.

Galli S.J., Tsai M., and Piliponsky A.M. 2008. The development of allergic inflammation. *Nature* 454(7203): 445–454.

Geldre E.V., Pauw I.D., Inze D., Montagu M.V., and Eeckhout E.V. 2000. Cloning and molecular analysis of two new sesquiterpene cyclases from *Artemisia annua* L. *Plant Sci.* 158(1-2): 163–171.

Gupta P.C., Dutta B., Pant D., Joshi P., and Lohar D.R. 2009. *In vitro* antibacterial activity of *Artemisia annua* Linn. growing in India. *Int. J. Green Pharm.* 3(3): 255–258.

Haghighian F., Sendi J.J., Aliakbar A., and Javaherdashti M. 2008. The growth regulatory, deterrency and ovicidal activity of worm wood (*Artemisia annua* L.) on *Tribolium confusum* duv. and identification of its chemical constituents by GC-MS. *Pestycydy* 1(2): 51–59.

Harfoush M.A., Hegazy A.M., Soliman A.H., and Amer S. 2010. Drug resistance evaluation of some commonly used anti-coccidial drugs in broiler chickens. *J. Egypt. Soc. Parasitol.* 40(2): 337–348.

Haynes R.K. 2006. From artemisinin to new artemisinin antimalarials: biosynthesis, extraction, old and new derivatives, stereochemistry and medicinal chemistry requirements. *Curr. Top. Med. Chem.* 6(5): 509–537.

Henderson D.A. 2009. *Smallpox: The Death of a Disease. The Inside Story of Eradicating a Worldwide Killer.* Amherst, New York: Prometheus Books, 334.

Holfels E.J., McAuley D., Mack W.K., and Milhous M.R. 1994. In vitro effects of artemisinin ether, cycloguanil hydrochloride (alone and in combination with sulfadiazine), quinine sulfate, mefloquine, primaquine phosphate, trifluoperazine hydrochloride, and verapamil on *Toxoplasma gondii. Antimicrob. Agents Chemother.* 38(6): 1392–1396.

Homewood C.A., Jewsbury J.M., and Chance M.L. 1972. The pigment formed during haemoglobin digestion by malarial and schiostosomal parasites. *Comp. Biochem. Physiol. B.* 43(3): 517–523.

Huang L., Liu J.F., Liu L.X., Li D.F., Zhang Y., Nui H.Z., Song H.Y., et al 1993. Antipyreticand anti-inflammatory effects of *Artemisia annua* L. *China Journal of Chinese Materia Medica* 18(1): 44–48, 63–64.

Huang L., Xie C., Duan B., and Chen S. 2010. Mapping the potential distribution of high artemisinin-yielding *Artemisia annua* L. (Qinghao) in China with a geographic information system. *Chin. Med.* 5: 18.

Iacob O.C., Duma V. 2009. Clinical, paraclinical and morphopathological aspects in cecal eimeriosis of broilers. *Sci. Parasitol.* 10: 43–50.

Islamuddin M., Chouhan G., Tyagi M., Abdin M. Z., Sahal D., and Afrin F. 2014. Leishmanicidal activities of *Artemisia annua* leaf essential oil against visceral leishmaniasis. *Front. Microbiol.* 5: 626.

Jelinek T. 2013. Artemisinin based combination therapy in travel medicine. *Travel Med. Infect. Dis.* 11(1): 23–28.

Jenkins M., Klopp S., Ritter D., Miska K.,and Fetterer R. 2010. Comparison of *Eimeria* species distribution and salinomycin resistance in commercial broiler operations utilizing different coccidiosis control strategies. *Avian Dis.* 54(3): 1002–1006.

Jung M., and Schinazi R. F. 1994. Synthesis and *in vitro* anti-human immunodeficiency virus activity of artemisinin (Qinghaosu)-related trioxanes. *Bioorg. Med. Chem. Lett.* 4(7): 931–934.

Juteau F., Masotti V., Bessiere J.M., Dherbomez M., Viano J. 2002. Antibacterial and antioxidant activities of *Artemisia annua* essential oil. *Fitoterapia* 73(6): 532–535.

Karamoddini M.K., Emami S.A., Ghannad M.S., Sani E.A., and Sahebkar A. 2011. Antiviral activities of aerial subsets of *Artemisia* species against herpes simplex virus type 1 (HSV1) *in vitro. Asian Biomed.* 5(1): 63–68.

Ke D.M.,Krug E.C., Marr J.J., and Berens R. 1990. Inhibition of growth of Toxoplasma gondii by qinghaosu and derivatives. *Antimicrob. Agents Chemother.* 34(10): 1961–1965.

Ke O.Y., Krug E.C., Marr J.J., Berens R.L. 1990. Inhibition of growth of *Toxoplasma gondii* by qinghaosu and derivatives. *Antimicrob. Agents Chemother.* 34: 1961–1965.

Keiser J., Veneziano V., Rinaldi L., Mezzino L., Duthaler U., and Cringoli G. 2010. Anthelmintic activity of artesunate against *Fasciola hepatica* in naturally infected sheep. *Res. Vet. Sci.* 88(1): 107–110.

Khan M.M.A.A., Jain D.C., Bhakuni R.S., Zaim M., Thakur R.S. 1991. Occurrence of some antiviral sterols in *Artemisia annua. Plant Sci.* 75(2): 161–165.

Khosravi R., Sendi J.J., Ghadamyari M., and Yezdani E. 2011. Effect of sweet wormwood *Artemisia annua* crude leaf extracts on some biological and physiological characteristics of the lesser mulberry pyralid, *Glyphodes pyloalis. J. Insect Sci.* 11(156): 1–13.

Kim M.H., Seo J.Y., Liu K.H., and Kim J.-S. 2014. Protective effect of *Artemisia annua* L. extract against galactose-induced oxidative stress in mice. *PLoS ONE* 9 (7): e101486.

Krishna S., Bustamante L., Haynes R.K., and Staines H.M. 2008. Artemisinins: Their growing importance in medicine. *Trends Pharmacol. Sci.* 29(10): 520–527.

Lai H., Sasaki T., Singh N.P., and Messay A. 2005. Effects of artemisinin-tagged holotransferrin on cancer cells. *Life Sci.* 76(11): 1267–1279.

Lai H., Singh N.P. 1995. Selective cancer cell cytotoxicity from exposure to dihydroartemisinin and holotransferrin. *Cancer Lett.* 91(1): 41–46.

Laughlin J.C. 2002. Post-harvest drying treatment effects on amtimalarial constituents of *Artemiasia annua* L. In: Bernáth J et al (eds) *Proceedings of the International Conference on MAP, Acta Hort,* 576, ISHS, Belgium.

Le W., You J., Yang Y., Mei J., Guo H., Yang H., and Zhang C. 1982. Studies on the efficacy of artemether in experimental schistosomiasis. *Acta Pharm. Sin.* 17: 187–193. (Article in Chinese with abstract in english).

Liao H.W., Wang D.Y., and Li X.M. 2006. Studies on the chemical constituents of essential oil of *Yao Xue Xue Bao.* 50(10): 366–370.

Lin J.Y., Chen T.S., Chen C.S. 1994. Flavonoids as nonpeptide angiotensin-converting enzyme inhibitors for hypertension treatment. Jpn-Kokai Tokkyo Koho Jp, Patent No. 06135830.

Liu C.H., Zou W.X., Lu H., and Tan R.X. 2001. Antifungal activity of *Artemisia annua* endophyte cultures against phytopathogenic fungi. *J. Biotechnol.* 88(3): 277–282.

Lubbe A., Seibert I., Klimkait T., van der Kooy F. 2012. Ethnopharmacology in overdrive: the remarkable anti-HIV activity of *Artemisia annua. J. Ethnopharmacol.* 141(3): 854–859.

Martínez M. J.A., del Olmo L. M. B., Ticona L.A., and Benito P.B. 2014. Chapter-5: Pharmacological potentials of artemisinin and related sesquiterpene lactones: Recent advances and trends. In Aftab, T., Ferreira, J.F.S., Khan, M.M.A., and Naeem, M. (eds.), *Artemisia annua – Pharmacology and Biotechnology,* pp. 75–93, Springer-Verlag, Berlin, Heidelberg.

Massiha A., Khoshkholgh-Pahlaviani M.M., Issazadeh K., Bidarigh S., and Zarrabi S. 2013. Antibacterial activity of essential oils and plant extracts of *Artemisia* (*Artemisia annua* L.) *in vitro. Zahedan J. Res. Med. Sci. (ZJRMS)* 15 (6): 14–18.

McDougald L.R. 2005. Blackhead disease (histomoniasis) in poultry: A critical review. *Avian Dis.* 49(4): 462–476.

McDougald L.R., and Reid W.M. 1991. Coccidiosis, In *Diseases of Poultry.* Edited by B.W. Calnek, H.J. Barnes, C.W. Beard, W.M. Reid, and H.W. Yoder, Ames: Iowa State University Press; pp. 780–797.

Merali S., Meshnick S.R. 1991. Susceptibility of *Pneumocystis carinii* to artemisinin *in vitro. Antimicrob. Agents Chemother.* 35(6): 1225–1227.

Miller L.H. and Su X. 2011. Artemisinin: Discovery from Chinese herbal garden. *Cell* 146 (6): 855–858.

Misra H., Mehta D., Mehta B.K., and Jain D.C. 2014. Extraction of artemisinin, an active antimalarial phytopharmaceutical from dried leaves of *Artemisia annua* L., using microwaves and a validated HPTLC-visible method for its quantitative determination. *Chromatogr. Res. Int.* Article I.D: 361405 (1–11).

Mouton J., Jansen O., Frédérich M., and Van der Kooy F. 2013. Is artemisinin the only antiplasmodial compound in the *Artemisia annua* tea infusion? An in vitro study, *Plant. Med.* 79(6): 468–470.

Mouton J. and Van der Kooy F. 2014. Identification of *cis-* and *trans-* Melilotoside within an *Artemisia annua* tea infusion. *Eur. J. Med. Plants,* 4(1): 52–63.

Mojarad T.B., Roghani M., and Zare N. 2005. Effect of subchronic administration of aqueous *Artemisia annua* extract on α_1-adrenoceptor agonist-induced contraction of isolated aorta in rat. *Iran. Biomed. J.* 9(2): 57–62.

Mustfa K., Landau I., Chabaud A.G., Chavatte J.M., Chandenier J., Duong T.H., and Richard-Lenoble D. 2011. Effects of antimalarial drugs ferroquine and artesunate on *Plasmodium yoelii* gametocytogenesis and vectorial transmission. *Sante* 21(3): 133–142.

Mueller M.S., Karhagomba I.B., Hirt H.M., and Wemakor E. 2000. The potential of *Artemisia annua* L. as a locally produced remedy for malaria in the tropics: Agricultural, chemical and clinical aspects. *J. Ethnopharmacol.* 73(3): 487–493.

Murray H.W., Berman J.D., Davies C.R., and Saravia N.G. 2005. Advances in leishmaniasis. *Lancet* 366(9496): 1561–1577.

Naidoo V., McGaw L.J., Bisschop S.P., Duncan N., and Eloff J.N. 2008. The value of plant extracts with antioxidant activity in attenuating coccidiosis in broiler chickens. *Vet. Parasitol.* 153(3-4): 214–219.

Nakase I., Gallis B., Takatani-Nakase T., Oh S., Lacoste E., Singh N.P., Goodlett D.R., et al. 2009. Transferrin receptor-dependent cytotoxicity of artemisinin-transferrin conjugates on prostate cancer cells and induction of apoptosis. *Cancer Lett.* 274(2): 290–298.

Nakase I., Lai H., Singh N.P., Sasaki T. 2008. Anticancer properties of artemisinin derivatives and their targeted delivery by transferrin conjugation. *Int. J. Pharm.* 354(1–2): 28–33.

Namdeo A.G., Mahadik K.R. and Kadam S.S. 2006. Antimalarial drug – *Artemisia annua*. *Pharmacogn. Mag.* 2(6): 106–111.

Nibret E., Wink M. 2010. Volatile components of four Ethiopian *Artemisia species* extracts and their *in vitro* antitrypanosomal and cytotoxic activities. *Phytomedicine* 17(5): 369–374.

Obeid S., Alen J., Nguyen V.H., Pham V.C., Meuleman P., Pannecouq C., Le T.N., Neyts J., Dehaen W., and Paeshuyse J. 2013. Artemisinin analogues as potent inhibitors of *in vitro* hepatitis C virus replication. *PLoS ONE* 8(12): e81783.

Ogutu B. 2013. Artemether and lumefantrine for the treatment of uncomplicated *Plasmodium falciparum* malaria in sub-Saharan Africa. *Expert Opin. Pharmacother.* 14(5): 643–654.

Ogwang P.E., Ogwal J.O., Kasasa S., Olila D., Ejobi F., Kabasa D., and Obua C. 2012. *Artemisia annua* L infusion consumed once a week reduces risk of multiple episodes of malaria: A randomised trial in a Ugandan community. *Trop. J. Pharmacol. Res.* 11(3): 445–453.

Oh S., Kim B.J., Singh N.P., Lai H., and Sasaki T. 2009. Synthesis and anti-cancer activity of covalent conjugates of artemisinin and a transferrin-receptor targeting peptide. *Cancer Lett.* 274(1): 33–39.

Okafor P.C., Ebiekpe V.E., Azike C.F., Egbung G.E., Brisibe E.A. and Ebenso E.E. 2012. Inhibitory action of *Artemisia annua* extracts and artemisinin on the corrosion of mild steel in H_2SO_4 solution. *Int. J. Corros.* 2012: 8.

Oriakpono O., Aduabobo H., Awi-Waadu G.D.B., and Nzeako S. 2012. Anti-parasitic effects of methanolic extracts of *Artemisia annua* L. against parasites of *Sarotherodon melanotheron. Int. J. Mod. Biol. Med.* 1(2): 108–116.

Pareek A., Chandurkar N., Srivastav V., Lakhani J., Karmakar P.S., Basu S., Ray A. et al. 2013. Comparative evaluation of efficacy and safety of artesunate-lumefantrine vs artemether-lumefantrine fixed-dose combination in the treatment of uncomplicated *Plasmodium falciparum* malaria. *Trop. Med. Int. Health* 18(5): 578–587.

Poiată A., Tuchiluş C., Ivănescu B., Ionescu A., and Lazăr M.I. 2009. Antibacterial activity of some Artemisia species extract. *Rev. Med. Chir. Soc. Med. Nat. Iasi.* 113(3): 911–914.

Preet S. 2010. Laboratory evaluation of molluscicidal and cercaricidal potential of *Artemisia annua* (fmily: Asteraceae). *Ann. Biol. Res.* 1(1): 47–52.

Price R.N. 2013. Potential of artemisinin-based combination therapies to block malaria transmission. *J. Infect. Dis.* 207(11): 1627–1629.

Rasooli I., Rezaee, M.B., Moosavi M.L., and Jaimand K. 2003. Microbial senstivity to and chemical properties of the essential oil of *Artemisia annua* L. *J. Essent. Oil Res.* 15(1): 59–62.

Ro D.K., Paradise E.M., Ouellet M., Fisher K.J., Newman K.L., Ndungu J.M., Ho K.A. et al. 2006. Production of the antimalarial drug precursor artemisinic acid in engineered yeast. *Nature* 440: 940–943.

Romero M.R., Efferth T., Serrano M.A., Castano B., Macias R.I.R., Briz O., and Marin J.G. 2005. Effect of artemisinin/artesunate as inhibitors of hepatitis B virus production in an *"in vitro"* replicative system. *Antiviral Res.* 68(2): 75–83.

Rosenthal P.J. 2008. Artesunate for the treatment of severe falciparum malaria. *New Engl. J. Med.* 358: 1829–1836.

Sadiq A., Hayat M.Q., and Ashraf M. 2014. Chapter-2: Ethnopharmacology of *Artemisia annua* L.: A review. Aftab T., Ferreira J.F. S., Khan M.M.A., and Naeem M. (eds.), *Artemisia annua – Pharmacology and Biotechnology*, pp. 9–25, ISBN: 978-3-642-41026-0, Springer-Verlag, Berlin, Heidelberg.

Sano M., Akyol C.V., Tungtrongchitr A., Ito M., and Ishih A. 1993. Studies on chemotherapy of parasitic helminths: Efficacy of artemether on Japanese strain of *Schistosoma japonicum* in mice. *Southeast Asian J. Trop. Med. Public Health* 24(1): 53–56.

Sarina S., Yagi Y., Nakano O., Hashimoto T., Kimura K., Asakawa Y., Zhong M., Narimatsu S., and Gohda E. 2013. Induction of nerite outgrowth in PC12 cells by artemisinin through activation of ERK and p38 MAPK signaling pathways. *Brain Res.* 1490: 61–71.

Sendi J. J. and Khosravi R. 2014. Chapter-13: Recent developments in controlling insect, acari, nematode, and plant pathogens of agricultural and medical importance by *Artemisia annua* L. (Asteraceae). In Aftab T., Ferreira J.F.S., Khan M.M.A., and Naeem M. (eds.), *Artemisia annua – Pharmacology and Biotechnology*, pp. 229–247, Springer-Verlag, Berlin, Heidelberg.

Sen R., Bandyopadhyay S., Dutta A., Mandal G., Ganguly S., Saha P., and Chatterjee M. 2007. Artemisinin triggers induction of cell-cycle arrest and apoptosis in *Leishmania donovani* promastigotes. *J. Med. Microbiol.* 56(Pt 9): 1213–1218.

Shahbazfar A. A., Zare P., Mohammadpour H., and Tayefi-Nasrabadi H. 2012. Effects of different concentrations of artemisinin and artemisinin-iron combination treatment on Madin Darby Canine Kidney (MDCK) cells. *Interdiscip. Toxicol.* 5(1): 30–37.

Shalaby H.A., El Namaky A.H., and Kamel R.O.A. 2009. *In vitro* effect of artemether and triclabendazole on adult *Fasciola gigantica*. *Vet. Parasitol.* 160(1-2): 76–82.

Shi J.Q., Zhang C.C., Sun X.L., Cheng X.X., Wang J. B., Zhang Y. D., Xu J., and Zou H.Q. 2013. Antimalarial drug artemisinin extenuates amyloidogenesis and neuroinflammation in APPswe/PS1dE9 transgenic mice via inhibition of nuclear factor-κB and NLRP3 inflammasome activation. *CNS Neurosci. Ther.* 19(4): 262–268.

Shukla A., Farooqi A.H.A., Shukla Y.N., and Sharma S. 1992. Effect of triacontanol and chlormequat on growth, plant hormonea and artemisinin yield in *Artemisia annua*. *Plant Growth Regul.* 11(2): 165.

Singh N.P., and Lai H. 2001. Selective toxicity of dihydroartemisinin and holotransferrin towards human breast cancer cells. *Life Sci* 70(1): 49–56.

Singh A. and Sarin R. 2010. *Artemisia scoparia*: A new source of artemisinin. *Bangladesh J. Pharmacol.* 5(1): 17–20.

Sullivan D.J.Jr. 2013. Timing is everything for artemisinin action. *Proc. Natl. Acad. Sci. USA* 110(13): 4866–4867.

Sundar S. and Chatterjee M. 2006. Visceral leishmaniasis: Current therapeutic modalities. *Indian J. Med. Res.* 123(3): 345–352.

Tang H.Q., Hu J., Yang L., and Tan R.X. 2000. Terpenoids and flavonoids from *Artemisia* species. *Planta Med.* 66(4): 391–393.

Tawfik A.F., Bishop S.J., Ayalp A., and el-Feraly F.S. 1990. Effects of artemisinin, dihydroartemisinin and arteether on immune responses of normal mice. *Int. J. Immunopharmacol.* 12(4): 385–389.

Tellez M.R., Canel C., Rimando A.M., and Duke S.O. 1999. Differential accumulation of isoprenoids in glanded and glandless *Artemisia annua* L. *Phytochemistry* 52(6): 1035–1040.

The Nobel Foundation. 2015. Available from URL: www.nobelprize.org/nobel_prizes/medicine/laureates/2015/ (accessed on 22 December 2016).

Thøfner I.C.N., Liebhart D., Hess M., Schou T.W., Hess C., Ivarsen E., Fretté X.C. et al. 2012. Antihistomonal effects of artemisinin and *Artemisia annua* extracts *in vitro* could not be confirmed by *in vivo* experiments in turkeys and chickens. *Avian Pathol.* 41(5): 487–496.

Tripathi A.K., Prajapati V., Aggarwal K.K., Khanuja S.P., and Kumar S. 2000. Repellency and toxicity of oil from *Artemisia annua* to certain store-products. *J. Econ. Entomol.* 93(1): 43–47.

Tripathi A.K., Prajapati V., Aggarwal K.K., and Kumar S. 2001. Toxicity, feeding deterrence, and effect of activity of 1, 8-cineol from *Artemisia annua* on progeny production of *Tribolium castanaeum* (Coleoptera: Tenebrionidae). *J. Econ. Entomol.* 94(4): 979–983.

Tzeng T.C., Lin Y.L., Jong T.T., and Chang C.M.J. 2007. Ethanol modified supercritical fluids extraction of scopoletin and artemisinin from *Artemisia annua* L. *Sep. Purif. Technol.* 56(1): 18–24.

Utizinger J., Shuhua X., Goran E.K., Bergquist R., and Tanner M. 2001. The potential of artemether for the control of schistosomiasis. *Int. J. Parasitol.* 31(14): 1549–1562.

Verdian-rizi M.R., Sadat-Ebrahimi E., Hadjiakhoondi A., Fazeli M.R., and Hamedani P.M. 2008. Chemical composition and antimicrobial activity of *Artemisia annua* L essential oil from Iran. *J. Med. Plants* 7(4): 58–62.

Vicidomini S. 2011. Alternative properties of *Artemisia* (Asteraceae) phyto–extracts to anti-malarian ones: preliminary bibliografic review on nemato-toxic effects. II Naturalista Compano 1–22. http://www.museonaturalistico.it/.

Wang B., Sui J., Yu Z., and Zhu, L. 2011. Screening the hemostatic active fraction of *Artemisia annua* L. In-vitro. *Iran. J. Pharm. Res.* 10(1): 57–62.

Weathers P.J., Arsenault P.R., Covello P.S., McMickle A., Teoh K.H., and Reed D.W. 2011. Artemisinin production in *Artemisia annua*: studies *in planta* and results of a novel delivery method for treating malaria and other neglected diseases. *Phytochem. Rev.* 10 (2): 173–183.

Weathers P.J., and Towler M.J. 2012. The flavonoids casticin and artemetin are poorly extracted and are unstable in an *Artemisia annua* tea infusion. *Planta Med.* 78(10): 1024–1026.

Wells S., Diap G., and Kiechel J.R. 2013. The story of artesunate-mefloquine (ASMQ), innovative partnerships in drug development: Case study. *Malaria J.* 12: 68–78.

White N.J. 2008a. Qinghaosu (artemisinin): The price of success. *Science* 320(5874): 330–334.

White N.J. 2008b. The role of anti-malarial drugs in eliminating malaria. *Malaria J.* 7 (Suppl.1): 8.

WHO 1993. The control of schistosomiasis. Second report of WHO expert committee. WHO Tech Rep Ser 830 WHO Geneva.

WHO 2003. Country representative office for Ethiopia proceedings for the planning workshop in determining the economic impact of epidemic malaria in East Africa, Addis Ababa, pp. 1–86. http://www.who.int/malaria/cmc_upload/0/000/016/663/Att2.pdf

WHO Library Cataloguing-in-Publication Data 2006 WHO monograph on good agricultural and collection practices (GACP) for *Artemisia annua* L. World Health Organization. www.who.int/medicines/publications/traditional/Artemisia Monograph.pdf

Willcox M. 2009. Artemisia species: From traditional medicines to modern antimalarials and back again. *J. Altern. Complement Med.* 15(2): 101–109.

Willcox M., Bodeker G., Bourdy G., Dhingra V., Falquet J., Ferreira J.F.S., Graz B. et al. 2004. *Artemisia annua* as a traditional herbal antimalarial. In: Wilcox M.L., Bodeker G., Rasoanaivo P. (eds) *Traditional Medicinal Plants and Malaria*, vol 4. CRC Press, Boca Raton, FL, pp 43–59.

Woerdenbag H.J., Moskal T.A., Pras N., Malingre T.M., el-Feraly F.S., Kampinga H.H., and Konings A.W. 1993. Cytotoxicity of artemisinin related endoperoxides to Ehrlich ascites tumor cells. *J. Nat. Prod.* 56(6): 849–856.

Wohlfarth C. and Efferth T. 2009. Natural products as promising drug candidates for the treatment of hepatitis B and C. *Acta. Pharmacol. Sin.* 30(1): 25–30.

Woodrow C.J., Haynes R.K., and Krishna S. 2005. Artemisinins. *Postgrad. Med. J.* 81(952): 71–78.

Worku N., Mossie A., Stich A., Daugschies A., Trettner S., Hemdan N.Y.A., and Birkenmeier G. 2013. Evaluation of the *in vitro* efficacy of *Artemisia annua*, *Rumex abyssinicus*, and *Catha edulis Forsk* extracts in cancer and *Trypanosoma brucei* cells. *ISRN Bioechemistry* 2013: 10.

Xiao S.H., Catto B.A. 1989. *In vitro* and *in vivo* studies of the effect of artemether on *Schistosoma mansoni. Antimicrob. Agents Chemother.* 33(9): 1557–1562.

Xiao S.H., Chollet J., Weiss N.A., Bergquist R.N., and Tanner M. 2000a. Preventive effect of artemether in experimental animals infected with *Schistosoma mansoni. Parasitol. Int.* 49(1): 19–24.

Xiao S.H., Utzinger J., Collet J., Endriss Y., N'Goran E.K., and Tanner M. 2000b. Effect of artemether against *Schistosoma haematobium* in experimentally infected hamsters. *Int. J. Parasitol.* 30(9): 1001–1006.

Yang D.M., Liew F.Y. 1993. Effects of qinghaosu (artemisinin) and its derivatives on experimental cutaneous leishmaniasis. *Parasitology* 106(Pt 1): 7–11.

Yang G.E., Bao L., Zhang X.Q., Wang Y., Li Q., Zhang W.K., and Ye W.C. 2009. Studies on flavonoids and their antioxidant activities of *Artemisia annua. Zhong Yao Cai* 32(11): 1683–1686.

Yao Q., Chen J., Lyu P.H., Zhang S.J., Ma F.C., and Fang J.G. 2012. Knowledge map of artemisinin research in SCI and medline database. *J. Vector Borne Dis.* 49(4): 205–216.

Yimer S. and Sahu O. 2014. Anti-mosquito repellent from *Artemisia annua. Int. J. Med. Clin. Sci.* 1(1): 1–8.

Youn H.J., and Noh J.W. 2001. Screening of the anticoccidial effects of herb extracts against *Eimeria tenella. Vet. Parasitol.* 96(4): 257–263.

Yu Z., Wang B., Yang F., Sun Q., Yang Z., and Zhu L. 2011. Chemical composition and anti-acetylcholinesterase activity of flower essential oils of *Artemisia annua* at different flowering stage. *Iran. J. Pharm. Res.* 10(2): 265–271.

Zanjani K.E., Rad A.S.H., Bitarafan Z., Aghdam A.M., Taherkhani T., and Khalili P. 2012. Physiological response of sweet wormwood to salt stress under salicylic acid application and non application conditions. *Life Sci. J.* 9(4): 1097–8135.

Zamani S., Sendi J. J., and Ghadamyari M. 2011. Effect of *Artemisia annua* L. (Asterales: Asteraceae) essential oil on mortality, development, reproduction and energy reserves of *Plodia interpunctella* (Hübner) (Lepidoptera: Pyralidae). *J. Biofertil. Biopestici.* 2(1): 105.

Zhang Y.Q., Ding W., and Luo J.X. 2012. Bioactivities of *Artemisia annua* against *Petrobia harti. The Proceedings of Chinese Society of Plant Protection in 2012*, Beijing, China, pp. 354–358.

Zhang Y.Q., Ding W., Zhao Z.M., Wu J., and Fan Y.H. 2008. Studies on acaricidal bioactivities of *Artemisia annua* L. extracts against *Tetranychus cinnabarinus* Bois. (Acari: Tetranychidae). *Agric. Sci. China* 7(5): 577–584.

Zheng G.Q., 1994. Cytotoxic terpenoids and flavonoids from *Artemisia annua. Plant Med.* 60(1): 54–7.

CHAPTER 10

Impact of Integrated Omics Technologies for Identification of Key Genes and Enhanced Artemisinin Production in *Artemisia annua* L.

Shashi Pandey-Rai, Neha Pandey, Anjana Kumari,
Deepika Tripathi, and Sanjay Kumar Rai

CONTENTS

10.1 INTRODUCTION

Medicinal plants are the source of an enormous variety of bioactive secondary metabolites and have potential synergistic effects against a broad range of human diseases. According to the World Health Organization (WHO), more than 80% of the world's population depend on medicinal plants for everyday healthcare. One of the most popular secondary metabolites with immense therapeutic potential is artemisinin (AN), present in a well-known Asteraceae family member, *Artemisia annua* L. The AN isolated from the leaves of *A. annua* by Chinese scientist You You Tu is acknowledged as an effective antimalarial compound (Barbacka and Baer-Dubowska 2011). AN and its bioactive derivatives isolated from *A. annua* are powerful medicines widely used for their ability to swiftly control Plasmodium malaria. AN-based combination therapies (ACTs), with their established safety record, are the first line of treatment recommended by WHO (2014) for malaria caused by *Plasmodium falciparum*. In addition to its antimalarial effects, AN has recently been evaluated for its potential antibacterial, antiviral, antitumor, antileishmanial, antischistosomiatic, anti-sleeping sickness, anticancer, and herbicidal properties (Efferth et al. 2011; Utzinger et al. 2001; Sen et al. 2007; Mishina et al. 2007). However, the low content of AN in plant tissue has resulted in poor yield/production of AN, which seems insufficient to fulfill the demand for 392 million courses of ACT each year (WHO 2014). Semisynthetic derivatives of AN, such as artemether and artesunate, are also commonly produced commercially, but they are not routinely available in remote rural areas. Moreover, these derivatives are very expensive, and low yields of AN result in relatively high costs for its extraction and purification. Further, *A. annua* requires a relatively long period of time for its agricultural cultivation, which results in wide swings in affordable, best-quality, robust supply of raw materials, and prices. Intensified efforts have been carried out to increase AN production (Liu et al. 2006). However, the routine metabolic engineering strategy, via overexpressing or downregulating key genes in AN biosynthetic pathways, has not proved very effective. Glandular secretory trichomes, sites of AN biosynthesis on the surface of *A. annua*, are the new target for increasing AN yield (Duke et al. 1994). In general, the population and morphology of glandular secretory trichomes in *A. annua* (AaGSTs) are positively correlated with AN content. Higher production of AN requires breeding of *A. annua* to optimize the biomass yield and trichome density. Various efforts have been made to breed high–trichome density cultivars of the plant for increased AN production. However, various approaches have been already taken into consideration for the semisynthesis of AN (Paddon et al. 2013). The production of AN is also challenging because *A. annua* remains relatively undeveloped as a crop. Therefore, there is a need to improve the varieties and cultivation strategies of *A. annua* for farmers in developing countries, because this would bring immediate benefits to the existing supply chain of AN. Major advancements in omics technologies such as genomics, proteomics, and metabolomics have enabled high-throughput monitoring of a variety of molecular and biochemical processes. These techniques have been widely applied to identify biological variants and complex biochemical pathways/systems. Many omics platforms target the comprehensive analysis of genes (genomics), mRNA (transcriptomics), proteins (proteomics), and metabolites (metabolomics). However, the interpretation of obtained data is challenging due to very complex biochemical and metabolic pathways.

10.2 OMICS STRATEGIES FOR ENHANCED AN PRODUCTION

Extensive efforts have already been made to improve AN production through many conventional and molecular approaches. The present chapter provides a compilation of recent advances in both conventional and modern omics approaches for AN yield enhancement, such as (1) various structural and functional genomics approaches, (2) proteomics approaches, (3) metabolomics approaches for identifying the players involved in regulation of AN biosynthesis, and (4) integrated (mutational and molecular) breeding of *A. annua* with the help of molecular markers.

In recent years, more emphasis has been placed on the integration of natural and artificial production systems, involving both biological and chemical synthesis, for AN synthesis/drug production. Success has been achieved with the completion of the "Semisynthetic Artemisinin Project," funded by the Bill & Melinda Gates Foundation, in a partnership between University of California (Berkeley, United States), Amyris, Inc., and the Institute for One World Health, which has achieved the production of 25 grams per liter artemisinic acid, a derivative of artemisinin, in engineered baker's yeast. Further, it has succeeded in the 40% chemical conversion of artemisinic acid into artemisinin (Paddon et al. 2013). Despite the successful AN production, research is still continuing to produce potent stable AN derivatives to meet the demand for ACTs. Therefore, it is still highly desirable to achieve low-cost, less labor-intensive, enhanced drug production at global level by other modern molecular techniques as well. Very recently, the consumption of dried *A. annua* leaves/leaf extracts has been demonstrated to overcome existing resistance to pure AN in *Plasmodium yoelii* (Elfawal et al. 2015). These recent reports give support to the novel idea of increasing AN synthesis through natural *in planta* yield enhancement with the help of omics approaches over the semisynthetic production of pure AN for effective malarial treatments (Pandey and Pandey-Rai 2014, 2016).

Scenarios of pathways from genes to metabolites with the help of genomics approaches in *A. annua* remain poorly investigated, and the use of an integrated approach combining metabolomics, transcriptomics, and gene function analyses to characterize gene-to-terpene and sesquiterpene pathways in this plant species is essential for understanding AN production. These "omic" technologies are primarily based on the universal detection of genes, their structure and functions (genomics), expression of mRNA (transcriptomics), protein structure and function (proteomics), and metabolite profiling (metabolomics), as represented in Figure 10.1. These omics approaches, when integrated together, are helpful in the data mining and interpretation of different biological samples/organisms.

10.2.1 Structural Genomics Studies in *A. annua*

Genomics is the systematic study of an organism's genome with the help of molecular tools. Traditionally, genes have been analyzed individually, but microarray technology has advanced substantially in recent years. Various steps of genome analysis

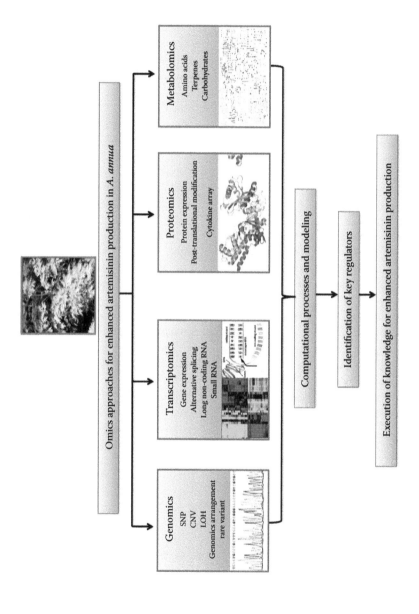

Figure 10.1 Schematic representation of omics approaches for enhanced artemisinin production in *A. annua* L.

involve (1) genome sequencing, (2) identification of repetitive as well as unique sequences, (3) gene prediction, (4) identification of functional expressed sequence tags (ESTs) and complementary DNA (cDNA) sequences, and (5) genome annotation and gene location/gene mapping. Recently, DNA microarray techniques have evolved as a powerful tool, which has the potential to measure differences in DNA sequences between individuals and the expression of thousands of genes simultaneously.

The mining of sequences generated from genomic libraries/cDNA libraries identified several key genes and selectable markers linked with AN biosynthesis in *A. annua*. Using the potential of various molecular markers, the preliminary genetic maps of *A. annua* with nine linkage groups have been reported by Graham et al (2010). For the construction of genetic maps, they used the Artemis (high-yielding variety of *A. annua*) pedigree lines and established genetic linkage groups and quantitative trait locus (QTL) maps. Positive QTLs related to AN biosynthesis and production were also independently validated with the help of linkage analysis. Further, with the help of these structural genetic markers, it would be interesting to screen genetically high-AN-yielding lines quickly. Screening of high AN producing varieties using sequence characterized amplified region (SCAR) marker has also been reported in *A. annua* (Pandey and Pandey-Rai 2016). Developed SCAR markers could be considered as complementary tools in identification of high-AN producer lines, and their use could greatly reduce the time and labor needed for screening, land usage, and other costs associated with the breeding of *A. annua* for high AN. Marker-based selection techniques also facilitate the screening of a large number of progenies together at an early stage of plant development. Efforts have also been made in large-scale sequencing of ESTs on the transcriptome of *A. annua* to identify novel genes and markers for fast-track breeding. Extensive genetic variation helped in the generation of a preliminary genetic map with nine linkage groups, and enrichment for positive QTLs confirms that the sequence knowledge and molecular tools are helpful in converting *A. annua* into a robust crop for molecular breeding. The use of Roche 454 pyrosequencing gave a good platform to produce EST databases from cDNA libraries derived from glandular trichomes, which serve as a home/factory for AN synthesis. In addition to the trichome-specific libraries, cDNA libraries and EST databases were also prepared from meristem tissues and cotyledons, which gave an idea about the structure and function of key genes associated with primary and secondary metabolic pathways, phenotypic traits, trichome development, plant architecture, and other tissue-specific traits that could affect AN yield. Further, these EST sequences can be used for *in silico* identification of single-nucleotide polymorphisms (SNPs), short sequence repeats (SSRs), and insertions/deletions (InDels) and as molecular markers for mapping and breeding. About 34,419 SNPs from DNA sequences have already been reported in different EST databases derived from the Artemis F1 hybrid material with a mean SNP frequency of 1 in 104 base pairs.

It was also revealed that predicted SSR–ESTs have potential for various molecular functions: regulation of transcription (mRNA synthesis), transport regulation, signaling, defense response, stress, and tRNA processing. Computational approaches help to provide an attractive alternative to conventional laboratory methods for the rapid and economical development of SSR markers by using freely available genomic sequences in public databases for *A. annua*. Microsatellite distributions are helpful in understanding mutational processes and evolutionary selection constraints as well as development in DNA repair mechanisms. In order to effectively use the studied information related to EST and to extend the utility of SSRs in *A. annua*, future work should be focused on both computational and molecular biology. Further sequence information can be used for the development of DNA chips with DNA sequences for the identification and screening of high-AN-producing lines (Harvey et al. 2015). DNA chip technology can provide a rapid, high-throughput screening and information-rich tool for genotyping, quality assurance, and species confirmation.

10.2.2 Functional Genomics Studies in *A. annua*

The transcriptome is the total mRNA in a cell, which reflects the genes that are actively expressed at any given moment. Gene expression microarrays measure packaged mRNA as a summary of gene activity. Recently, next-generation sequencing (NGS) technology for transcriptome (RNA-seq) has provided new insight for both obtaining gene sequences and quantifying the transcriptome of any organ/organism. RNA-seq has also been an influential method for distinguishing genes involved in important metabolic pathways, such as the synthesis of secondary metabolites. Several key genes and ESTs of AN biosynthetic pathways have already been reported on different databases. The names of the genes, their functions, and their primer sequences for amplification are summarized in Table 10.1.

This information is very helpful in microarrays and quantitative real-time reverse transcriptase-polymerase chain reaction (Q-RT-PCR) analysis for gene expression profiling. Based on the target sequences, a significant change in mRNA profile can be estimated for thousands of genes with the help of this technique. Gene expression changes and Q-RT-PCR-based approaches focused on specific genes have also been developed for this plant (Soetaert et al. 2013). The latter is a highly recommended confirmatory tool for quantifying transcript level/gene expression with improved sensitivity and specificity.

As for other terpenes, AN biosynthesis involves two pathways: the cytosolic "mevalonate (MVA) pathway" and the plastidic "non-mevalonate/MEP pathway." These two pathways are involved in the formation of two isoprenoid precursors: isopentenyl diphosphate (IPP) and dimethylallyl diphosphate (DMAPP). Two stages of AN biosynthesis have now been completely elucidated: the formation of farnesyl diphosphate (FPP) and the cyclization of FPP to form amorpha-4,11-diene, which ultimately leads to the synthesis of AN and its various other derivatives. All the biosynthetic pathway genes and their sequences and functions have already been reported in *A. annua* (Bouwmeester et al. 1999). The transcriptomes of leaves and glandular trichomes of *A. annua* have also been well characterized, as, later, has the site of AN biosynthesis (Arsenault et al. 2010; Olofsson et al. 2011). The transcriptome data are very helpful in manipulating the biosynthetic pathways to channelize the carbon flux for enhanced AN synthesis by either upregulation of the desired

Table 10.1 Identified Genes of Secondary Metabolism and Their Function in *Artemisia annua*

Gene ID	Gene Name	Functions
ADS	Amorpha 4,11-diene synthase	Catalyzes the reaction of farnesyl diphosphate (FPP) to amorpha-4,11-diene
ALDH1	Aldehyde dehydrogenase	Catalyzes the NAD(P)-dependent oxidation of the putative artemisinin precursors, artemisinic and dihydroartemisinic aldehydes
BAS	β-amyrin synthase	Catalyzes the generation of β-amyrin from squalene
BFS	β-farnesene synthase	Catalyzes the reaction of farnesyl diphosphate (FPP) to β-farnesene
BPS	β-pinene synthase	Catalyzes the reaction of geranyl diphosphate to β-pinene
CPR	Cytochrome P450 reductase	Required for electron transfer from NADPH to cytochrome P450
CYP71AV1	Cytochrome P450 dependent monooxygenase/hydroxylase	Catalyzes the oxidation of amorpha-4,11-diene, artemisinic alcohol, and artemisinic aldehyde
DBR2	Artemisinin aldehyde delta 11(13) double bond reductase	Catalyzes the reaction of artemisinic aldehyde to di-artemisinic aldehyde
DXR	1-deoxyxylulose 5-phosphate reductoisomerase	Catalyzes the first committed step of the 2-C-methyl-d-erythritol 4-phosphate pathway for isoprenoid biosynthesis
DXS	1-deoxyxylulose 5-phosphate synthase	Catalyzes the first, rate-limiting step in the methylerythritol phosphate pathway of isoprenoid biosynthesis
ECS	8-epicedrol synthase	Catalyzes the synthesis of epicedrol from farnesyl diphosphate
FDS	Farnesyl diphosphate synthase	Catalyzes the production of geranyl pyrophosphate and farnesyl pyrophosphate from isopentenyl pyrophosphate and dimethylallyl pyrophosphate
GAS	Germacrene A synthase	Catalyzes the reaction of farnesyl diphosphate to germacrene
GGPRS	Geranylgeranyl pyrophosphate synthase	Catalyzes the synthesis of GGPP from farnesyl diphosphate and isopentenyl diphosphate
HDR	4-hydroxy-3-methylbut-2-enyl diphosphate reductase	Acts upon (E)-4-hydroxy-3-methyl-but-2-enyl pyrophosphate
HMGR	3-hydroxy-3-methylglutaryl-CoA reductase	Catalyzes the conversion of HMG-CoA to mevalonic acid
IPPi	Isopentenyl pyrophosphate isomerase	Catalyzes the isomerization of IPP to DMAPP by an antarafacial transposition of hydrogen
LS	Limonene synthase	Catalyzes the generation of limonene from geranyl diphosphate
QSH	β-caryophyllene synthase	Catalyzes the generation of β-caryophyllene from farnesyl diphosphate
RED1	Dihydroartemisinic aldehyde reductase	Catalyzes the reaction of dihydroartemisinic aldehyde to dihydroartemisinic alcohol
SQC	Sesquiterpene cyclase	A branch point enzyme in the general isoprenoid pathway for the synthesis of phytoalexin capsidiol
SQS	Squalene synthase	Converts two molecules of FPP into squalene via an intermediate: presqualene diphosphate

pathway or downregulation of competing pathways through overexpression and/or suppression of their respective pathway genes.

10.2.2.1 Overexpression of Key Biosynthetic Genes

Overexpression of key biosynthetic genes is a promising approach for greater synthesis of product AN, and great success has been achieved in this direction via engineering almost all the key genes of the AN biosynthetic pathway. Two important key enzymes are HMG-CoA reductase (HMGR) and DXP reductoisomerase (DXR), which regulate the upstream pathway of AN biosynthesis (as represented in Figure 10.2) and act as initiators for isoprenoid biosynthesis. HMGR shunts HMG-CoA into the mevalonate pathway, whereas DXR, in the plastidic non-mevalonate pathway, catalyzes the first step of isoprenoid (C5) synthesis (Kiran et al. 2010). These two important key genes have been overexpressed in *A. annua* by many workers. Transfer of the *Catharanthus roseus HMGR* gene to *A. annua*, first reported by Aquil et al. (2009), resulted in 38.9% higher AN content in one of the transgenic lines.

Co-expression of *HMGR* with the *ADS* gene enhanced AN production up to 7.65-fold in a transgenic line (Alam and Abdin 2011). Recently, CaMV 35S promoter-driven *DXR* gene overexpression in *A. annua* also produced higher AN in transgenic plants (Xiang et al. 2012), while farnesyl diphosphate synthase (FPS) gene overexpression resulted in 2- to 3-fold higher AN content in transgenic plants (Chen et al. 2000; Banyai et al. 2010). Further, the FPS gene co-expressed with the HMGR gene boosted AN content 1.8-fold in transgenic plants. Overexpression of ADS is also a promising approach for upregulated AN biosynthesis in *A. annua*. In addition, Chen et al. (2012) tried to co-overexpress CYP71AV1 and CRP with the FPS gene, and in a similar approach, CYP71AV1 and CPR were co-overexpressed with ADS and resulted in 2.4-fold higher AN. In addition to this, DBR2-overexpressing lines also produced more arteannuin B and its direct precursor, artemisinic acid (Yuan et al. 2015).

10.2.2.2 Regulating the Expression of Transcription Factors

The biosynthetic pathway of AN is almost completely elucidated, but its regulatory mechanism remains largely unknown. In recent years, a set of transcription factors have been reported to participate in regulating the biosynthesis and accumulation of AN, and several of them were already being used in the production of transgenic *A. annua* plants to obtain a higher AN content. The recent investigations related to transcription factors were summarized by Kexuan Tang's research group, who have been involved in the study of regulatory mechanisms related to AN to improve the AN content in *A. annua* plants. Transcription factors (TFs) and transporters are receiving extensive interest, as the former may regulate secondary metabolic pathways and the latter may function to transfer natural products (Borevitz et al. 2000). Overexpression of TFs also offers an alternative/complementary strategy to increase the transcript level of biosynthetic pathway genes (Figure 10.3). Three key genes

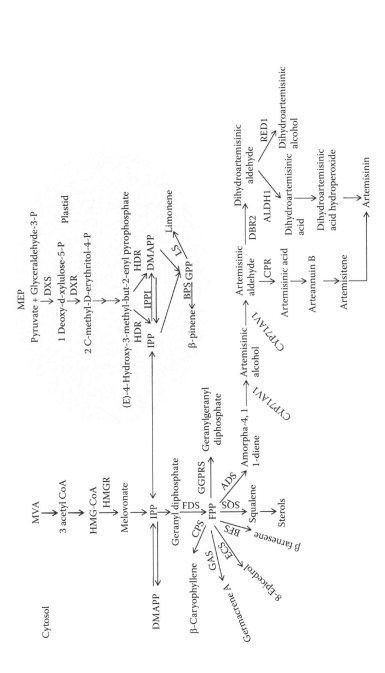

Figure 10.2 Artemisinin biosynthetic pathway in *Artemisia annua* L. MEP pathway genes: *DXS*, 1-deoxy-D-xylulose-5-phosphate synthase; *DXR*, 1-deoxy-D-xylulose 5-phosphate reductase; *HDR*, 4-hydroxy-3-methylbut-2-enyl diphosphate reductase; *IPPI*, isopentenyl pyrophosphate isomerase. MVA pathway genes: *HMGR*, 3-hydroxy-3-methylglutaryl coenzyme A reductase. Artemisinin biosynthesis pathway genes: *FDS*, farnesyl diphosphate; *ADS*, amorpha-4,11-diene synthase; *CPR*, cytochrome P450 reductase; *CYP71AV1*, cytochrome P450 monooxygenase; *DBR2*, artemisinic aldehyde D-11(13)-double bond reductase; *ALDH1*, aldehyde dehydrogenase 1; *RED1*, dihydroartemisinic aldehyde reductase. Genes involved in other terpene biosynthesis pathways: *BFS*, beta-farnesene synthase; *CPS*, beta-caryophyllene synthase; *GAS*, germacrene A synthase; *ECS*, 8-epicedrol synthase; *SQS*, squalene synthase; *BAS*, beta-amyrin synthase; *LS*, limonene synthase; *BPS*, beta-pinene synthase; *GGPPS*, geranylgeranyl pyrophosphate synthase.

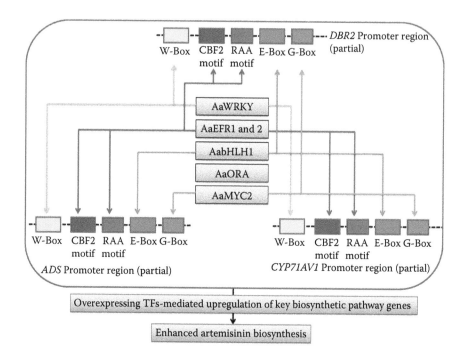

Figure 10.3   Schematic representation of overexpressing transcription factor–mediated regulation of key artemisinin biosynthetic pathway genes for enhanced artemisinin biosynthesis.

of AN biosynthesis promoters, ADS, CYP71AV1, and DBR2, have been characterized to date. The first characterized transcription factor of *A. annua* is AaWRKY1 (a WRKY type transcription factor), which binds to the W-boxes in ADS and CYP71AV1 promoters. Transient expression of AaWRKY cDNA in *A. annua* was shown to increase the transcript level of several AN biosynthetic genes (Ma et al. 2009), and AaWRKY-overexpressing transgenic plants produced higher AN as compared with the control (Tang et al. 2012; Han et al. 2014). Very recently, Chen et al. (2016) succeeded in cloning the GLANDULAR TRICHOME-SPECIFIC WRKY1 (AaGSW1) in *A. annua* and found that AaGSW1 positively regulates CYP71AV1 and AaORA expression by directly binding to the W-box motifs in their promoters, which ultimately leads to increased AN biosynthesis. They have also reported the direct regulation of AaGSW1 by AaMYC2 and AabZIP transcription factors. AabZIP transcription factors have a direct positive regulatory effect on AN biosynthesis (Zhang et al. 2015). In another recent study by Shen et al. (2016), overexpression of AaMYC2, which binds to G-box-like motifs within the promoters of *CYP71AV1* and *DBR2*, significantly enhanced the transcript level of *CYP71AV1* and *DBR2* genes, which resulted in higher AN content. In contrast, the AN level was reduced in RNA interference (RNAi) transgenic lines of AaMYC2, which supports the positive role of AaMYC2 in AN biosynthesis. Further, CRTDREHVCBF2 (CBF2) and RAV1AAT (RAA) motifs of both ADS and CYP71AV1 promoters have been proposed as the binding sites of two TFs, AaERF1 and AaERF2, which belong to the JA-responsive AP2 family of TFs. Another TF that has been cloned and overexpressed is AaORA, which is a trichome-specific APETALA2/ethylene-response factor (AP2/ERF) family TF in *A. annua* (Yu et al. 2012; Lu et al. 2013). Its overexpression in *A. annua* significantly boosted the transcription of several key genes of the biosynthetic pathway. Recently, AabHLH1, a bHLH TF, has been also successfully isolated from glandular secretary trichomes (GSTs) of *A. annua*, and transient expression of AabHLH1 in the leaves of *A. annua* significantly increased the transcript levels of HMGR, ADS, and CYP71AV1 (Ji et al. 2014).

In addition to TF engineering, blocking/silencing AN competitive pathway/genes with the help of RNAi-mediated post-transcriptional gene silencing is also helpful in enhancing AN production. So far, among the competitive pathway enzymes, squalene synthase (SQS) and β-caryophyllene synthase (CPS) have been identified in *A. annua*. Zhang et al. (2009) applied RNAi technology for the first time to knock down the expression of SQS, which catalyzes the first step in the sterol biosynthetic pathway by converting FPP into squalene. The suppression of SQS expression in *A. annua* with the help of an RNAi technique was reported to result in about a 3-fold higher AN content in transgenic plants as compared with control. In another study using antisense RNA technology, knocking down the expression of the CPS gene by an *Agrobacterium*-mediated antisense fragment of CPS cDNA into *A. annua*, cloned in pBI121 plant expression vector, resulted in higher AN. Antisense strand expression lowered the endogenous CPS expression and elevated the AN content. Some recent reports determined that the activity of the DBR2 enzyme is also involved in the enhancement of AN biosynthesis in certain chemotypes (Yang et al. 2015). Fortunately, the release of the entire genome of *A. annua* will resolve this challenge and provide benefits in terms of better understanding of the regulation of the genetic machinery with the help of TF regulation for AN production.

10.2.3 Proteomic Approaches Used for Identifying AN Biosynthetic Proteins

Proteomics provides a powerful tool for the study of proteins on a genome-wide scale. The *A. annua* genome sequence information available to date and the improvement in methods of protein characterization are driving this technology forward. The field of proteo-genomics is important due to its potential to support the annotation of DNA sequence data by exploiting the information obtained through proteomic studies for the identification/characterization of the actual products of gene expression. Currently, two approaches, based on mass spectrometry, used for global quantitative protein profiling are (1) two-dimensional electrophoresis (2DE) followed by staining, selection, and identification with the help of mass spectrometry and (2) isotope tags to label proteins and separation by multidimensional liquid chromatography followed by mass spectrometry analysis (Ansong et al. 2008). Both these basic proteomic approaches are supplemented with valuable information provided by molecular imaging. Most importantly, trichome-specific proteome data are now available for *A. annua*, including proteins that are related to the biosynthesis of AN as well as other highly abundant proteins, which suggest additional enzymatic processes within the trichomes. In particular, the trichome-specific expression of peroxidases represents strong oxidation activity in trichomes, which helps in effective oxidative reactions at the final phase of the biosynthesis of AN, which had earlier been presumed to be non-enzymatic in nature (Bryant et al. 2015). There is also a developing interest in applying genomics and proteomics data in *A. annua*, but one of the biggest handicaps is the lack of availability of a well-annotated database for this plant. The *in silico* comparison of *A. annua* databases, including the EST trichome library of *A. annua* trichome, the Trinity contig database, Uni/Prot/*A. annua*, and UniProt/*viridiplantae*, has been a useful tool for identifying important enzymes and TFs. However, these important tools also have significant differences in their utility for genomic and proteomic analyses. Despite these differences, the EST trichome library has allowed the identification of essential proteins, enriched in the *A. annua* trichomes, that are involved in the biosynthesis/regulation of AN (Bryant et al. 2015). Table 10.2 summarizes several identified proteins in *A. annua* and their categorization according to their role and function. Proteomic studies from trichome-enriched samples as represented in Figure 10.4 display a rough molecular functional classification of the UNIProtKB (taxonomy; *Viridiplantae*) proteins. The secondary metabolic pathway of AN and its synthesis occur in glandular secretory trichomes of *A. annua*, which have been relatively well characterized in terms of transcriptomics and proteomics, and significant efforts have also been made using plant molecular genetic engineering to increase the synthesis/production of this compound in tissues as well as at the whole plant level. Proteomics has also been useful in revealing an abiotic stress tolerance mechanism in *A. annua* by redirecting carbon metabolism (Rai et al. 2014). Results look promising; however, further efforts should be addressed toward the optimization of the most cost-effective approaches by combining proteomic studies for the synthesis/production of secondary metabolites (monoterpenes, diterpenes, and sesquiterpenes) as medicines against the malarial parasite (Lange and Ahkami 2013).

10.2.4 Metabolomic Studies for Metabolite Profiling in *A. annua*

The metabolome is highly responsive to and interactive with both environmental/stress and biological regulatory mechanisms such as epigenetics, transcription, and post-translational modification. The analysis of metabolites presents a unique approach to characterizing the phenotypic and genotypic functions. A key aspect of metabolite profiling is that it can be used in high-throughput operation and provides a valuable combination of high performance and low unit cost per sample. However, metabolomics by itself may not be sufficient for the full characterization of complex biological systems. In various other plant systems, numerous genes have been interpreted on the basis of relationships between transcript and metabolite levels (Nakabayashi and Saito 2013).

Table 10.2 List of Some Identified Proteins Responsible for Different Functions in *Artemisia annua*

Electron Transport Chain	Translation and Transcription	Metabolism	Protease	Detoxification Defense and Stress Response	Others
ATP synthase CF1 alpha subunit	Calreticulin	4-aminobutyrate aminotransferase	Cell division protein FtsH-like protein	ABC_NikE_OppD_ transporters	Ef-hand calcium binding protein, putative
ATP synthase CF1 epsilon subunit	Chloroplast heat shock 70-1	Artemisinic aldehyde delta-11(13) reductase	Cell division protein Chloroplast FtsH protease	Ascorbate peroxidase-2-like protein	F5I10.22 gene product
ATP ase epsilon chain	Eukaryotic translation initiation factor eIF5A	Carbonic anhydrase	Cysteine protease	Dehydration stress-induced protein)	Hypothetical protein LOC100193724
ATP synthase CF1 beta subunit	LEAFY like protein	Chalcone synthase	Mitochondrial-processing peptidase subunit alpha	Dehydro ascorbate reductase	Pine globulin-1
ATP synthase subunit alpha	Maturase K	Carbonic anhydrase 3	Peptidases_S8_Tripeptidyl_ Aminopeptidase_II	Ferritin	Os04g0550400
ATP synthase beta subunit	Nitrogen regulatory protein P-II	ATP dependent Chloroplast protease putative	Predicted protein, proteasome	Glycine-rich RNA-binding protein RGP-1c	Rubber elongation factor protein
ATP synthase gamma subunit	GLUTAMINE	Fructose-bisphosphate aldolase	Putative zinc dependent protease	Hairpin binding protein 1	
ATP synthase delta chain, chloroplastic	NAD(P)-rossaman binding protein	Glyceraldehyde-3-phosphate dehydrogenase		Peroxiredoxin	
ATP synthase β chain	Ribosomal 50S protein L12	Glycine cleavage system H protein		Rhodanese	
Cytochrome C oxidase polypeptide	RNA polymerase beta' chain	Malate dehydrogenase		Rhodanese Homology Domain (RHOD)	
Chlorophyll a, b, binding protein	Ribosomal L12 1a	Nitrilase/cyanide hydratase			
Mitochondrial F0 ATP synthase D chain	SGRP glycine rich binding protein	Nucleoside diphosphate kinase II			
NAD(P)H-quinone oxidoreductase subunit M	Transcription factor APFI-like	Nucleoside diphosphate kinase B			
PSII PsP protein		Predicted protein (PRK)			
Photosynthetic electron transfer-like protein		Plastidic aldolase			
Rieske-FeS protein		NPALDP1			
SDH1-1; ATP binding/ succinate dehydrogenase		Phosphoglycerate mutase 1			
V-type (H+)-ATPase V1, A subunit		Plastidic aldolases			
		Phosphopyruvate hydratase			
		Ribulose-1,5-bisphosphate carboxylase activase			
		Ribose-5-phosphate isomerase			
		S-adenosyl methionine synthase			
		Transketolase			
		Triosephosphate isomerase			

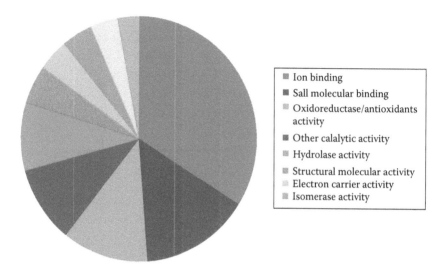

Figure 10.4   Categorization of identified proteins in trichome-enriched *Artemisia annua* L.

Integrated genomics and transcriptomics has recently included metabolite profiling–based approaches. These approaches are very useful in the systematic characterization of biological processes and in the characterization of *A. annua* chemotypes for complete complementation related to metabolites (Liu et al. 2011). Metabolite characterization, in contrast to genome sequencing, involves comprehensive metabolomics: (1) the *chemical diversity* of a typical metabolome is analyzed to accommodate variation in chemico-physical properties, (2) there is a *dynamic range* of metabolite concentrations, (3) the *spatial distribution* includes specific organs, (4) the *temporal distribution* is similarly large, with variations, and (5) unlike proteomics and transcriptional studies, *genomic information is not a constraint* in the identification of molecular species. Recently, metabolomics has become a popular approach for identifying the molecular underpinnings of biological phenomena under varying environmental conditions (Dixon and Paiva 1995).

In contrast to high-throughput technology for the study of DNA, RNA, and proteins, metabolite characterization still currently faces significant obstacles due to the high degree of chemical diversity among metabolite pools in *A. annua*. Metabolite analysis with multivariate tools showed an evolutionary relationship of artemisinic metabolites with two distinct sub-groups of the metabolites (Bedair and Sumner 2008), and their characterization suggests that on the one hand, AN and dihydroartemisinin were clearly associated with each other, and on the other hand, there was a correlation between artemisinic acid, arteannuin B, and artemisitene (Suberu et al. 2016). Some of the identified secondary metabolites, such as terpenes (monoterpenes, diterpenes, and sesquiterpenes) and fatty acids, are listed in Table 10.3.

Table 10.3   List of Identified Volatile Compounds in *A. annua*

| Monoterpenes | | Sesquiterpenes | | Diterpenes and Others | |
Compounds	Groups	Compounds	Groups	Compounds	Groups
		Acetyl cedrene	Alkene	Lignocero	Saturated fatty acid
				Methyl hexadecanoate	—
				Hexadecanoic acid	Fatty acid
				Methyl octadecanoate	—
				Octadecanoic acid	Saturated fatty acid
				Ethyl 2-methyl butyrate	—
				Octanal	Aldehyde
				n-Hexanol	Aldehyde
				Hexyl 3-methyl butyrate	Alcohol
				(Z)-3-Hexenol	Alcohol
				1-Octen-3-ol	Alcohol
				(Z)-3-Hexenyl isovalerate	—
				Eicosane	
				Heneicosane	
				Sanderacopamira-8(14),15-diene	
				Tricosane	
				Tetracosane	
				Pentacosane	
				Hexcosane	
1,8-Cineole	Ether	Aristolene epoxide	Alkene	Larixol	Alcohol
α-Terpineol	Alcohol	Aristolone	Ketone	Phytol	Alcohol
Camphene	Alkene	α-Bisabolol	Alcohol	3-Decanone	Ketone
Camphor	Ketone	γ-Cadinene	Ketone	Butanoate <2 methyl-,(3Z)(hexenyl)	—
trans-Carveol	Alcohol	β-Cadinene	Alkene	Butyrate <2-methylbutyl-,2-methyl-1,1'-Biphenyl, 4 hydroxyacetyl	—
cis-Carveol	Alcohol	β-Caryophyllene	Alkene	Capillene	Benzene
Carvone	Ketone	IsoCaryophyllene	Alkene	Decanoic acid	Saturated fatty acid
Citronellol	Alcohol	β-Copaene	Alkene	4 Caprylorpholine	Aliphatic carboxylic acid
β-Cymene	Alkene	α-Copaene	Alkene	Hexadecanoic	Aliphatic carboxylic acids

(*Continued*)

Table 10.3 (Continued)   List of Identified Volatile Compounds in *A. annua*

Monoterpenes		Sesquiterpenes		Diterpenes and Others	
Compounds	Groups	Compounds	Groups	Compounds	Groups
Isoborneol	Alcohol	Corymbolone	Ketone	Caprylmorpholine	Aliphatic carboxylic acids
Jasmone	Ketone	Cubebanol	Alcohol	n-Butyro-morpholine	Aliphatic hydrocarbon
Lavandulylacetate	Ester	Cyclocolorenone	Ketone	Simvastatin	Polycyclic hydrocarbon
Limonene oxide	Alkene	Dihydromayurone	Ketone		Alkane
Myrcene	Alkene	β-Farnesene	Alkene		Aldehyde
Myrtenol	Alcohol	(E, E)-α Farnesene	Alkene		Alkane
α-Pinene	Alkene	Farnesol	Alcohol		Unsaturated fatty acid
β-Pinene	Alkene	γ-Elemene	Alkene		Saturated fatty acid
Pinocarvone	Ketone	Germacrene A	Alkene		Alkane
Sabinene	Alkene	Himachalol	Alcohol		Alkane
γ-Terpinene	Alkene	Humulene epoxide II	Epoxide		
4-Terpineol	Alcohol	α-Humulene	Alkene		
Thujopsanone	Ketone	Intermedeol	Alcohol		
Verbenol	Alcohol	Isogeranial	—		
Cymene	Alkene	Isocomene	Alkene		
Isoeugenol	Alcohol	10-epi-γ-Eudesmol	Alcohol		
Sesquicineole	Alcohol	β-Isocomene	Alkene		
α-Phellandrene	Alkene	γ-Eudesmol	Alcohol		
Dehydro-1,8 cineole	Alcohol	Lanceol acetate	Ester		
(Z)-β-Ocimene	Alkene	Longifolol	Alcohol		
(E)-β-Ocimene	Alkene	β-Longipinene	Alkene		
δ-3-Carene	Alkene	Longipinocarvone	Ketone		
Terpinolene	Alkene	β-Malliene	Alkene	Linyl acetate	Acetate
Santolina alcohol	Alcohol	γ-Selinene	Alkene	Hexyl tiglate	Ester
trans-Sabinene hydrate	—	α-Selinene	Alkene	9-Decen-1-ol	Alcohol

(Continued)

Table 10.3 (Continued) List of Identified volatile Compounds in A. annua

Monoterpenes		Sesquiterpenes		Diterpenes and Others	
Compounds	Groups	Compounds	Groups	Compounds	Groups
cis-p-Mentha-28-dien1ol	Alcohol	β-Selinene	Alkene	Methyl salicylate	Ester
Chrysanthenone	Ketone	Spathulenol	Alcohol	Para-Mentha-1,4(8)-dien-3-ol	Alcohol
Myrtenal	Aldehyde	Silphinene	Alkene	Octadecane	Alkane
trans-Pinocarveol	Alcohol	Valerenyl acetate	Ester	Benzyl isovalorate	–
trans-Sabinol	Alcohol	Arteannuic alcohol	Alcohol	Tetradecane	Alkane
γ–Cadinone	Ketone	α-Agarofuran	–	Tridecanal	Alkane
Cuminaldehyde	Aldehyde	β-Bergamotene	Alkene	Docosane -n	Alkene
Para-Cymen-8-ol	Alcohol	β-Caryophyllene oxide	Oxide	Linoleic acid	Alkane
cis-Sabinol	Alcohol	Isocaryophyllene oxide	Oxide	Octadecanoic acid	Alkane
(E)-Anethol	Alcohol	β-Ionone	Ketone	Heptadecane	Alkane
Eugenol	Alcohol	β-Selinene	Alkene	Nonadecane	Alkane
Carvacrol	Alcohol	Occidentalol	Alcohol		
Tricyclene	Alkene	Occidentalol acetate	Ester		
		Occidol	Alcohol		
		(E, E)-Farnesyl acetate	Ester		
		Δ-Elemene	Alkene		
		β-Guaiene	Alkene		
		Bicyclogermacrene	Alkene		
		δ-Cadinene	Alkene		
		cis-Calamenene	Alkene		
		Dihydroartemisinic acid	Acid		
		Cedrol	Alcohol		
		Thymol	Alcohol		
		α-Bisabolol	Alcohol		
		Caryophylla-1(12),8(15)-diene-14-ol	Alcohol		
		Caryophylladienol	Alcohol		

10.3 MOLECULAR BREEDING APPROACHES
FOR ENHANCED AN PRODUCTION

The integration of different molecular breeding techniques in *A. annua* may be a viable, cost-effective, and less labor-intensive alternative to produce more AN, thereby aiding in the overall enhanced production of AN globally *in planta*. Both conventional and new biotechnological plant breeding techniques using omic approaches can be applied in *A. annua* at the genetic level for improving yield and uniformity with stable high-performing phenotypes across adverse/variable environments and to modify desired valuable characters/compounds by metabolite profiling assessment. Conventional plant breeding practices when combined with molecular markers can facilitate fast-track breeding through the selection/creation of a population or germplasm with useful desired genetic variation, identification of superior individuals, and development of improved varieties from selected individuals. The success of conventional breeding is dependent on the selection process, and numerous selection methods can be adopted for *A. annua*. Mass selection is dependent mainly on selection of plants according to their phenotypes and performance and can be used to improve the overall population of *A. annua* by positive or negative mass selection. One drawback of mass selection is the influence of the environment on the development, phenotype, and performance of single plants (Pandey and Pandey-Rai 2015).

Another method is recurrent selection, which is more suitable for *A. annua*, as it involves cross-pollinating species. Through different selection methods, a few high-yielding varieties/cultivars of *A. annua*, such as "CIM-Arogya," "Jeevan Raksha," and Asha, were developed by the Central Institute for Medicinal and Aromatic Plants (CIMAP), India, as superior lines rich in AN (Patra and Kumar 2005). Recently, Townsend et al. (2013) have also reported that selection of material for breeding using combining ability analysis of a diallele cross can be used for the selection/identification of elite parents to produce high-AN-producing *A. annua* hybrids. This selection method was found to be consistent with advanced QTL-based molecular breeding approaches. The advancement of plant breeding using different biotechnological tools is now opening a new platform for crop improvement. Among these tools is molecular/mutational breeding, which involves induced mutational changes through chemicals/radiation or by site-directed mutagenesis with the advantage of improving one or two yield-related characters without modifying the rest of the genetic constitution. A successful effort was made by Mediplant (a Swiss not-for-profit organization) by developing a hybrid of *A. annua* known as "Artemis" (F1 hybrid) with a high mean annual AN production of about 32 kg/ha. Further, they also created a new high-yielding hybrid with 40.5–52.0 kg/ha AN production (Simonnet et al. 2008).

There has been significant progress in molecular plant breeding techniques using various molecular tools such as DNA markers (e.g., restriction fragment length polymorphism [RFLP], RAPD, amplified fragment length polymorphism [AFLP], SSRs, SNPs, sequence tagged microsatellite sites [STMS], SCAR, etc.) and functional markers (ESTs, microarray, qRT-PCR, etc.), which can be variously used to speed up the selection/recognition of desired genotypes for high-yielding traits at an early stage of development (Figure 10.5). Although these marker-assisted

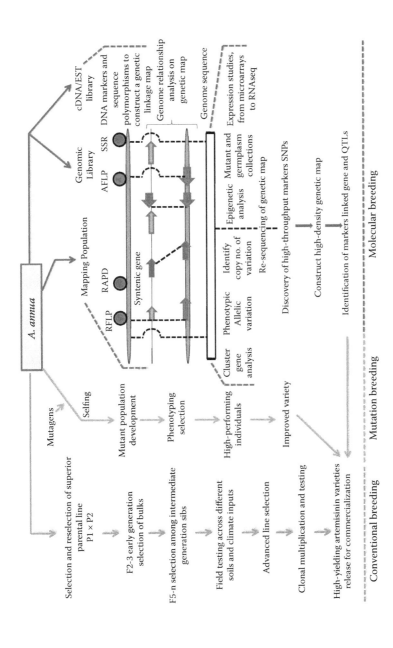

Figure 10.5 Possible breeding approaches for enhanced artemisinin production in *A. annua* L.

molecular breeding techniques are being applied in various crops, there is relatively little information on molecular marker–based approaches in medicinal plants, in which secondary metabolism is of great importance. One such effort was made in *A. annua* by CIMAP (India) for developing the high-AN variety CIM-Arogya through marker-assisted selection breeding. For the production of high-yield varieties of *A. annua*, a fast-track molecular breeding project, led by CNAP's Director Dianna Bowles and Deputy Director Ian Graham, was funded by the Bill & Melinda Gates Foundation. With the aim of producing a better non-GM variety of *A. annua*, about 23,000 parental lines were screened for desired high-yielding traits, and several hybrid crosses were made. After a rigorous selection procedure aided by molecular tools, the two best-performing hybrids, Hyb1209r (Shennong) and Hyb8001r (Zenith), with enhanced AN production have been commercially released.

10.4 CONCLUSION AND FUTURE DIRECTIONS

Various molecular and conventional approaches have been adopted to improve the yield of AN. However, it is still a challenge for any individual approach alone to meet the global demand for AN. *In planta* yield enhancement through various strategies, as discussed in this chapter, remains the prime focus for a sustainable drug supply of AN. It is worth mentioning that all transgenic/non-transgenic tools for the enhancement of AN biosynthesis can be more fruitful if implemented on better chemotypes/hybrids of *A. annua* when integrated with metabolite profiling. The integration of conventional methods, such as breeding and selection, with advanced techniques will serve to enhance the continued production and supply of the drug. Deficiency in the continuous supply chain of AN can be remediated by integrating agri-based technologies with molecular tools for affordable drug production. For example, in India, intercropping of high-yielding varieties of *A. annua* (developed through molecular/mutational breeding) with staple food crops can be a better option for its cultivation without engaging additional cultivable fertile land, and meanwhile, the execution of multiharvesting techniques and advanced drug extraction practices may result in greater AN production with a lower production cost. Metabolomics integrated with transcriptomics has been successfully applied as a smart analytical and screening tool for quality control of AN, and it also offers efficient, comprehensive quantitative and qualitative analysis of metabolites. On the other hand, metabolomics can be coupled with bioactivity assays to identify the components responsible for bioactivity, investigate synergistic effects, and even predict the bioactivity of different metabolites/bioactive products. Similar approaches may also be applied in other countries, which can give a further boost to AN production while at the same time bettering the living standards of their farmers.

REFERENCES

Alam P. and Abdin M.Z. 2011. Over-expression of HMG-CoA reductase and amorpha-4, 11-diene synthase genes in *Artemisia annua* L. and its influence on artemisinin content. *Plant Cell Reports* 30, no. 10: 1919–28.

Ansong C., Purvine S.O., Adkins J.N., Lipton M.S. and Smith R.D. 2008. Proteogenomics: Needs and roles to be filled by proteomics in genome annotation. *Briefings in Functional Genomics & Proteomics* 7, no. 1: 50–62.

Aquil S., Husaini A.M., Abdin M.Z. and Rather G.M., 2009. Overexpression of the HMG-CoA reductase gene leads to enhanced artemisinin biosynthesis in transgenic *Artemisia annua* plants. *Planta Medica* 75, no. 13: 1453–8.

Arsenault P.R., Vail D., Wobbe K.K., Erickson K. and Weathers P.J. 2010. Reproductive development modulates gene expression and metabolite levels with possible feedback inhibition of artemisinin in *Artemisia annua*. *Plant Physiology* 154, no. 2: 958–68.

Banyai W., Kirdmanee C., Mii M. and Supaibulwatana K. 2010. Overexpression of farnesyl pyrophosphate synthase (FPS) gene affected artemisinin content and growth of *Artemisia annua* L. *Plant Cell, Tissue and Organ Culture (PCTOC)* 103, no. 2: 255–65.

Barbacka K. and Baer-Dubowska W. 2011. Searching for artemisinin production improvement in plants and microorganisms. *Current Pharmaceutical Biotechnology* 12, no. 11: 1743–51.

Bedair M. and Sumner L.W. 2008. Current and emerging mass-spectrometry technologies for metabolomics. *TrAC Trends in Analytical Chemistry* 27, no. 3: 238–50.

Borevitz J.O., Xia Y., Blount J., Dixon R.A. and Lamb C. 2000. Activation tagging identifies a conserved MYB regulator of phenylpropanoid biosynthesis. *The Plant Cell* 12, no. 12: 2383–93.

Bouwmeester H.J., Wallaart T.E., Janssen M.H., et al. 1999. Amorpha-4, 11-diene synthase catalyses the first probable step in artemisinin biosynthesis. *Phytochemistry* 52, no. 5: 843–54.

Bryant L., Flatley B., Patole C., Brown G.D. and Cramer R. 2015. Proteomic analysis of *Artemisia annua*—Towards elucidating the biosynthetic pathways of the antimalarial pro-drug artemisinin. *BMC Plant Biology* 15, no. 1: 1.

Chen D.H., Ye H.C. and Li G.F. 2000. Expression of a chimeric farnesyl diphosphate synthase gene in *Artemisia annua* L. transgenic plants via *Agrobacterium tumefaciens*-mediated transformation. *Plant Science* 155, no. 2: 179–85.

Chen M., Yan T., Shen Q., Lu X., Pan Q., Huang Y., Tang Y., et al. 2016. GLANDULAR TRICHOME-SPECIFIC WRKY 1 promotes artemisinin biosynthesis in *Artemisia annua*. *The New Phytologist*, doi:10.1111/nph.14373.

Chen Y., Shen Q., Wang Y., Wang T. et al. 2012. The stacked overexpression of FPS, CYP71AV1 and CPR genes leads to the increase of artemisinin level in *Artemisia annua* L. *Plant Biotechnology Reports* 7: 287–95.

Dixon R.A. and Paiva N.L. 1995. Stress-induced phenylpropanoid metabolism. *The Plant Cell* 7, no. 7: 1085.

Duke M.V., Paul R.N., Elsohly H.N., Sturtz G. and Duke S.O. 1994. Localization of artemisinin and artemisitene in foliar tissues of glanded and glandless biotypes of *Artemisia annua* L. *International Journal of Plant Sciences*: 365–72.

Efferth T., Herrmann F., Tahrani A. and Wink M. 2011. Cytotoxic activity of secondary metabolites derived from *Artemisia annua* L. towards cancer cells in comparison to its designated active constituent artemisinin. *Phytomedicine* 18, no. 11: 959–69.

Elfawal M.A., Towler M.J. Reich N.G., Weathers P.J. and Rich S.M. 2015. Dried whole-plant *Artemisia annua* slows evolution of malaria drug resistance and overcomes resistance to artemisinin. *Proceedings of the National Academy of Sciences* 112, no. 3: 821–6.

Graham I.A., Besser K., Blumer S., Branigan C.A., Czechowski T., Elias L., Guterman I., et al. 2010. The genetic map of *Artemisia annua* L. identifies loci affecting yield of the antimalarial drug artemisinin. *Science* 327, no. 5963: 328–31.

Han J., Wang H., Lundgren A. and Brodelius P.E. 2014. Effects of overexpression of AaWRKY1 on artemisinin biosynthesis in transgenic *Artemisia annua* plants. *Phytochemistry* 102: 89–96.

Harvey A.L., Edrada-Ebel R. and Quinn R.J. 2015. The re-emergence of natural products for drug discovery in the genomics era. *Nature Reviews Drug Discovery* 14, no. 2: 111–29.

Ji Y., Xiao J., Shen Y. et al. 2014. Cloning and characterization of AabHLH1, a bHLH transcription factor that positively regulates artemisinin biosynthesis in *Artemisia annua*. *Plant and Cell Physiology* 55, no. 9: 1592–604.

Kiran U., Ram M., Mather Ali Khan S.K., Jha P., Alam A. and Abdin M.Z. 2010. Structural and functional characterization of HMG-COA reductase from *Artemisia annua*. *Bioinformation* 5, no. 4: 146.

Lange B.M. and Ahkami A. 2013. Metabolic engineering of plant monoterpenes, sesquiterpenes and diterpenes—Current status and future opportunities. *Plant Biotechnology Journal* 11, no. 2: 169–96.

Liu B., Wang H., Du Z., Li G. and Ye H. 2011. Metabolic engineering of artemisinin biosynthesis in *Artemisia annua* L. *Plant Cell Reports* 30, no. 5: 689–94.

Liu C., Zhao Y. and Wang Y. 2006. Artemisinin: Current state and perspectives for biotechnological production of an antimalarial drug. *Applied Microbiology and Biotechnology* 72, no. 1: 11–20.

Lu X., Zhang L., Zhang F. et al. 2013. AaORA, a trichome-specific AP2/ERF transcription factor of *Artemisia annua*, is a positive regulator in the artemisinin biosynthetic pathway and in disease resistance to *Botrytis cinerea*. *New Phytologist* 198, no. 4: 1191–202.

Ma D., Pu G., Lei C. et al. 2009. Isolation and characterization of AaWRKY1, an *Artemisia annua* transcription factor that regulates the amorpha-4, 11-diene synthase gene, a key gene of artemisinin biosynthesis. *Plant and Cell Physiology* 50, no. 12: 2146–61.

Mishina Y.V., Krishna S., Haynes R.K. and Meade J.C. 2007. Artemisinins inhibit *Trypanosoma cruzi* and *Trypanosoma brucei rhodesiense* in vitro growth. *Antimicrobial Agents and Chemotherapy* 51, no. 5: 1852–4.

Nakabayashi R. and Saito K. 2013. Metabolomics for unknown plant metabolites. *Analytical and Bioanalytical Chemistry* 405, no. 15: 5005–11.

Olofsson L., Engström A., Lundgren A. and Brodelius P.E. 2011. Relative expression of genes of terpene metabolism in different tissues of *Artemisia annua* L. *BMC Plant Biology* 11, no. 1: 1.

Paddon C.J., Westfall P.J., Pitera D.J. et al. 2013. High-level semi-synthetic production of the potent antimalarial artemisinin. *Nature* 496, no. 7446: 528–32.

Pandey N. and Pandey-Rai S. 2014. Modulations of physiological responses and possible involvement of defense-related secondary metabolites in acclimation of *Artemisia annua* L. against short-term UV-B radiation. *Planta* 240, no. 3: 611–27.

Pandey N. and Pandey-Rai S., 2016. Updates on artemisinin: An insight to mode of actions and strategies for enhanced global production. *Protoplasma* 253, no. 1: 15–30.

Patra N.K. and Kumar B. 2005. Improved varieties and genetic research in medicinal and aromatic plants (MAPs). In: Kumar A., Mathur A.K., Sharma A., Singh A.K., Khanuja S.P.S. (eds) *Proceeding of Second National Interactive Meet on Medicinal and Aromatic Plants.* CSIR-CIMAP, Lucknow: 53–61.

Rai R., Pandey S., Shrivastava A.K. and Pandey Rai S. 2014. Enhanced photosynthesis and carbon metabolism favor arsenic tolerance in *Artemisia annua*, a medicinal plant as revealed by homology-based proteomics. *International Journal of Proteomics*, 21.

Sen R., Bandyopadhyay S., Dutta A. et al. 2007. Artemisinin triggers induction of cell-cycle arrest and apoptosis in Leishmania donovani promastigotes. *Journal of Medical Microbiology* 56, no. 9: 1213–18.

Shen Q., Lu X., Yan T., Fu X., Lv Z., Zhang F., Pan Q., Wang G., Sun X. and Tang K. 2016. The jasmonate-responsive AaMYC2 transcription factor positively regulates artemisinin biosynthesis in *Artemisia annua*. *The New Phytologist* 210 no. 4: 1269–81.

Simonnet X., Quennoz M. and Carlen C. 2008. New *Artemisia annua* hybrids with high artemisinin content. *Acta Horticulturae* 769: 371–3.

Soetaert S.S., Van Neste C.M., Vandewoestyne M.L. et al. 2013. Differential transcriptome analysis of glandular and filamentous trichomes in *Artemisia annua*. *BMC Plant Biology* 13, no. 1: 1.

Suberu J., Gromski P.S., Nordon A. and Lapkin A. 2016. Multivariate data analysis and metabolic profiling of artemisinin and related compounds in high yielding varieties of *Artemisia annua* field-grown in Madagascar. *Journal of Pharmaceutical and Biomedical Analysis* 117: 522–31.

Tang K.X., Jiang W.M., Lu X., Qiu B. and Wang G.F. 2012. Overexpression AaWRYK1 gene increased artemisinin content in *Artemisia annua* L. Shanghai Jiao Tong University, China. Patent CN201210249469. X 14.

Townsend T., Segura V., Chigeza G. et al. 2013. The use of combining ability analysis to identify elite parents for *Artemisia annua* F1 hybrid production. *PLoS ONE* 8, no. 4: e61989.

Utzinger J., Shuhua X., N'Goran E.K., Bergquist R. and Tanner M. 2001. The potential of artemether for the control of schistosomiasis. *International Journal for Parasitology* 31, no. 14: 1549–62.

WHO (World Health Organization). 2014. Status Report on Artemisinin Resistance. www.who.int/malaria/publications/atoz/status-rep-artemisinin-resistance Sep2014. pdf.

Xiang L., Zeng L., Yuan Y. et al. 2012. Enhancement of artemisinin biosynthesis by overexpressing dxr, cyp71av1 and cpr in the plants of *Artemisia annua* L. *Plant Omics* 5, no. 6: 503–7.

Yang K., Monafared R.S., Wang H., Lundgren A. and Brodelius P.E. 2015. The activity of the artemisinic aldehyde Δ11 (13) reductase promoter is important for artemisinin yield in different chemotypes of *Artemisia annua* L. *Plant Molecular Biology* 88, no. 4–5: 325–40.

Yu Z.X., Li J.X., Yang C.Q., Hu W.L., Wang L.J. and Chen X.Y. 2012. The jasmonate-responsive AP2/ERF transcription factors AaERF1 and AaERF2 positively regulate artemisinin biosynthesis in *Artemisia annua* L. *Molecular Plant* 5, no. 2: 353–65.

Yuan Y., Liu W., Zhang Q. et al. 2015. Overexpression of artemisinic aldehyde Δ11 (13) reductase gene–enhanced artemisinin and its relative metabolite biosynthesis in transgenic *Artemisia annua* L. *Biotechnology and Applied Biochemistry* 62, no. 1: 17–23.

Zhang F., Fu X., Lv Z. et al. 2015. A basic leucine zipper transcription factor, AabZIP1, connects abscisic acid signaling with artemisinin biosynthesis in *Artemisia annua*. *Molecular Plant* 8, no. 1: 163–75.

Zhang L., Jing F., Li F. et al. 2009. Development of transgenic *Artemisia annua* (Chinese wormwood) plants with an enhanced content of artemisinin, an effective anti-malarial drug, by hairpin-RNA-mediated gene silencing. *Biotechnology and Applied Biochemistry* 52, no. 3: 199–207.

CHAPTER 11

Engineering the Plant Cell Factory for Artemisinin Production

Mauji Ram, Himanshu Misra, Ashish Bharillya, and Dharam Chand Jain

CONTENTS

11.1 INTRODUCTION

Malaria causes more than a million deaths a year and has over 500 million clinical cases annually. Despite tremendous efforts toward the control of malaria, the global morbidity and mortality rates have not been significantly changed in the last 50 years (Riley, 1995). The key problem is the failure to find effective medicines against malaria. Artemisinin, a sesquiterpene lactone containing an endoperoxide bridge obtained from a Chinese medicinal plant *Artemisia annua* L., has been demonstrated as an effective and safe alternative therapy against malaria (Luo and Shen, 1987). Artemisinin and its derivatives are found to be effective against multidrug-resistant *Plasmodium* sp., which is especially prevalent in Southeast Asia, South America, and more recently in Africa (Mohapatra et al., 1996; Newton and White, 1999; Krishana et al., 2004). It has also been found to be effective against other infectious diseases such as schistosomiasis, HIV, hepatitis B, and leishmaniasis (Borrmann et al., 2001; Jung and Schinazi, 1994; Utzinger et al., 2001; Romero et al., 2005, and Sen et al., 2007), and a variety of cancer cell lines including breast cancer, human leukemia, colon cancer, and small-cell lung carcinomas (Moore et al., 1995; Efferth et al., 2001; Singh and Lai, 2001). Unfortunately, in some parts of the Cambodia–Thailand border, the malaria parasite (*Plasmodium falciparum*) has developed resistance against artemisinin monotherapies (Noedl et al., 2008). To combat this problem, the World Health Organization (WHO) has now recommended use of an artemisinin-based combination therapy (ACT). The exponential increase in the number of countries adopting ACTs has led to a rapid increase in demand for artemisinin and its derivatives. However, its global production (120 metric tons year^{-1}) is far behind its global demand (180 metric tons year^{-1}) (Kindermans et al., 2007).

The relatively low yield of artemisinin in *A. annua* L. leaves (0.01%–1.2%) and unavailability of economically viable biotechnological or synthetic protocols are the major limitations to commercialization of artemisinin-derived drugs (Laughlin, 1994; Van Agtmael et al., 1999; Kumar et al., 1999; Abdin et al., 2003; Alam and Abdin, 2011). It is imperative to enhance the production of artemisinin in order to bring down the cost of ACT treatment, making it affordable to developing countries. To overcome this problem, efforts are being made worldwide to enhance its production, employing various approaches such as conventional breeding, and biochemical, physiological, molecular, and hairy root culture techniques (Dong and Thuang, 2003; Ro et al., 2006; Zeng et al., 2007; Newman et al., 2006; Zhang et al., 2009; Aquil et al., 2009; Weathers et al., 2005). These approaches show potential for future development, but improvements delivered by them so far have not met the global demand.

Genetic engineering aims to (1) modify cellular metabolite composition in order to produce new compounds, (2) enhance the production of existing compounds, and (3) eliminate the undesirable compounds. Plant metabolism is modified either by introducing novel genes or pathways, or enhancing the expression of endogenous pathways, for example, by up-regulating transcription factors. Further, down-regulation of endogenous genes to suppress or block the production of undesirable metabolites is accomplished by silencing target genes with anti-sense expression or RNA interference (RNAi) (Dixon, 2005). The advancements and limitations of genetic modification of plants have been regularly overviewed in numerous reviews and commentaries (Dixon and Arntzen, 1997; Ohlrogge, 1999; DellaPenna, 2001; Broun and Somerville, 2001; Capell and Christou, 2004). In addition, altering metabolic enzymes or pathways has become an important approach for investigating cell physiology (Farmer and Liao, 1996).

The application of genetic engineering for the production of artemisinin and its precursors, particularly in *A. annua* L., *Cichorium intybus* L., and microbes, has been adapted very recently (Mercke et al., 2000; Martin et al., 2003; Han et al., 2006; Ro et al., 2006; Newman et al., 2006; Zhang et al., 2009; Aquil et al., 2009; Paddon et al., 2013; Alam and Abdin, 2011). The production of precursors of artemisinin such as amorpha-4,11-diene and artemisinic acid, especially in *Escherichia coli* and yeast, has become a prime example of the capabilities of this parts-list and systems-design approach to microbial genetic engineering. Initially, the production of the farnesyl pyrophosphate (FPP) precursor was optimized, and then an enzyme catalyzing the first committed step in the artemisinin pathway, that is, amorpha-4,11-diene synthase (ADS) was overexpressed in *E. coli* (Martin et al., 2003). Several enzymes involved in the early steps of artemisinin biosynthesis have been discovered including HMG-CoA reductase (*HMGR*), *FPS*, *ADS*, *DXS*, *DXR*, *SQS*, *CYP71AV1*, etc. (Table 11.1).

Keeping in view the importance of artemisinin as a novel therapeutic agent for the treatment of drug-resistant malaria and its potential to treat other infectious diseases including cancer, we have made an attempt to critically analyze and summarize recent developments related to genetic engineering of the artemisinin biosynthetic pathway in this chapter.

11.2 BIOSYNTHETIC PATHWAY OF ARTEMISININ

The biosynthetic pathway of artemisinin belongs to the isoprenoid metabolite pathway (Figure 11.1). Based on the experimental evidence related to its biosynthesis, artemisinin is suggested to be derived from two common precursors, isopentenyl pyrophosphate (IPP) and its isomer, dimethylallyl diphosphate (DMAPP). It has been established that higher plants have two independent biosynthetic pathways leading to the formation of IPP: the cytosolic mevalonate pathway and the plastid-localized mevalonate-independent (MEP/Rohmer) pathway (Liu et al., 2005). As a result, the mevalonate pathway is no longer considered as the sole route to the synthesis of artemisinin in *A. annua* L. This was further supported by isolation of two clones encoding deoxy-D-xylulose-5-phosphate synthase (DXPS) and deoxy-D-xylulose-5-phosphate reductoisomerase (DXPR) from transformed hairy roots of *A. annua* L. (Souret et al., 2002; Krushkal et al., 2003). The partial carbon supply to the synthesis of artemisinin was reported to be made by the MEP pathway operating in plastids and DXR catalyzing the rate-limiting step (Towler and Weathers, 2007). Recently, the relative contribution of these pathways toward carbon supply in artemisinin production was evaluated by Ram et al. (2010). They demonstrated that the mevalonate pathway is the major contributor of carbon, supplying 80% of the carbon to artemisinin biosynthesis, whereas the MEP pathway supplies only 20% of the carbon.

In mevalonate pathways, three molecules of acetyl-coenzyme A condense together to yield 3-hydroxy-3-methylglutaryl CoA (HMG-CoA), which is subsequently reduced by the enzyme HMGR to yield mevalonic acid (MVA). Then, under the catalysis of mevalonate kinase, mevalonate 5-diphosphate is formed which is subsequently decarboxylated to yield IPP (Newman and Chappell, 1999). The synthesis of IPP and DMAPP by either MVA or DXP pathways is followed by chain elongation. The carbonium ion is a potent alkylating agent that reacts with IPP, giving geranyl diphosphate (GPP). GPP has the active allylic phosphate group and further reacts with IPP to produce FPP. FPP takes part in a cyclization reaction catalyzed by cyclases to produce various final products of isoprenoids including artemisinin (Barkovich and Liao, 2001).

Table 11.1   Genes Related to Artemisinin Biosynthesis in *A. annua* L

Enzyme	Gene	Function	Location	Gene Bank Accession No.	References
Deoxyxylulose synthase	dxs	1-Deoxy-D-xylulose-5-phosphate synthase activity	Plastid	AF182286	Souret et al., 2002
Deoxyxylulose reductase	dxr	Isomerase and oxidoreductase activity	Plastid	AF182287	Souret et al., 2002
3-Hydroxy-3-methyl glutaryl coenzyme A	hmgr	Catalyse the two step reduction of S-HMG-CoA into R-mevalonate	Cytosol	AF142473	Souret et al., 2002
Farnesyl diphosphate synthase	fps	Synthesis of FPP	Cytosol	AF112881	Chen et al., 2000; Souret et al., 2002
Sesquitermene cyclases		Catalyzes cyclization of FPP to:	All likely in cytosol		
(i) Epicedrol synthase	eps	8-Epicedrol		AJ001539	Mercke et al., 1999; Hua and Matsuda, 1999
(ii) Amorphadiene synthase	ads	Amorpha-4,11 diene		AJ251751	Mercke et al., 2000; Chang et al., 2000
(iii) β-caryophyllene synthase	cs (qhs1)	β-Caryophyllene		AF472361	Cai et al., 2002
(iv) β-farnesene synthase	fs	β-Farnesene		AY835398	Picaud et al., 2005a
Putative sesquiterpene cyclases	casc125	Isoprenoid biosynthesis and lyase activity	Isolated from: Flowers	AJ271792	Van geldre, 2000
	Casc34	Isoprenoid biosynthesis and lyase activity	Leaves and flowers	AJ271793	Van geldre, 2000
	ses	Reduction product not determined	Young leaves	AAD39832	Lie et al., 2002
Squalene synthase	aasqs	Farnesyl-diphosphate farnesyltransferase activity	Endoplasmic reticulum	AY445506	Liu et al., 2003
Squalene synthase fragment	sqsl	Transferase activity	Cytosol	AF182286	Souret et al., 2003
CYP71AV1	na	Catalyzes 3 steps post ADS	Trichomes	DQ315671	Teoh et al., 2006
PsbA (Fragment)	psba	Act as barcode for flowering plants	Chloroplast	DQ006143	Kress et al., 2005
Ribulose-1,5-biphosphate Carboxylase/oxygenase	rbcl	Carbon dioxide fixation; barcoding for flowering plants	Chloroplast	DQ006057	Kress et al., 2005
Peroxidase 1	pod1	Favored the bioconversion of artemisinic acid to artemisinin	Root, stems and leaves	AY208699	Zhang et al., 2004
Beta-pinene synthase	gh6	Circadian pattern of expression	Juvenile leaves	AF276072	Liu et al., 2002
(3R)-linalool synthase	gh1	Lyase activity	Leaves and flowers	AF154125	Jia et al., 1999
Isopentenyl transferase	ipt	Biosynthesis of cytokinines phytohormones	Transferred into A. annua L. via A. tumefaciens	M91610	Sa et al., 2001

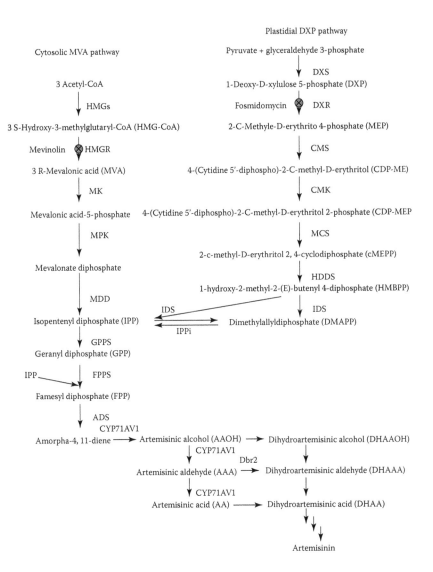

Figure 11.1   Proposed artemisinin biosynthesis pathway in *A. annua* L. CMK, 4-(Cytidine 5'-diphospho)-2-C-methyl-D-erythritol kinase; CMS, 2-C-methyl-D-erythritol 4-phosphate cytidyl transferase; DXR, 1-deoxy-D-xylulose 5-phosphate reductoisomerase; DXS, 1-deoxy-D-xylulose 5-phosphate synthase; FPPS, farnesyl diphosphate synthetase; GPPS, geranyl diphosphate synthase; HMGR, 3-hydroxy-3-methylglutaryl coenzyme A(HMGCoA) reductase; HMGS, HMG-CoA synthase; IDS, isopentenyl diphosphate synthase; MCS, 2-C-methyl-D-erythritol 2,4-cclodiphosphate synthase; MDD, mevalonate diphosphate decarboxylase; MK, mevalonate kinase; MPK, mevalonate-5-phosphate kinase; SES, sesquiterpene synthase; CYP71AV1, cytochrome P450 monooxygenase; Dbr2, artemisinic aldehyde reductase.

All the steps of mevalonate and MEP pathways have been fully characterized, but post-FPP production of artemisinin is not yet completely elucidated. The formation of the sesquiterpene carbon skeleton, amorpha-4,11-diene is catalyzed by ADS (Bouwmeester et al., 1999), for which corresponding cDNAs have been cloned (Chang et al., 2000; Mercke et al., 2000; Wallaart et al., 2001). The non-descript arrangement of the amorphadiene product belies the unique structural features that ultimately allow for the formation of the 1-, 2-, and 4-trioxane moiety (Sy and Brown, 2002) (Figure 11.2). An expression analysis of *CYP71AV1* in *A. annua* L. tissues indicates that it is most highly expressed in glandular secretory trichomes (GSTs) (Teoh et al., 2006). The moderate expression observed in flower buds presumably reflects their high density of GSTs. Low but detectable levels of reverse transcription polymerase chain reaction (RT-PCR) products could be observed in leaves. The role of *CYP71AV1* in the hydroxylation of amorpha-4,11-diene is undoubtedly important in artemisinin biosynthesis. The subsequent route to artemisinin is less clear. Most evidence implicates dihydroartemisinic acid as a late precursor to artemisinin biosynthesis, which is derived from artemisinic alcohol by oxidation at C12 and reduction of the C11-C13 double bond. This is based on *in vitro* biochemical evidence (Bertea et al., 2005), as well as the conversion of dihydroartemisinic acid to artemisinin both *in vivo* (Brown and Sy, 2004) and *in vitro* in an oxygen-dependent non-enzymatic fashion (Sy and Brown, 2002).

Bertea et al. (2005) showed that *A. annua* L. leaf microsomes convert amorphadiene to artemisinic alcohol in the presence of NADPH. The route from artemisinic alcohol to artemisinin is still not entirely clear, which is evident from the published data

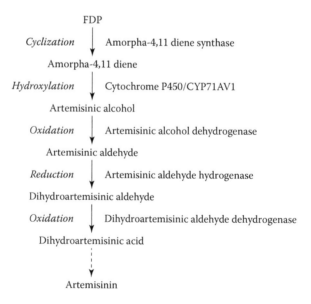

Figure 11.2 Proposed biosynthetic pathway of artemisinin starting from farnesyl diphosphate. On the left is the type of reaction, on the right is the enzyme for each known enzymatic action. Broken arrow indicates multiple steps.

reviewed by Liu et al. (2006) . In this regard, it is useful to consider the possible route(s) to artemisinin among the pathways shown in Figure 11.2. These pathways are based on a few conversions whose order may vary. These conversions include the oxidation of C12 from alcohol to aldehyde as well as aldehyde to acid, the reduction of the double bond at C11-13 and the formation of the 1-, 2-, and 4-trioxane moiety. The later steps in artemisinin biosynthesis remain controversial and theories differ mainly in the identification of either artemisinic acid or dihydroartemisinic acid as the later precursor. The evidence for artemisinic acid has been reviewed by Li et al. (2005). This includes the suggestion that C11-13 double bond reductions occur at the level of an intermediate beyond artemisinic acid, such as arteannuin B or artemisitene. On the other hand, the co-occurrence of dihydroartemisinic acid with high artemisinin levels suggests that even if double bond reduction could occur at a very late step, it also occurs in less oxidized precursors. The double bond reduction at C11-13 is of general interest biochemically, given the relative rarity of enzymes catalyzing double bond reductions (Kasahara et al., 2006). The dihydroartemisinic acid is also being considered as a late precursor of artemisinin biosynthesis. Dihydroartemisinic acid is incorporated into artemisinin *in vivo*, a sequence which can occur in the absence of enzymes (Brown and Sy, 2004; Haynes et al., 2006; Sy and Brown, 2002). Upstream of dihydroartemisinic acid, the order of oxidations and reduction of artemisinic alcohol *en route* to dihydroartemisinic acid is still not settled. Bertea et al. (2005) provided biochemical evidence for the fate of artemisinic alcohol in *A. annua* L. using GST cell-free extracts.

11.3 COMBINATORIAL GENETIC ENGINEERING OF ARTEMISININ BIOSYNTHEIC PATHWAY

Naturally occurring terpenoids are produced in small quantities, and thus, their purification results in low yields. Further, the complex structures of these molecules make their chemical synthesis challenging and often uneconomical due to poor yields. Transferring metabolic pathways in genetically traceable industrial biological hosts (*E. coli* and *Saccharomyces cerevisiae*) offers an attractive alternative, allowing production of large quantities of these complex molecules. To accomplish the production of artemisinin in a microbial host, the altering of MVA and MEP pathways along with the addition of very specialized enzymes, for example ADS, is required. Based on preliminary work on the engineering of the MEP pathway to increase isoprenoid precursors for high-level production of carotenoids (Kajiwara et al., 1997; Harker and Bramley, 1999; Farmer and Liao, 2001; Kim and Keasling, 2001; Abdin et al., 2003), Keasling's group further developed a base technology for the production of amorpha-4,11-diene in *E. coli* (Martin et al., 2003). Bacterium contains the MEP pathway, which allows production of IPP/DMAPP, but lacks the MVA pathway. Keasling's group postulated that the MEP pathway is subjected to unknown control elements in bacteria and direct alteration might impair growth. They, therefore, added a truncated MVA pathway from *S. cerevisiae* that was coupled to ADS in *E. coli* (Figure 11.3). It resulted in good bacterial growth and high-level production of amorpha-4,11-diene, estimated at up to 100 µg/L in 12 h. Thus,

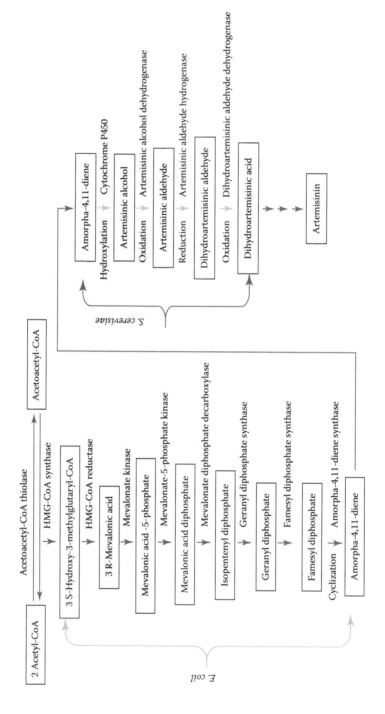

Figure 11.3   Combinatorial biosynthesis of artemisinin starting from acetyl coenzyme A.

these engineered *E. coli* strains can serve as platform hosts for the production of essentially any terpenoid for which the biosynthetic genes are available, since IPP and DMAPP produced by either arm of the terpenoid pathway are universal precursors to all terpenoids. This strategy has been combined with engineering of genes from the mevalonate-dependent isoprenoid pathway (Figure 11.3), which resulted in an *E. coli* strain producing 24 µg/ml amorpha-4,11-diene (calculated as a caryophyllene equivalent) from acetyl-CoA after supplementation of 0.8% glycerol (Martin et al., 2003). Tsuruta et al. (2009) successfully achieved up to 27.4 g/L amorpha-4,11-diene through an *E. coli* fermentation system. Teoh et al. (2006) have isolated the next enzyme in the artemisinin biosynthetic pathway, that is, cytochrome P450 enzyme (CYP71AV1). This enzyme appears to catalyze the next three steps in artemisinin biosynthesis, an enzymatic function which has also been confirmed by Keasling's group.

The Bill and Melinda Gates Foundation awarded a five-year grant of $42.6 million in December 2004 to the Institute for One World Health and a non-profit pharmaceutical company (Amyris Biotechnologies), to fund the research and development partnership between Amyris and the University of California Berkeley (U.C. Berkeley). The research used synthetic biology to develop a stable and scalable, low-cost technology platform for producing artemisinin and its derivatives. The goal of the collaboration is to create a consistent, high-quality, and affordable new source of artemisinin, a key ingredient for making the life-saving antimalarial drugs known as ACTs. In this case, the project team is using synthetic biology to insert genes from the plant *A. annua* L. into *E. coli*, a bacterium. Professor Jay Keasling's laboratory in the Centre for Synthetic Biology at U.C. Berkeley has completed the synthetic biological process to produce artemisinic acid, a precursor to artemisinin (Figure 11.4). In another study, attempts have been made to use *S. cerevisiae* to produce artemisinin

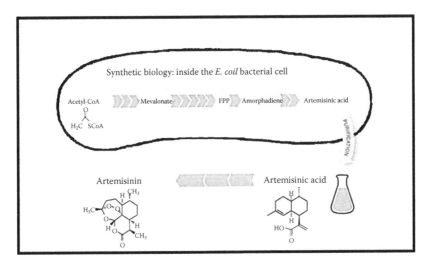

Figure 11.4 Semisynthetic approach to produce artemisinin and its precursors using *E. coli* fermentation system.

precursors. The expression of the ADS gene in yeast using plasmids and chromosomal integration led to the production of 600 and 100 µg L^{-1} amorpha-4,11-diene, after 16 days of batch cultivation (Lindahl et al., 2006). Ro et al. (2006) have reported the production of 100 mg/L artemisinic acid in a *S. cerevisiae* strain containing an engineered MVA pathway coupled with the genes encoding ADS and CYP71AV1. This strain transported artemisinic acid, the artemisinin precursor, outside the yeast cell, which makes purification of the product less complex. Paddon et al. (2013) provided a major breakthrough using strains of *S. cerevisiae* (baker's yeast) and achieved up to 25 g/L of artemisinic acid with fermentation and also achieved 40%–45% conversion rate of artemisinic acid into artemisinin. Artemisinic acid can be used for the semisynthesis of artemisinin, but to lower the cost of production of the drug, bioprocessing must be optimized (Liu et al., 1998b).

11.4 GENETIC ENGINEERING OF *ARTEMISIA* SPECIES FOR ENHANCED PRODUCTION OF ARTEMISININ

The ability to produce high-artemisinin-yielding transgenic strains of *A. annua* L. plants is envisioned; this will ensure a constant high production of artemisinin by overexpressing the key enzymes in the terpene and artemisinin biosynthetic pathways, or by inhibiting enzyme(s) of another pathway competing for artemisinin precursors. In recent years, remarkable progress has been made in the understanding of molecular biology of artemisinin biosynthesis and its regulation (Bouwmeester et al., 1999; Weathers et al., 2006). The genes of the key enzymes involved in the biosynthesis of artemisinin, such as HMGR, farnesyl pyrophosphate synthase (This chapter: FPPS), ADS, and the genes of the enzymes involved in the pathway competing for artemisinin precursors, such as squalene synthase (SQS), which is involved in sterol biosynthesis, have been cloned from *A. annua* L. (Matsushita et al., 1996; Mercke et al., 2000; Wallaart et al., 2001; Liu et al., 2003, Abdin et al., 2003). On the other hand, Weathers et al. (1994) and Qin et al. (1994) induced hairy roots in *A. annua* L. employing *Agrobacterium rhizogenes*. Further, the factors influencing transformation efficiency of *A. rhizogenes* were explored to optimize the transformation system by Liu et al. (1998a). Xie et al. (2001) induced hairy root in *A. annua* L. leaf blade pieces and petiole segments infected with *A. rhizogenes* strain-1601 and obtained a clone with a high content of artemisinin (1.195 mg/g DW).

To develop transgenic *A. annua* L. strains with a high content of artemisinin by modulating the expression of the previously mentioned genes, an efficient system of genetic transformation as well as regeneration of explants of *A. annua* L. should be in place. Vergauwe et al. (1996) developed an *Agrobacterium tumefaciens*-mediated transformation system for *A. annua* L. plants with high transformation rates (75% regenerants harboring the foreign gene). Artemisinin content in the leaves of regenerated plants was 0.17%, which is higher than that present in the leaves of normally cultured plants (0.11% DW). They further investigated the factors *viz.*, the age of explants, the *A. tumefaciens* strain, and plant genotype influencing the transformation efficiency (Vergauwe et al., 1998). Later, Han et al. (2005) established a high-efficiency genetic transformation and regeneration system for *A. annua* L. via *A. tumefaciens*.

Artemisinic acid is one of the precursors of biosynthesis of artemisinin, which has a cadinene structure. Chen et al. (1998) transformed a cotton cadinene synthase cDNA into the leaf explants of *A. annua* L. using *A. rhizogenes*. In the isoprenoid biosynthesis pathway, FPS catalyzes the two sequential 1–4 condensations of IPP with DMAPP to produce GPP and with GPP to give FPP, which is then utilized by the isoprenoid pathway and artemisinin biosynthetic pathway to produce isoprenoids and artemisinin, respectively (Cane, 1990). The cDNAs encoding FPS have been isolated from a number of plant species, including *Arabidopsis thaliana* (Delourme et al., 1994) and *Lupinus albus* (Attucci et al., 1995). Since 15-carbon FPP can be catalyzed by sesquiterpene cyclases, such as, ADS to form cyclic sesquiterpenoids (amorpha-4,11-diene in *A. annua* L.), overexpressing foreign FPS gene into *A. annua* L. plants holds the possibility of affecting the accumulation of artemisinin. A cDNA encoding cotton FPS placed under a CaMV 35S promoter was hence transferred into *A. annua* L. plants *via A. tumefaciens* strain LBA-4404- or *A. rhizogenes* strain ATCC-15834-mediated genetic transformation (Chen et al., 1999 and 2000). In the transgenic plants, the concentration of artemisinin was approximately 8–10 mg/g DW, which was 2- to 3-fold higher than that in the control plants. Han et al. (2006) achieved about a 34.4% increase in artemisinin content by overexpressing FPS. We have overexpressed one of the key regulatory enzymes of MVA pathway (*HMGR*) in *A. annua* L. plants via *A. tumefaciens* mediated transformation and achieved 39% enhancement in artemisinin contents as compared to control plants (Tazyeen et al., 2010). Jing et al. (2008) simultaneously overexpressed *CYP71AV1* and *CPR* genes in *A. annua* L. and recorded a 2.4-fold enhancement in artemisinin content. When *HMGR* was co-expressed with *ADS*, a 7-fold change was measured (Alam and Abdin, 2011).

The cytokinin biosynthetic gene codes for the enzyme isopentenyl transferase (ipt), which catalyzes the condensation of isopentenyl pyrophosphate and adenosine monophosphate (AMP) to yield isopentenyl AMP, is believed to represent the rate-limiting step in cytokinin biosynthesis in tumorous plant tissue (Akiyoshi et al., 1983 and 1984). The influence of overexpression of the ipt gene on the physiological and biochemical characteristics of *A. annua* L. plant was studied by Geng et al. (2001b). The transgenic *A. annua* L. plants were found to accumulate more cytokinins (2- to 3-fold), chlorophyll (20%–60%), and artemisinin (30%–70%), when compared with control plants (Geng et al., 2001b). Previous studies indicated that capitate glands on the leaf surface (Duke and Paul, 1994) and specialized chloroplasts of the capitate gland appeared to play a very important role in artemisinin biosynthesis (Duke and Paul, 1993). Light affects terpene biosynthesis in general and artemisinin biosynthesis in particular, by modulating carbon flux through regulation of HMGR, a key regulatory enzyme in the mevalonate pathway. In the case of potatoes, it has been reported that light regulates *HMGR* at both transcriptional and translational level (Korth et al., 2000). In *A. annua* L., β-pinene synthase was found to have a circadian pattern of gene expression, accompanied by a similar temporal pattern of β-pinene emission under light exerting a stimulatory effect (Liu et al., 2002). Analysis of root cultures of *A. annua* L. suggested that light also positively regulates artemisinin biosynthesis because the root cultures exhibited a substantial decrease in artemisinin content when moved from light to dark (Liu et al., 2002; Guo et al., 2004). Hong et al. (2009), hence, overexpressed *Arabidopsis* blue light receptor CRY1 in *A. annua* L. to evaluate its effect on artemisinin synthesis and accumulation. They found that overexpression of the *CYP1* gene had resulted in increased accumulation of both artemisinin (30%–40%) and anthocyanins (2-fold) as compared to control plants. Xiang et al. (2012) co-overexpressed *DXR*, *CYP71AV1*, and *CPR* in *A. annua* L. plants and achieved three-times higher levels of artemisinin in comparison to wild-type *A. annua* line. In a recent study carried out by Dilshad et al. (2015a), *A. annua* L. plants were transformed through *A. tumefaciens* harboring *rol* B and C genes. Transgenic of *rol* B gene showed 2- to 9-fold increase in artemisinin, 4- to 12-fold increase in artesunate and 1.2- to 3-fold increase in dihydroartemisinin. Whereas in the case of *rol* C gene transformants, a 4-fold increase in artemisinin, 4- to 9-fold increase in artesunate, and 1- to 2-fold increase in dihydroartemisinin concentration was observed. In another study (Dilshad et al., 2015b), genetic transformation of *Artemisia carvifolia* was carried out with *A. tumefaciens* GV3101 harboring the rol B and rol C genes. Artemisinin content increased 3- to 7-fold in transgenics bearing the rol B gene, and 2.3- to 6-fold in those with the rol C gene. A similar pattern was observed for artemisinin analogues. The dynamics of artemisinin content in transgenics and wild-type *A. carvifolia* was also correlated with the expression of genes involved in its biosynthesis.

11.5 GENETIC ENGINEERING OF *C. INTYBUS* FOR ARTEMISININ PRODUCTION

Dafra Pharma International NV and Plant Research International (PRI) have initiated new research to produce artemisinin *via* genetically modified chicory plants. In studies carried out at Wageningen, the complete biosynthetic pathway of artemisinin was resolved (de Kraker et al., 2003; Bertea et al., 2005; Figure 11.5). In addition, the Wageningen group, headed by Prof. Harro Bouwmeester and Dr. Maurice Franssen, demonstrated that chicory enzyme(s) normally involved in the biosynthesis of the bitter sesquiterpene lactones in chicory, were capable of performing reactions required for the biosynthesis of artemisinin (de Kraker et al., 2003). Prof. Bouwmeester's group have tried to produce the chemical precursor for artemisinin (dihydroartemisinic acid) in the roots of chicory *via* a diversion of the biosynthesis of bitter compounds. Prof. Bouwmeester has also shown, in a wide range of plant species, that diversion of the biosynthesis of terpenes can be carried out very efficiently (Kappers et al., 2005). Moreover, they also demonstrated that up to 40 kg ha^{-1} dihydroartemisinic acid can be produced using genetically modified chicory.

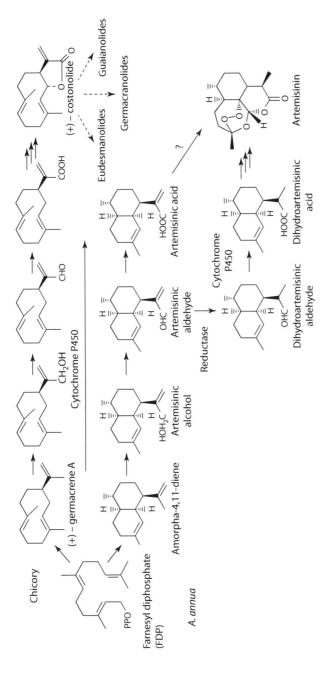

Figure 11.5 Biosynthetic routes of bitter sesquiterpene lactones in chicory and artemisinin in *A. annua* L.

11.6 GENETIC ENGINEERING OF *NICOTIANA TABACUM* FOR ARTEMISININ PRODUCTION

Successful production of amorphadiene in *Nicotiana tabacum* and identification of the genes encoding the subsequent enzymes in the artemisinin biosynthetic pathway facilitated Zhang, Nowak, Reed, and Covello's (2011) attempt to generate the late pathway molecules in plants. They used a multigene construct for the stable transformation of tobacco and expressed *A. annua* ADS and FPS genes fused with a plastidial targeting sequence and under the control of 35S promoters and NOS terminators (2011). This allowed for amorphadiene production at an intermediate level of ca. 0.25 µg/g FW. The addition of CYP71AV1 using a similar vector led to an average reduction in the hydrocarbon's level and detection of artemisinic alcohol. Interestingly, neither of the expected successive metabolites–artemisinic aldehyde and artemisinic acid–which are produced by the action of CYP71AV1 *in vitro* and in yeast, could be detected (Zhang et al., 2008). As with mitochondria, localized production of amorphadiene, the hydrocarbon sesquiterpene synthesized by the plastid-localized ADS, was still accessible to the cytosolic CYP71AV1. This is similar to the case of native cytochrome P450-dependent modifications of terpenoids produced in plastids (Croteau, Davis, Ringer, and Wildung, 2005). In other tobacco lines generated with a construct in which DBR2 was added to CYP71AV1, the plastid targeted FPS and ADS, in addition to amorphadiene and artemisinic alcohol, small amounts of molecules with a reduction in the $\Delta^{11(13)}$ double bond were detected, but again only dihydroartemisinic alcohol was detected and not the more oxidized successive pathway compounds, that is, the aldehyde or acid forms. Speculating that CYP71AV1 activity did lead to the formation of aldehydes in the transgenic tobacco but endogenous reducing activity prevented their accumulation, the authors attempted to enhance flux toward dihydroartemisinic acid by adding ALDH1 to the expression cassette and generating new transgenes. However, this did not lead to the accumulation of amorphane acids or artemisinin, but for some reason elevated the total yield of amorphane sesquiterpenes from ~0.25 to 0.75 µg/g FW. It seems that some native reducing action prevented the successive oxidizing metabolism of the alcohols in the transgenic tobacco, or that in these experiments, the oxidation activity of the cytochrome P450 was not high enough to drive the formation of the aldehydes and acids (Zhang et al., 2011). It should be noted that general eukaryotic ER-bound cytochrome P450 enzymes require an interaction with CPR and available NADPH (which acts as co-substrate). For example, routine functional characterization of cytochrome P450 enzymes using microsomes is performed in lines expressing or overexpressing a suitable reductase, for example, the *S. cerevisiae* WAT11 strain expressing the *Arabidopsis* reductase ATR1 (Rontein et al., 2008; Urban, et al., 1997).

11.7 RNAi (RNA INTERFERENCE) MEDIATED GENE SILENCING

The mechanism of artemisinin biosynthesis has recently become much clearer (Abdin et al., 2003; Bertea et al., 2005; Liu et al., 2006). It has been shown that artemisinin belongs to the isoprenoid group of compounds, which are derived from two common precursors, namely, IPP and its isomer DMAPP. GPP is formed by chain elongation from IPP and DMAPP when they react with a carbonium ion, and GPP can then further react with IPP to produce FPP. FPP can be converted through enzymic catalysis to produce various isoprenoid final products, such as artemisinin and sterols. SQS is reported to be the key enzyme catalyzing the first step of the sterol biosynthetic pathway, a pathway competing with artemisinin biosynthesis (Figure 11.6). The gene for SQS has been cloned from *A. annua* L. (Liu et al., 2003). Previous studies have shown that inhibition of the sterol biosynthetic pathway by chemical methods could improve the artemisinin content of *A. annua* L. (Kudakasseri et al., 1987). Woerdenbag et al. (1993) showed in their study that artemisinin production could be enhanced by the addition of naftiphine, an inhibitor of the enzyme squalene epoxidase, to the medium. Kudakasseril et al. (1987) also demonstrated that the application of many sterol inhibitors, including miconazole or chlorocholine, resulted in an increase in artemisinin in shoot cultures of *A. annua* L. Ro et al. (2006) confirmed that down-regulation of *ERG9* (ergosterol biosynthesis-pathway gene 9), a gene that encodes SQS in yeast, using a methionine-repressible promoter (PMET3), increased the production of amorpha-4,11-diene 2-fold in a yeast

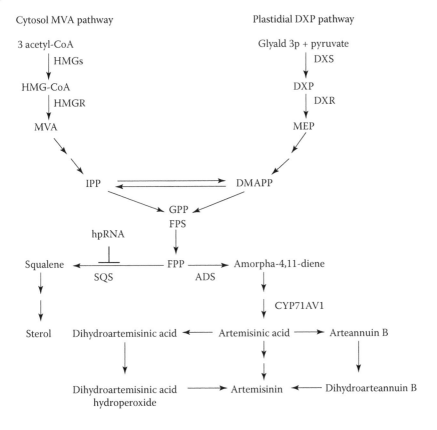

Figure 11.6   RNAi-mediated suppression of SQS gene.

strain into which FPP-synthetic-pathway genes and the *ADS* gene from *A. annua* L. had been incorporated. RNAi mediated by hpRNA has been used in gene silencing in many species of plants. Liu et al. (2002) reported that hpRNA mediated downreg-ulation of ghSAD-1 and ghFAD2-1, two key enzymes in the fatty acid biosynthesis pathway in cotton (*Gossypium hirsutum*), it elevated the stearic acid content (44% compared to a normal level of 2%) and oleic acid content (77% compared with a nor-mal level of 15%) in cotton seeds. It was also reported that the suppression of one key enzyme, *CaMXMT1*, involved in the caffeine biosynthetic pathway through hpRNA mediated interference in coffee (*Coffea* spp.) decreased obromine and caffeine accu-mulation efficiently: 30%–50% of that normally found in the species (Ogita et al., 2004). In a study with opium poppy (*Papaver somniferum*), a hpRNA construct con-taining sequences from multiple cDNAs of genes in the codeine reductase gene fam-ily was used to silence several enzymes in the pathway. In the developed transgenic plants, the non-narcotic alkaloid (*S*)-reticuline, which occurs upstream of codeine in the pathway, accumulated at the expense of morphine, codeine, opium, and theba-ine (Allen et al., 2004). In tomato (*Solanum lycopersicum*), a hp (hairpin) construct was used to suppress an endogenous photomorphogenesis regulatory gene, *DET1*, driven by a fruit-specific promoter. *DET1* was degraded and the carotenoid and fla-vonoid content of tomato fruits were increased, while all other traits for fruit quality remained unchanged in transgenic plants compared with that in wild-type tomatoes (Davuluri et al., 2005). In another study, suppression of an arsenic reductase gene *ACR2* in *Arabidopsis* (thale cress) using hp constructs improved the arsenic con-tent significantly in transgenic shoots 10- to 16-fold compared with that in the wild type (Dhankher et al., 2006). Following the similar strategy, artemisinin content was enhanced up to 2- to 3-fold in transgenic *A. annua* L. plants ($\approx$31.4 mg g^{-1}dw as compared to 8–10 mg g^{-1}dw in control plants) by suppressing the expression of *sqs* (squalene synthase gene), encoding SQS, a key enzyme of sterol pathway (a pathway competitive with that of artemisinin biosynthesis) by means of hp-RNA-mediated RNAi (RNA interference) (Zhang et al., 2009) (Figure 11.6). The sterol content of transgenic plants was also reduced to 37%–58% as compared to that of wild type plants, but it had not affected their growth and development. This study, along with others, demonstrates that the metabolic engineering strategy of suppressing sterol biosynthesis using RNAi could become an effective and suitable means for increas-ing the artemisinin content of plants.

11.8 ROLE OF TRANSCRIPTION FACTORS IN ARTEMISININ BIOSYNTHESIS

The functions of an increasing number of plant transcription factors are being elucidated, and many of these factors have been found to impact flux through meta-bolic pathways. Since transcription factors, as opposed to most structural genes, tend to control multiple steps of pathways, they have emerged as powerful tools for the manipulation of complex metabolic pathways in plants. The importance of transcrip-tion factors in the regulation of the flavonoid pathway suggests that they may play an equally important role in regulating other pathways of plant secondary metabolism (Broun et al., 2006). In several species, a relevant observation is the fact that terpenoid accumulation is preceded by the coordinated induction of several pathway genes. It is likely that several aspects of terpenoid metabolism are regulated at the level of gene expression, while relatively little is known of the transcription factors involved. The first direct evidence of transcription factor control over terpenoid pathway gene expression was observed in *Catharanthus* cells overexpressing (mono) terpenoid indole alkaloid (TIA) pathway activator *ORCA3* (Van der Fits and Memelink, 2000). Analysis of the transcript levels showed that, in addition to TIA pathway genes, the gene encoding *DXS* was also induced. Although induction was found to be significant, it was also fairly limited. *G10H* encoding geraniol 10-hydroxylase, another gene in the monoterpene branch of the pathway, which was monitored in this experiment, was not affected. These observations suggest a role for *ORCA3* in regulating terpenoid as well as TIA biosynthesis, although additional factors are likely to be involved (van der Fits and Memelink 2000). Recently, Ma et al. (2009) isolated and characterized *AaWRKY1*, an *A. annua* L. transcription factor that regulated the ADS gene, a key gene of artemisinin biosynthesis. Promoters of *ADS* contain two reverse-oriented TTGACC W-box cis-acting elements, which are binding sites of WRKY transcription factors. A full-length cDNA (*AaWRKY1*) was isolated from a cDNA library of the GSTs in which artemisinin is synthesized and sequestered. *AaWRKY1* and *ADS* genes were highly expressed in GSTs and both were strongly induced by methyl jasmonate and chitosan. Ma et al. (ibid) also demonstrated that *AaWRKY1* has a similar propensity. Transient expression of *AaWRKY1* activated the expression of *HMGR*, *ADS*, *CYP71AV1*, and *DBR2* of the artemisinin biosynthesis pathway. It is possible that the W-box also existed in the promoters of *CYP71AV1*, *HMGR*, and *DBR2* in *A. annua* L. Indeed, the W-box has been found in the promoter of cytochrome P450 genes of many other plants such as *Arabidopsis* (Narusaka et al., 2004), *Camellia japonica* (Kato et al., 2007), and cotton (Xu et al., 2004), while two W-box elements were also found in the HMGR1 promoter of *Camptotheca acuminate* (Burnett et al., 1993).

11.9 CONCLUSIONS

A. annua L. is the main source of artemisinin, which is the most potent and efficacious antimalarial drug after quinine. Artemisinin has also been demonstrated as a selective anticancer drug. Currently, the limited availability of artemisinin and the lack of real competition among producers of the raw material seem to be the major barriers to scaling-up production and are partially responsible for its high price (World Bank, 2003). The relatively low yields of artemisinin in *A. annua* L. and the non-availability of economically viable synthetic protocols have been the major obstacles for its commercial production and clinical use. However, multipoint genetic engineering is now beginning to supersede single-point engineering as the best way to manipulate metabolic flux to enhance artemisinin synthesis in transgenic plants. It is evident from the foregoing discussion that several points in a given metabolic pathway can be controlled simultaneously either by overexpressing and/or suppressing several enzymes or through the use of transcriptional regulators to control several endogenous genes. Moreover, applied genomics, proteomics, and metabolomics are continuing to expand our knowledge of metabolic pathways, while advances in systems biology help us to model the impact of different modifications made at the gene level with greater accuracy. Heterologous production of artemisinin in an alternative plant is now feasible and might enable the development of a designer production platform based on plants with high yields and biomass production rooted in defined agricultural practices.

11.10 FUTURE PROSPECTS

Genetic engineering of biosynthetic pathways of artemisinin has shown promising results. Because of the intricate and highly complicated metabolic networks of the artemisinin pathway in *A. annua* L., there are multiple rate-limiting steps in the pathway. The fluxes through the pathway are controlled to a great extent at the level of genes, enzymes, compartmentation, transport, and accumulation (Verpoorte and Memelink, 2002). It has been established that the enzyme(s) HMGR, IPT, FPS, ADS, CYP71AV1, DXR, and CPR are catalyzing for the rate-limiting steps in isoprenoid and artemisinin biosynthesis (Aquil et al., 2009; Geng et al., 2001b; Banyai et al., 2010; Han et al., 2006; Ping et al., 2008; Lulu et al., 2008; Xiang et al., 2012). The overexpression of genes for these enzymes in transgenic *A. annua* L. plants has resulted in a considerable increase in artemisinin content. Artemisinin production in metabolically engineered plants can be boosted by further modifications to the flux of the FPP precursor's supply to elevate amorphadiene accumulation and by combining several subcellular compartments for production. For example, flux through the MVA pathway toward FPP synthesis can be increased by expressing more efficient variants of the eukaryotic pathway (Wang et al., 2012). Expression of growth promoting genes *rol* B and C in *A. annua* L. plants were also found to be key factors in enhancing artemisinin yield (Dilshad et al., 2015a). The down-regulation of sterol biosynthesis by suppressing genes encoding key enzymes, such as SQS, has also been shown to enhance artemisinin content (Zhang et al., 2009). These genetic engineering steps, when used together, could be able to enhance artemisinin content in *A. annua* L. The other way to enhance artemisinin biosynthesis in *A. annua* L. could be the use of transcription factors (AaWRKY1 and ORCA3), which are involved in up-regulation of artemisinin biosynthetic pathway and down-regulation of sterol biosynthesis (Ma et al., 2009). Isolation and characterization of the promoters of HMGR, CYP71AV1, DBR2, and other genes in the artemisinin biosynthetic pathway would be beneficial for further evaluation of the function of WRKY transcription factors in artemisinin metabolism.

To this end, tobacco is the only plant, apart from *A. annua*, in which artemisinin production has been demonstrated; this opens up new opportunities for molecular farming of artemisinin in plants with better biomass generation and more established fast-growth practices than *A. annua*. Tobacco itself is a good production platform since there is ample knowledge on its genetics and growing practices to allow for rapid and high biomass yield (Zhang et al., 2011). Furthermore, tobacco can be efficiently grown in areas such as China and East Africa where *A. annua* cultivation and extraction facilities are already operational. Other engineered plants may be capable of accumulating higher levels of artemisinin. For example, an interesting collaboration between Dafra Pharma International NV and Plant Research International of the University of Wageningen relies on the observation that *C. intybus* (chicory) produces large amounts of sesquiterpenes lactones, similar to artemisinin (de Kraker et al., 2003 and Bertea et al., 2005). The assumption is that this host may serve as an efficient platform for dihydroartemisinic acid production due to its biochemical properties, which are similar to those of *A. annua* (http://www.dafra.be/content.chicory-project). The outcome of this initiative should be quite interesting.

Moreover, combinatorial biosynthesis has been utilized for the production of important classes of natural products, including alkaloids (vinblastine and vincristine), terpenoids (artemisinin and paclitaxel) and flavanoids. It is expected, therefore, that combinatorial biosynthetic strategies will yield interesting alternatives in the near future.

WHO recommends ACTs as a first-line treatment for malaria. However, supplies of plant-derived artemisinin are subject to the seasonality and volatility common to many plant-based commodities, leading to fluctuations in the price of artemisinin. Commercial-scale production of semisynthetic artemisinin would have the potential to stabilize supply and supplement existing plant-derived materials to create a consistent, high-quality, and affordable new source of artemisinin and can help to meet the projected world-wide demand for ACTs. Although progress has been made in metabolic engineering of artemisinin biosynthesis in *A. annua* L., more needs to be done in order to further increase artemisinin production to a level of practical application through metabolic regulation.

REFERENCES

Abdin M.Z., Israr M., Rehman R.U., and Jain S.K., 2003. Artemisinin, a novel antimalarial drug: Biochemical and molecular approaches for enhanced production. *Planta Med.* 69: 289–299.

Akiyoshi D.E., Klee H., Amasino R., Nester E.W., and Gordon M.P., 1984. T-DNA of *Agrobacterium tumefaciens* encodes an enzyme of cytokinin biosynthesis. *Proc. Natl. Acad. Sci. U.S.A* 81: 5994–5998.

Akiyoshi D.E., Morris R.O., Mischke B.S., Kosuge T., Garfinkel D., Gordon M.P., and Nester E.W., 1983. Cytokinin auxin balance in crown gall tumors is regulated by specific loci in the T-DNA. *Proc. Natl. Acad. Sci. USA* 80: 407–411.

Alam P., Abdin M.Z., 2011. Over-expression of HMG-CoA reductase and amorpha-4,11-diene synthase genes in *Artemisia annua* L. and its influence on artemisinin content. *Plant Cell Rep.* 10: 1919–1928.

Allen R.S., Millgte A.G., Chitty J.A., Thisleton J., Miller J.A., Fist A.J., Gerlach W.L., et al, 2004. RNAi-mediated replacement of morphine with the nonnarcotic alkaloid reticuline in Opium poppy. *Nat. Biotechnol* 22: 1559–1566.

Aquil S., Husaini A.M., Abdin M.Z., and Rather G.N., 2009. Overexpression of HMG-CoA reductase gene leads to enhanced artemisinin biosynthesis in transgenic *A. annua* L. plants. *Planta Med.* 75(13): 1–6.

Attucci S., Aitken S.M., Ibrahim R.K., and Gulick P.J.A., 1995. cDNA encoding farnesyl pyrophosphate synthesis in white lupine. *Plant Physiol*. 108(2): 835–836.

Banyai W., Kirdmanee C., Mii M., and Supaibulwatana K., 2010. Overexpression of farnesyl pyrophosphate synthase (FPS) gene affected artemisinin content and growth of *Artemisia annua* L. *Plant Cell Tiss. Organ Cult*. 103(2): 255–265.

Barkovich R. and Liao J.C., 2001. Metabolic engineering of isoprenoids. *Metab. Eng*. 3(1): 27–39.

Bertea C.M., Freije J.R., van der Woude H., Verstappen F.W., Perk L., and Marquez V., 2005. Identification of intermediates and enzymes involved in the early steps of artemisinin biosynthesis in *Artemisia annua*. *Planta Med*. 71(1): 40–47.

Borrmann S., Szlezak N., Faucher J.F., Matsiegui P.B., Neubauer R., Biner R.K., Lell B., et al. 2001. Artesunate and praziquantel for the treatment of *Shistosoma haematobium* infections: A doubleblind, randomized, placebo-controlled study. *J. Infect. Dis*. 184(10): 1363–1366.

Boumeester H.J., Wallaart T.E., Janssen M.H., van Loo B., Jansen B.J., Posthumus M.A., Schmidt C.O., et al. 1999. Amorpha-4,11-diene synthase catalyze the first probable step in artemisinin biosynthesis. *Phytochemisty* 52(5): 843–854.

Broun P., Liu Y., Queen E., Schwarz Y., and Leibman A.M., 2006. Importance of transcription factors in the regulation of plant secondary metabolism and their relevance to the control of terpenoid accumulation. *Phytochem. Rev*. 5(1): 27–38.

Broun P. and Somerville C., 2001. Progress in plant metabolic engineering. *Proc. Nat. Acad. Sci*. 98: 8925–8927.

Brown G.D. and Sy L.-K., 2004. *In vivo* transformations of dihydroartemisinic acid in *Artemisia annua* plants. *Tetrahedron* 60(5): 1139–1159.

Burnett R.J., Maldonado-Mendoza I.E., McKnight T.D., and Nessler C.L., 1993. Expression of a 3-hydroxy-3-methylglutaryl coenzyme A reductase gene from *Camptotheca acuminata* is differentially regulated by wounding and methyl jasmonate. *Plant Physiol*. 103(1): 41–48.

Cai Y., Jia J.W., Crock J., Lin Z.X., Chen X.Y., and Croteau R.A., 2002. cDNA clone for b-caryophyllene synthase from *Artemisia annua*. *Phytochem* 61: 523–529.

Cane D.E., 1990. Enzymatic formation of sesquiterpenes. *Chem. Rev*. 90(7): 1089–1103.

Capell T., and Christou P., 2004. Progress in plant metabolic engineering. *Curr. Opin. Biotechnol*. 15: 148–154.

Chang Y.J., Song S.H., Park S.H., and Kim S.U., 2000. Amorpha-4,11-diene synthase of *Artemisia annua*: cDNA isolation and bacterial expression of a terpene synthase involved in artemisinin biosynthesis. *Arch. Biochem. Biophys*. 383(2): 178–184.

Chen D., Liu C., Ye H., Li G., Liu B., Meng Y., and Chen X., 1999. Ri-mediated transformation of *Artemisia annua* with a recombinant farnesyl diphosphate synthase gene for artemisinin production. *Plant Cell Tissue Organ Cult*. 57: 157–162.

Chen D., Ye H., and Li G., 2000. Expression of a chimeric farnesyl diphosphate synthase gene in *Artemisia annua* L. transgenic plants via *Agrobacterium tumefaciens*-mediated transformation. *Plant Sci*. 155: 179–185.

Chen D.H., Meng Y., Ye H.C., Li G.F., and Chen X.Y., 1998. Cultures of transgenic *Artemisia annua* hairy root with cadinene synthase gene. *Acta Bot. Sin*. 40: 711–714.

Croteau R., Davis E., Ringer K., and Wildung M. 2005. (-)-Menthol biosynthesis and molecular genetics. *Naturwissenschaften* 92: 562–577.

Davuluri G.R., van Tuinen A., Fraser P.D., Manfredonia A., Newman R., Burgess D., Brummell D.A., King S.R., Palys J., and Uhlig J., 2005. Fruit-specific RNAi-mediated suppression of DET1 enhances carotenoid and flavonoid content in tomatoes *Nat. Biotechnol*. 23: 890–895.

de Kraker J., Schurink M., Franssen M.C.R., König W.A., Groot A., and Bouwmeester H.J., 2003. Hydroxylation of sesquiterpenes by enzymes from chicory (*Cichorium intybus* L.) roots. *Tetrahedron* 59: 409–418.

DellaPenna D., 2001. Plant metabolic engineering. *Plant Physiol.* 125: 160–163.

Delourme D., Lacroute F., and Karst F., 1994. Cloning of an *Arabidopsis thaliana* cDNA coding for farnesyl diphosphate synthase by function complementation in yeast. *Plant Mol. Biol.* 26: 1867–1873.

Dhankher O.P., Rosen B.P., McKinney E.C., and Meagher R.B., 2006. Hyperaccumulation of arsenic in the shoots of *Arabidopsis* silenced for arsenate reductase (ACR2) *Proc. Natl. Acad. Sci. U.S.A.* 103(14): 5413–5418.

Dilshad E., Cusido R.M., Palazon J., Estrada K.R., Bonfill M., and Mirza B., 2015a. Enhanced artemisinin yield by expression of *rol* genes in *Artemisia annua*. *Malaria J.* DOI: 10.11.1186/s12936-015-0951-5.

Dilshad E., Cusido R.M., Palazon J., Estrada K.R., Bonfill M., and Mirza B., 2015b. Genetic transformation of *Artemisia carvifolia* Buch with *rol* genes enhances artemisinin accumulation. *PLOS One*. DOI: 10.1371/journal.pone.0140266.

Dixon R.A., 2005. Engineering of plant natural product pathways. *Curr. Opin. Plant Biol.* 8: 329–336.

Dixon R.A. and Arntzen C.J., 1997. Transgenic plant technology is entering the era of metabolic engineering. *Trends in Biotechnol.* 15: 441–444.

Dong N.H. and Thuang N.V., 2003. Breeding of high leaf and artemisinin yielding *Artemisia annua* variety. *Paper presented at the International Conference on Malaria: Current Status and Future Trends*, Chulabhorn Research Institute, Bangkok, Thailand, Feb. 16–19.

Duke S.O. and Paul R.N., 1993. Development and fine structure of the glandular trichomes of *Artemisia annua* L. *Int. J. Plant Sci.* 154: 107–118.

Duke M.V. and Paul R.N., 1994. Localization of artemisinin and artemisitene in foliar tissue of glanded and glandless biotypes of *Artemisia annua*. *Int. J. Plant Sci.* 155: 365–372.

Efferth T., Dunstan H., Sauerbrey A., Miyachi H., and Chitambar C.R., 2001. The anti-malarial artesunate is also active against cancer. *Int. J. Oncol.* 18: 767–773.

Farmer W.R., and Liao J.C., 1996. Progress in metabolic engineering. *Curr. Opin. Biotechnol.* 7: 198–204.

Farmer W.R. and Liao J.C., 2001. Precursor balancing for metabolic engineering of lycopene production in *Escherichia coli*. *Biotechnol. Prog.* 17: 57–61.

Geng S., Ma M., Ye H.C., Liu B.Y., Li G.F., and Kang C., 2001b. Effect of ipt gene expression on the physiological and chemical characteristics of *Artemisia annua* L. *Plant Sci.* 160: 691–698.

Guo C., Liu C., Ye H., and Li G., 2004. Effect of temperature on growth and artemisinin biosynthesis in hairy root cultures of Artemisia annua. *Acta Bot. Sin.* 24: 1828–1831.

Han J.L., Liu B.Y., Ye H.C., Wang H., Li Z.Q., and Li G.F., 2006. Effects of overexpression of the edogenous farnesyl diphosphate synthesis on the artemisinin content in *Artemisia annua* L. *J. Integr. Plant Biol.* 48 (4): 482–487.

Han J.L., Wang H., Ye H.C., Liu Y., Li Z.Q., Zhang Y., Zhang Y.S., Yan F., and Li G.F., 2005. High efficiency of genetic transformation and regeneration of *Artemisia annua* L. via *Agrobacterium tumefaciens*-mediated procedure. *Plant Sci* 168: 73–80.

Harker M. and Bramley P.M., 1999. Expression of prokaryotic 1-deoxy-D-xylulose-5-phosphatases in *Escherichia coli* increases carotenoid and ubiquinone biosynthesis. *FEBS Lett.* 448: 115–119.

Haynes K.A., Caudy A.A., Collins L., and Elgin S.C.R., 2006. Element 1360 and RNAi components contribute to HP1-dependent silencing of a pericentric reporter. *Curr. Biol.* 16(22): 2222–2227.

Hong G.J., Hu W.L., Li J.X., Chen X.Y., and Wang L.J., 2009. Increased accumulation of arte-misinin and anthocyanins in *Artemisia annua* expressing the *Arabidopsis* blue light receptor CRY1. *Plant Mol. Biol. Rep.*, DOI 10.1007/s11105-008-0088-6.

Hua L., and Matsuda S. P., 1999. The molecular cloning of 8-epicedrol synthase from *Artemisia annua*. *Arch. Biochem. Biophys.* 369: 208–212.

Jia J., Crock J., Lu S., Croteau R., and Chen X., 1999. (3R)-Linalool synthase from *Artemisia annua* L.: cDNA isolation, characterization, and wound induction. *Arch. Biochem. Biophys.* 372: 143–149.

Jing F., Zhang L., Li M., and Tang K., 2008. Over-expressing cyp71av1 and cpr genes enhances artemisinin content in *Artemisia annua* L. *J. Agric. Sci. Technol.* 10: 64–70.

Jung M., ElSohly H.N., and Mc Chesney J.D., 1990. Artemisinic acid: A versatile chiral syn-thon and bioprecursor to natural products. *Planta Med.* 56: 624.

Jung M., Schinazi R.F., 1994. Synthesis and in vitro anti-human immunodeficiency virus aciv-ity of artemisinin (Qinghaousu) related trioxanes. *Bioorg. Med. Chem. Lett.* 4: 941–934.

Kajiwara S., Fraser P.D., Kondo K., and Misawa N. 1997. Expression of an exogenous iso-pentenyl diphosphate isomerase gene enhances isoprenoid biosynthesis in *Escherichia coli*. *Biochem. J.* 324: 421–426.

Kappers I.F., Aharoni A., Vanherpen T.W.J.M., Luckerhoff L.L.P., Dick, M., and Bouwmeester, H.J., 2005. Genetic engineering of terpenoid metabolism attracts body-guards to Arabidopsis. *Science.* 309: 2070–2072.

Kasahara H., Jiao Y., Bedgar D.L., Kim S.J., Patten A.M., Xia Z.Q., Davin L.B., et al, 2006. Pinus taeda phenylpropenal double-bond reductase: purification, cDNA clon-ing, heterologous expression in Escheria Ccoli, and subcellular localization in P. taeda. *Phytochem.* 67: 1765–1780.

Kato N., Dubouzet E., Kokabu Y., Yoshida S., Taniguchi Y., and Dubouzet J.G., 2007. Identification of a WRKY protein as a transcriptional regulator of benzylisoquinoline alkaloid biosynthesis in *Coptis japonica*. *Plant Cell Physiol.* 48: 8–18.

Kim S.W. and Keasling J.D., 2001. Metabolic engineering of the nonmevalonate isopente-nyl diphosphate synthesis pathway in *Escherichia coli* enhances lycopene production. *Biotechnol. Bioeng.* 72: 408–415.

Kindermans J.M., Pilloy J., Olliaro P., and Gomes M., 2007. Ensuring sustained ACT produc-tion and reliable artemisinin supply. *Malar. J.* 6: 1–6.

Korth K.L., Jaggard D.A.W., and Dixon R.A., 2000. Development and light regulated post translational control of 3-hydroxy 3-methylglutary coenzyme A reductase levels in potato. *The Plant J.* 23: 507–516.

Kress W.J., Wurdack K.J., Zimmer E.A., Weigt L.A., and Janzen D.H., 2005. Use of DNA barcodes to identify flowering plants. *Proc. Natl. Acad. Sci. USA.* 102: 8369–8374.

Krishna S., Uhlemann A.C., and Haynes R.K., 2004. Artemisinins: Mechanisms of action and potential for resistance. *Drug Resist. Updat.* 7: 233–244.

Krushkal J., Pistilli M., Ferrell K.M., Souret F.F., and Weathers P.J. 2003. Computational analysis of the evolution of the structure and function of 1-deoxy-D-xylulose-5-phos-phate synthase, a key regulation of the mevalonate-independent pathway in plants. *Gene* 313: 127–138.

Kudakasseril G.J., Lukem L., and Stabam E.J., 1987. Effect of sterol inhibitors on incorpo-ration of ^{14}C-isopentenyl pyrophosphate into artemisinin by a cell free system from *Artemisia annua* tissue culture and plants. *Planta Med.* 53: 280–284.

Kumar S., Khanuja S.P.S., Shasany A.K., and Darokar M.P., 1999. "Jeevan Raksha" from an isolated population containing high artemisinin in foliage (0.5%–1.0%). *J. Med. Arom. Plant Sic.* 21: 47–48.

Laughlin J.C., l994. Agricultural production of artemisinin: A review. *Trans Royal Soc. Trop. Med. Hyg*. 88 (Suppl.1): 21–22.

Lindahl A.L., Olsson M.E., Mercke P., Tollbom O., Schelin J., Brodelius M., and Brodelius, P.E., 2006. Production of the artemisinin precursor amorpha- 4,11-diene by engineered *Saccharomyces cerevisiae*. *Biotechnol. Lett*. 28: 571–580.

Liu B.Y., Ye H.C., and Li G.F., 1998a. Studies on dynamic of growth and artemisinin biosynthesis of hairy root of *Artemisia annua* L. *Chin. J. Biotechnol*. 14: 401–404.

Liu C., Zhao Y., and Wang Y., 2006. Artemisinin: Current state and perspectives for biotechnological production of an antimalarial drug. *Appl. Microbiol. Biotechnol*. 72: 11–20.

Liu C.Z., Wang Y.C., Guo C., Ouyang F., Ye H.C., and Li G.F., 1998b. Enhanced production of artemisinin by *Artemisia annua* L. hairy root cultures in a modified inner-loop airlift bioreactor. *Biopr. Eng*. 19: 389–392.

Liu Q., Singh S.P., and Green A.G., 2002. High-steric and high-oleic cotton seed oils produced by hairpin RNA-mediated post-transcriptional gene silencing. *Plant Physiol*. 129: 1732–1743.

Liu Y., Wang H., Ye H.C., and Li G.F., 2005. Advances in the plant isoprenoid biosynthesis pathway and its metabolic engineering. *J. Integr. Plant Biol*. 47: 769–782.

Liu Y., Ye H.C., Wang H., and Li G.F., 2003. Molecular cloning, *Escherichia coli* expression and genomic organization of squalene synthase gene from *Artemisia annua*. *Acta Bot. Sin*. 45: 608–613.

Lulu Y., Chang Z., Ying H., Ruiyi Y., and Qingping Z., 2008. Abiotic stress-induced expression of artemisinin biosynthesis genes in *Artemisia annua* L. *Chin. J. Appl. Environ. Biol*. 14: 1–5.

Luo X.D. and Shen C.C., 1987. The chemistry, pharmacology and clinical application of (qinghaosu) artemisinin and its derivatives. *Med. Res. Rev*. 7(1): 29–52.

Ma D., Pu G., Lei C., Ma L., Wang H., Guo Y., Chen J., Du Z., et al. 2009. Isolation and characterization of AaWRKY1, an *Artemisia annua* transcription factor that regulate the Amorpha-4,11-diene synthase gene, a key gene of artemisinin biosynthesis. *Plant Cell. Physiol*. 50(12): 1246–2161.

Martin V.J.J., Pitera D.J., Withers S.T., Newman J.D., and Keasling J.D., 2003. Engineering a mevalonate pathway in *Escherichia coli* for production of terpenoids. *Nat. Biotechnol*. 21(7): 796–802.

Matsushita Y., Kang W.Y., Charlwood B.V., 1996. Cloning and analysis of a cDNA encoding farnesyl diphosphate synthase from *Artemisia annua*. *Gene* 172: 207–209.

Mercke P., Bengtsson M., Bouwmeester H.J., Posthumus M.A., and Brodelius P.E., 2000. Molecular cloning, expression and characterization of amorpha-4,11-diene synthase, a key enzyme of artemisinin biosynthesis in *Artemisia annua* L. *Arch. Biochem. Biophys*. 381(2): 173–180.

Mercke P., Crock J., Croteau R., and Brodelius P.E., 1999. Cloning, expression, and characterization of epi-cedrol synthase, a sesquiterpene cyclase from *Artemisia annua* L. *Arch. Biochem. Biophys*. 369: 213–222.

Mohapatra P.K., Khan A.M., Prakash A., Mahanta J., and Srivastava V.K., 1996. Effect of arteether alpha/beta on uncomplicated falciparum malaria cases in Upper Assam. *Ind. J. Med. Res*. 104: 284–287.

Moore J.C., Lai H., Li J.R., Ren R.L., McDougall J.A., Singh N.P., and Chou C.K., 1995. Oral administration of dihydroartemisinin and ferrous sulphate retarded implanted fibrosarcoma growth in the rat. *Canc. Lett*. 98(1): 83–87.

Narusaka Y., Narusaka M., Seki M., Umezawa T., Ishida J., Nakajima M., 2004. Crosstalk in the responses to abiotic and biotic stresses in Arabidopsis: analysis of gene expression in cytochrome P450 gene superfamily by cDNA microarray. *Plant Mol. Biol*. 55(3): 327–342.

Newman J.D., and Chappell. J., 1999. Isoprenoid biosynthesis in plants: carbon partitioning within the cytoplasmic pathway. *Crit. Rev. Biochem. Mol. Biol.* 34: 95–106.

Newman J.D., Marshall J., Chang M.C.Y., Nowroozi F., Paradise E., Pitera D., Newman K.L., et al. 2006. High-level production of amorpha-4,11-diene in a two phase partitioning bioreactor of metabolically engineered *Escherichia coli. Biotechnol. Bioeng.* 95(4): 684–691.

Newton P. and White N., 1999. Malaria: New development in treatment and prevention. *Ann. Rev. Med.* 50: 179–192.

Noedl H., Se Y., Socheat D., and Fukuda M.M., 2008. Evidence of artemisinin-resistant malaria in Western Cambodia. *N. Eng. J. Med.* 359: 2619–2620.

Ogita S., Uefuji H., Morimoto M., and Sano H., 2004. Application of RNAi to confirm theobromine as the major intermediate for caffeine biosynthesis in coffee plants with potential for construction of decaffeinated varieties. *Plant Mol. Biol.* 54(6): 931–941.

Ohlrogge J. 1999. Plant metabolic engineering: Are we ready for phase two? *Curr. Opin. Plant Biol.* 2(2): 121–122.

Paddon C.J., Westfall P.J., Pitera D.J., Benjamin K., Fisher K., and McPhee D. 2013. High-level semi-synthetic production of the potent antimalarial artemisinin. *Nature.* 496: 528–532.

Ping Z.Q., Chang Z., Lulu Y., Ruiyi Y., Xiao M.Z., Ying H., Li F., and XueQin Y., 2008. Cloning of artemisinin biosynthetic cDNAs and novel EST and quantification of low temperature-induced gene overexpression. *Sci. China Ser. C-Life Sci.* 51(3): 232–244.

Qin M.B., Li G.Z., Yun Y., Ye H.C., and Li G.F., 1994. Induction of hairy root from *Artemisia annua* with *Agrobacterium rhizogenes* and its culture in vitro. *Acta Bot. Sin.* 36: 165–170.

Ram M., Khan M.A., Jha P., Khan S., Kiran U., Ahmad M.M., Javed S., and Abdin M.Z., 2010. HMG-CoA reductase limits artemisinin biosynthesis and accumulation in *Artemisia annua* L. *Plants. Acta Physiol. Plant.* 32(5): 859–866.

Riley E.M., 1995. The London School of Tropical Medicine: A new century of malarial research. *Mem. Inst. Oswaldo Cruz* 95: 25–32.

Ro D.K., Paradise E.M., Ouellet M., Fisher K.J., Newman K.L., Ndungu J.M., Ho K.A. et al., 2006. Production of the antimalarial drug precursor artemisinic acid in engineered yeast. *Nature.* 440: 940–943.

Romero M.R., Effeth T., Serrano M.A., Castano B., Macias R.I., Briz O., and Martin J.J., 2005. Effect of artemisinin/artesunate as inhibitors of hepatitis B virus production in an "in vitro" system. *Antiviral Res.* 68(2): 75–83.

Rontein D., Onillon S., Herbette G., Lesot A., Werck-Reichhart D., Sallaud C., and Tissier A., 2008. CYP725A4 from yew catalyzes complex structural rearrangement of taxa-4(5),11(12)-diene into the cyclic ether 5(12)-oxa-3(11)-cyclotaxane. *J. Biol. Chem.* 283(10): 6067–6075.

Sa G., Mi M., He-chun Y., Ben-ye L., Guo-feng L., and Kang C., 2001. Effect of ipt gene expression on the physiological and chemical characteristices of *Artemisia annua* L. *Plant Sci.* 160: 691–698.

Sen R., Bandyopadhyay S., Dutta A., Mandal G., Ganguly S., and Saha P., 2007. Artemisinin triggers induction of cell-cycle arrest and apoptosis in *Leishmania dovani* promastigotes. *J. Med. Microbiol.* 56(Pt 9): 1213–1218.

Singh N.P. and Lai H., 2001. Selective toxicity of dehydroartemisinin and holotransferrin on human breast cancer cells. *Life Sci.* 70(1): 49–56.

Souret F.F., Kim Y., Wyslouzil B.E., Wobbe K.K., and Weathers P.J., 2003. Scale-up of *Artemisia annua* L. hairy root cultures produces complex patterns of terpenoid gene expression. *Biotechnol. Bioeng.* 83: 653–667.

Souret F.F., Weathers P.J., and Wobbe K.K., 2002. The mevalonate independent pathway is expressed in transformed roots of *Artemisia annua* and regulated by light and culture age. *In Vitro cell. Dev. Biol. Plant* 38(6): 581–588.

Sy L.K., and Brown G.D., 2002. The mechanism of the spontaneous autoxidation of dihydro-artemisinic acid. *Tetrahedron* 58(5): 897–908.

Tazyeen N., Akmal M., Ram M., Alam P., Ahlawat S., Anis M., and Abdin M.Z., 2010. Enhancement of artemisinin content by constitutive expression of HMG-CoA reductase gene in high yielding strain of *Artemisia annua* L. DOI:10.1007/s11816-010-0156-x.

Teoh K.H., Polichuk D.R., Reed D.W., Nowak G., and Covello P.S., 2006. *Artemisia annua* L. (Asteraceae) trichome-specific cDNAs reveal CYP71AV1, a cytochrome P450 with a key role in the biosynthesis of the antimalarial sesquiterpene lactone artemisinin. *FEBS Lett.* 580(5): 1411–1416.

Towler M.J., and Weathers P.J., 2007. Evidence of artemisinin production from IPP stemming from both the mevalonate and the nonmevalonate pathways. *Plant Cell Rep.* 26(12): 2129–2136.

Tsuruta H., Paddon C.J., Eng D., Lenihan J.R., Horning T., Anthony L.C., Regentin R., et al. 2009. High-level production of amorpha-4,11-diene, a precursor of the antimalarial agent artemisinin, in *Escherichia coli. PLoS One* 4(2): e4489. doi:10.1371/journal.pone.0004489.

Urban P., Mignotte C., Kazmaier M., Delorme F., and Pompon D. 1997. Cloning, yeast expression, and characterization of the coupling of two distantly related *Arabidopsis thaliana* NADPH-cytochrome P450 reductases with P450 CYP73A5. *J. Biol. Chem.*, 272(31): 19176–19186.

Utzinger J., Xiao S., N'Goran E.K., Berquist R., and Tanner M., 2001. The potential of arte-mether for the control of shistosomiasis. *Int. J. Parasitol.* 31(14): 1549–1562.

Van Agtmael M.A., Eggetle T.A., and Van Boxtel C.J., 1999. Artemisinin drugs in the treat-ment of malaria: From medicinal herb to registered medication. *Trends Pharmacol. Sci.* 20(5): 199–204.

Van Geldre E., De Pauw I., Inze D., Van Montagu M., and Van den Eeckhout E., 2000. Cloning and molecular analysis of two new sesquiterpene cyclises from *Artemisia annua* L. *Plant Sci.* 158: 163–171.

Van der Fits L., Memelink J., 2000. ORCA3, a jasmonate responsive transcriptional regulator of plant primary and secondary metabolism. *Science.* 289(5477): 295–297.

Vergauwe A., Cammaert R., Vandenberghe D., Genetello C., van Montagu M., and Vanden Eeckhout E., 1996. *Agrobacterium tumefaciens*-mediated transformation of *Artemisia annua* L. and regeneration of transgenic plant. *Plant Cell Rep.* 15: 929–937.

Vergauwe A., Van G.E., Inze D., and Van D.E.E., 1998. Factor influencing *Agrobacterium tumefaciens*-mediated transformation of *Artemisia annua* L. *Plant Cell Rep.* 18: 105–110.

Verpoorte R. and Memelink J., 2002. Engineering secondary metabolite production in plants. *Curr. Opin. Biotechnol.* 13(2): 181–187.

Wallaart T.E., van Uden W., Lubberink H.G.M., Woerdenbag H.J., Pras N., and Quax W.J., 1999b. Isolation and identification of dihydroartemisinic acid from *Artemisia annua* and its possible role in the biosynthesis of artemisinin. *J. Nat. Prod.* 62(3): 40–42.

Wang H., Nagegowda D.A., Rawat R., Bouvier-Nave P., Guo D., Bach T.J., and Chye M.L., 2012. Overexpression of *Brassica juncea* wild type and mutant HMG-CoA synthase 1 in *Arabidopsis* up- regulates genes in sterol biosynthesis and enhance sterol production and stress tolerance. *Plant Biotechnol. J.* 10(1): 31–42.

Weathers P.J., Bunk G., and Mccoy M.C., 2005. The effect of phytohormones on growth and artemisinin production in *Artemisia annua* hairy roots. *In Vitro Cell. Dev. Biol. Plant* 41(1): 47–53.

Weathers P.J., Cheetham R.D., Follansbee E., and Tesh K., 1994. Artemisinin production by transformed roots of *Artemisia annua*. *Biotech. Lett.* 16(12): 1281–1286.

Weathers P., Elkholy S., and WOBBE K.K., 2006. Invited review: Artemisinin: The biosynthetic pathway and its regulation in *Artemisia annua*, a terpenoid-rich species. *In Vitro Cell. Dev. Biol.—Plant* 42(4): 309–317.

WHO 2001. Antimalarial drug combination therapy: Report of a WHO technical consultation. WHO/CDS/ RBM/2001/35, reiterated in 2003.

Woerdenbag H.J., Luers J.F.J., van Uden W., Pras N., Malingre T.H., and Alfermann A.W., 1993. Production of the new antimalarial drug artemisin in shoot cultures of *Artemisia annua* L. plant cell. *Plant Cell Tissue Organ Cult.* 32(2): 247–257.

Woerdenbag H.J., Moskal T.A., Pras N., Malingre T.M., el-Feraly F.S., Kampinga H.H., and Konings A.W., 1993. Cytotoxicity of artemisininrelated endoperoxides to Ehrlich ascites tumor cells. *J. Nat. Prod.* 56: 849–856.

World Bank 2003. Expert consultation on the procurement & financing of antimalarial Drugs. Production of the new antimalarial drug artemisin in shoot cultures of *Artemisia annua* L. plant cell. Meeting Report, Draft 3, 7 November 2003. Washington, DC: World Bank.

Xiang L., Zeng L., Yuan Y., Chen M., Wang F., Liu X., Zeng L., et al. 2012. Enhancement of artemisinin biosynthesis by overexpressing dxr, cyp71av1 and cpr in the plants of *Artemisia annua* L. *Plant Omics J.* 5(6): 503–507.

Xie D.Y., Zou Z.R., Ye H.C., Li G.F., and Guo Z.C., 2001. Selection of hairy root clones of *Artemisia annua* L. for artemisinin production. *Isr. J. Plant Sci.* 49(2): 129–134.

Xu Y.H., Wang J.W., Wang S., Wang J.Y., and Chen X.Y., 2004. Characterization of GaWRKY1, a cotton transcription factor that regulates the sesquiterpene synthase gene (+)-delta-cadinene synthase-A. *Plant Physiol.* 135(1): 507–515.

Zeng Q.P., Qiu F., Yuan L., 2007. Production of artemisinin by genetically modified microbes. *Biotech. Let.* DOI 10.1007/s10529-007-9596-y.

Zhang L., Jing F., Li F., Li M., Wang Y., Wang G., Sun X., et al. 2009. Development of transgenic *Artemisia annua* (Chinese wormwood) plants with an enhanced content of artemisinin, an effective anti-malarial drug, by hairpin-RNA mediated gene silencing. *Biotechnol. Appl. Biochem.* 52(Pt 3): 199–207.

Zhang Y., Nowak G., Reed D.W., and Covello P.S. 2011. The production of artemisinin precursors in tobacco. *Plant Biotechnol. J.* 9(4): 445–454.

Zhang Y., Teoh K.H., Reed D.W., Maes L., Goossens A., Olson D.J. H., and Covello P.S. 2008. The molecular cloning of artemisinic aldehyde delta 11(13) reductase and its role in glandular trichome-dependent biosynthesis of artemisinin in *Artemisia annua*. *J. Biol. Chem.* 283(31): 21501–21508.

Zhang Y.S., Liu B. Y., Li Z.Q., Ye H.C., Wang H.L., and Guo-Feng H.J.L., 2004. Molecular cloning of a classical plant peroxidase from *Artemisia annua* and its effect on the biosynthesis of artemisinin in vitro. *Acta. Bot. Sin.* 46: 1338–1346.

CHAPTER 12

Use of Nanocarriers to Enhance Artemisinin Activity

Anna Rita Bilia

CONTENTS

12.1 INTRODUCTION

By 1972, researchers at the Beijing Pharmaceutical Institute had identified the active ingredient of *Artemisia annua* L. It was called *qinghaosu* ("the active principle from qinghao") and is nowadays known as *artemisinin* (van Agtmael et al., 1999). Chemically, artemisinin (1) is a sesquiterpene trioxane lactone containing an endoperoxide bridge that is essential for its activity (Figure 12.1).

Immediately after its discovery, clinical evaluations focused on the pure isolated molecule and the efficacy of artemisinin against malaria was amply demonstrated. The first clinical trials of artemisinins began in China in 1975, and large-scale studies of these invaluable antimalarials continued more than a quarter of a century later (Haynes and Krishna, 2004). Several interesting features characterized the new molecule: high efficacy, low toxicity, no resistance, rapidity of action, and first-order pharmacokinetics (de Vries and Dien, 1996) but a very limiting aspect of this molecule is its poor bioavailability.

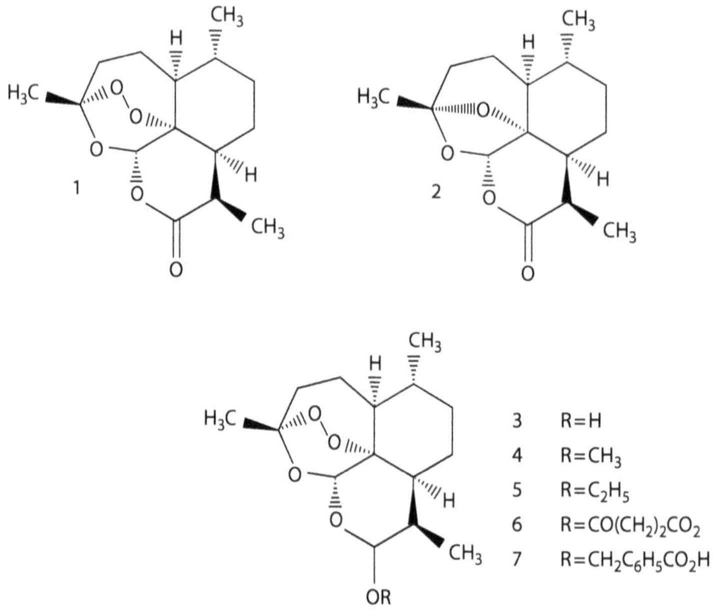

Figure 12.1 Structures of artemisinin (1), deoxyartemisinin (2), dihydroartemisinin (3), artemether (4), arteether (5), artesunic acid (6), artelinic acid (7).

Accordingly, when the structure of artemisinin was established, work started at once on the study of the relationship between the molecule's chemical structure and its antimalarial activity. At first, deoxyartemisinin (2) (Figure 12.1) was found to be inactive (Li et al., 1981). The results clearly showed that the peroxyl group was necessary for antimalarial activity. Derivatization of artemisinin was carried out in order to improve some of the parent's properties. During the studies, it was also discovered that the lactone of artemisinin could be reduced by sodium borohydride to yield dihydroartemisinin (DHA) (3) (Figure 12.1), which has even more antimalarial activity *in vitro* than artemisinin itself (Li et al., 1981; van Agtmael et al., 1999). However, the solubility problem remained. Consequently, three series of DHA derivatives were designed and synthesized: ethers, esters, and carbonates. In early 1977, 25 derivatives were tested against chloroquine-resistant strains of *Plasmodium berghei* in mice. The majority of these were proven to be much more active than artemisinin (Gu et al., 1980). In the ether series (lipid-soluble relatives), artemether (4) and arteether (5) were the most stable (Figure 12.1).

During the same period, a water-soluble derivative, the hemi-succinate of DHA (artesunic acid, compound (6), Figure 12.1), was also synthesized with significant pharmacokinetic advantages over its lipid-soluble relatives (Liu, 1980). More recently, another water-soluble compound, artelinic acid (7) has been synthetized (see Figure 12.1) (Balint, 2001).

In 1986 and 1987, after successful large-scale clinical trials, artemisinin, artemether, and artesunate were approved as new antimalarial drugs by the

Chinese authorities; DHA was approved in 1991 (Li and Wu, 1998). Their outstanding antimalarial efficacy has also been confirmed in clinical trials around the world. It is worth mentioning that research related to artemisinin and its derivatives has been supported by the World Health Organization (WHO) since 1979.

Artemether and artesunate are now used as oral formulations and the latter is also available in suppository form. It is WHO policy to promote the use of these drugs intra-rectally as an emergency treatment in primary healthcare situations in developing countries. Delay in treatment, often due to considerable distance to the hospital, is regarded as a significant factor in the high mortality among children. This delay would be substantially reduced if the mother could administer a suppository to her ill child at home (van Agtmael et al., 1999).

DHA, the lactol form of artemisinin, and the ethers and esters obtained from DHA, such as artemether, artheether, artesunic acid, and artenilic acid, have yielded superior results than artemisinin but they still suffer from the same chemical, bio-pharmaceutical, and treatment issues (Table 12.1).

In the last 25 years, hundreds of new ether and ester derivatives have been synthetized and tested against malaria and other illnesses including tumors. In addition, various lactam derivatives, 12 β-allyl deoxo derivatives and modifications at the C -9 position of artemisinin, modifications of B or C rings, ring-contracted derivatives, dimers, and trimers have also been reported. However, only a few molecules are actually in preclinical or clinical studies because they have demonstrated clear superiority to artemisinin and its existing derivatives (Jung et al., 2004; Sriram et al., 2004).

12.2 HOW TO INCREASE ARTEMISININ BIOAVAILABILITY: THE PHARMACEUTICAL TECHNOLOGY APPROACH

Key features of artemisinin and its derivatives are the rapid onset of action, killing of all asexual stages of the parasite *Plasmodium falciparum*, efficacy against severe malaria, and a rapid clearance rate (probably the reason for the slow development of artemisinin resistance, but probably also the reason for the frequent recurrence of infections). By contrast, artemisinin has been reported as having low bioavailability and, in recent decades, studies to enhance this property have increased exponentially in number. Bioavailability is defined by the European Medicines Agency as "the rate and extent to which the active ingredient or active moiety is absorbed from a drug product and becomes available at the site of action." Bioavailability is usually assessed by determining the area under the curve (AUC; plasma drug concentration–time relationship after administration of a hypothetical drug). After intravenous administration, it is assumed that the given dose of the drug is 100% bioavailable, since the drug is introduced directly into the systemic circulation. All other forms of systemic administrations, including oral, intramuscular, and subcutaneous routes, generally present a bioavailability of less than 100% (CPMP, 2010). In particular, after oral administration, the solubility, dissolution rate, gastrointestinal permeability,

Table 12.1   Some Pharmacokinetic Data of Artemisinin and Its Derivatives

Drug	Absorption	Distribution volume (L/kg)	Eliminationhalf-life (h)	Bioavailability(%)	Peak plasma concentration(h)	First-passeffect
Artemisinin	Rapid and incomplete		2–5	8–10	<2	Extensive
Artemether	Rapid and incomplete	0.67	3–11	54	3	Extensive
Arteether		0.72	>20	34		
Artesunate	Rapid and incomplete	0.14	<1	82		Extensive
Dihydroartemisinin		0.90	3.1	85	0.65	

first-pass metabolism, and susceptibility to efflux mechanisms are fundamental parameters that control the rate and extent of drug absorption and its bioavailability.

Formulation factors and technological processes can strongly affect drug dissolution or release from the dosage form, drug absorption and stability (at the site of administration), or metabolic processes, resulting in the modification of bioavailability and bioequivalence of formulated drugs (Bilia et al., 2017).

Up until 1990, thousands of patients were successfully treated with artemisinin and its derivatives (Qinghaosu Antimalaria Coordinating Research Group, 1979; Luo and Shen, 1987; Myint et al., 1989) but reliable pharmacokinetic data from humans were not available. A few pharmacokinetic studies in animals have been carried out (Niu et al., 1985; Zhao et al., 1986). The first report on use in humans was the study of Zhao (1987). He described the use of HPLC assay for the analysis of human plasma and saliva after administration of a capsule or a suppository of artemisinin to a volunteer. The first study of pharmacokinetics in humans was published in 1990 by Titulaer and coworkers (Titulaer et al., 1990). The authors used the HPLC assay reported by Zhao (1987) to evaluate the pharmacokinetics after oral, intramuscular, and rectal administration of artemisinin to healthy volunteers (n = 10, 21–30 years of age). The study was set up as a four-way cross-over design with a wash-out period of 1 week between the test days. The data obtained highlighted that artemisinin had been dosed most frequently by intramuscular injection. The suspension in oil resulted in rapid absorption and relatively high peak concentrations. An aqueous suspension resulted in a more sustained release with a slow onset and low concentrations. Artemisinin is completely absorbed, in a strict sense, after oral administration, but a high first-pass metabolism results in relatively low bioavailability. Absorption after rectal administration as an aqueous suspension or in suppositories is incomplete. Artemisinin is rapidly distributed through the tissues, passing the erythrocyte wall, the blood–brain barrier, the salivary glands, and the placenta. The relatively high recrudescence rate is probably dependent on the formulation characteristics, and might be reduced by better products and better dosage schemes.

The relative bioavailability in comparison with an intramuscular injection of a suspension in oil is estimated to be about 32% (Navaratnam et al., 2000). A study carried out by Ashton et al. (1998) showed high interindividual variability in plasma concentrations after both oral and rectal administration. The low bioavailability after oral intake can be due to (1) low transepithelial transport across the intestinal mucosa or (2) to the poor dissolution characteristics of artemisinin in the intestinal fluids. In any case, due to the facility of administration and improved patient compliance, artemisinin solid oral dosage forms, such as tablets and capsules, represented the most widely used formulations against malaria in China, Vietnam, and Thailand. However, no data have been reported on their comparability. During the development of a new oral dosage form, the pharmaceutical or *in vitro* bioavailability is generally measured for the investigation of drug bioavailability. It is evaluated by dissolution testing, considered to be sensitive, reliable, and rational for predicting *in vivo* drug bioavailability behavior, and is now one of the most important quality control tests performed on drugs and drug products.

In 1996, Ngo et al. (1996) evaluated dissolution profiles as the first rationale in the determination of quality of the different products on the market. Six commercial formulations (M1–M6, three tablets and three capsules) and an innovative one, M7 (developed to give fast dissolution), were investigated.

The dissolution behavior of the different formulations such as tablets, capsules, and M7 proved that drugs with large differences in dissolution characteristics could reach the market. As an example, tablet M4 showed poor disintegration and no released artemisinin could be measured during the first 6 h of the experimental process. Tablet M3 had the best dissolution profile, and after 6 h had released about 45% of its artemisinin. No other formulations achieved a release of more than 30%. These results demonstrated the indispensability of dissolution control in drug production to ensure drug quality.

A further investigation, reported by the same researchers (Ngo et al., 1997a), concerned an interesting study on the influence of some formulation variables on increasing the dissolution of artemisinin oral dosage forms in order to improve the bioavailability. Thus, in the case of tablets and capsules, the excipients added to the formulations and the technological manufacturing processes accounted for the variability in dissolution testing.

The reason for recrudescence after treatment with artemisinin and its derivatives is not known. The observation that the recrudescence rate is higher with tablets than after parenteral administration (Hien and White, 1993; Luo and Shen, 1987) clearly shows that fast and complete dissolution of the active compound from oral dosage forms is one of the most important factors in preventing recrudescence.

In the study by Ngo et al. (1997a), five excipients were selected to develop an innovative formulation with a fast dissolution rate. Lactose, sodium dioctyl sulfosuccinate, sodium starch glycolate, gelatine, and magnesium stearate were chosen for the investigation.

It was shown that sodium dioctyl sulfosuccinate gave the largest positive effect, an amount of magnesium stearate exceeding 1% gave a negative effect, and the effects of the other compounds followed the order: sodium starch glycolate > gelatine > lactose.

The same group of researchers (Ngo et al., 1997b) investigated the bioavailability of three selected formulations in rabbits, in order to verify the correlation between *in vivo* data and data obtained by the *in vitro* dissolution method. The formulations were a commercial product (M2) and two innovative tablets FD4 (60 mg lactose, 20 mg sodium dioctyl sulfosuccinate, 5 mg sodium starch glycolate, 1 mg magnesium stearate and 0.7% gelatine) and FD6 (60 mg lactose, 5 mg sodium dioctyl sulfosuccinate, 25 mg sodium starch glycolate, 1 mg magnesium stearate and 1.8% gelatin).

Differences were observed during the first 60 min. and the concentration of artemisinin in plasma, after reaching the maximum value, rapidly dropped to a level around 50 ng/mL, before slowly decreasing to about 20 ng/mL over 1 week.

The dissolution rate was obtained by using a two-phase partition–dissolution method, and a good correlation between the bioavailability results in rabbits and the dissolution data obtained from the tablets was evident.

Van Nijlen et al. (2003) investigated the improvement of the dissolution rate of artemisinin by the marinization and formation of solid dispersions with PVPK25 in order to improve the intestinal absorption characteristics of artemisinin.

The use of a hydrophilic carrier such as PVPK25 leads to improved wettability of the drug substances, and the formation of solid dispersions with PVPK25 as a carrier appears to be a promising method to improve the intestinal absorption characteristics of artemisinin.

The effect of these two approaches was evaluated in an *in vitro* dissolution system. Micronization by means of supercritical fluid technology resulted in a significant decrease in particle size as compared with untreated artemisinin. All micronized powders appeared to be crystalline. The dissolution rate of the micronized forms was improved in comparison to the untreated form, but showed no difference in comparison to mechanically ground artemisinin. Solid dispersions of artemisinin with PVPK25 as a carrier were prepared by the solvent method. Both powder x-ray diffraction and differential scanning calorimetry showed that the amorphous state was reached when the amount of PVPK25 was increased to 67%. The dissolution rate of solid dispersions containing at least 67% of PVPK25 was significantly improved in comparison to untreated and mechanically ground artemisinin. Improvement of the dissolution rate of artemisinin was obtained by both particle size reduction and formation of solid dispersions. The effect of particle size reduction on the dissolution rate was limited. Solid dispersions could be prepared by using a relatively small amount of PVPK25. The formation of solid dispersions with this carrier appears to be a promising method to improve the intestinal absorption characteristics of artemisinin.

12.3 ARTEMISININ NANOCARRIERS: IMPROVED BIOAVAILABILITY AND EFFICACY

The therapeutic value of artemisinins is limited due to poor water and oil solubility, a low bioavailability after oral administration because of rapid degradation by the liver, and a short (ca. 2.5 h) half-life (Isacchi et al., 2011). Therefore, there is an urgent need to develop new artemisinin formulations to increase its bioavailability, selectivity, and therapeutic application.

A simple and affordable technique to enhance the solubility of a drug is to modify its physical characteristics by reducing particle size and/or modifying its crystal habit. Apart from conventional micronizing techniques, particle technology now deals with various nanoparticle-engineering processes as promising methods to improve drug solubility. Nanopowders in the solid state (either amorphous or crystalline) have a typical size range of 10–1000 nm. Nanoscale materials have far greater surface areas than similar masses of larger-scale materials, resulting in an increased dissolution. Nanopowders, both in the form of nanocrystals and amorphous systems, can, at least theoretically, improve all common drug administration routes (oral, parenteral, transdermal, transmucosal, ocular, pulmonary), controlling the rate and extent of drug absorption and, ultimately, bioavailability (Bilia et al., 2017). Indeed, the formulation of "nanosized" vectors offers additional advantages that are not achievable with simple nanopowders. Nanosized drug delivery systems (10–400 nm diameter) hold great interest in research because of their advantages in modifying pharmacokinetics and biodistribution. They improve the therapeutic index of drugs by increasing their localization to specific tissues, organs, or cells, decreasing potential side effects, and having extreme versatility in the routes of administration that can be used (Bilia et al., 2014, 2017). The size of nanosystems is directly related to the cellular uptake rate (the optimum is 200–300 nm in diameter) and to the time they remain in the blood circulation (those having a diameter less than 10 nm are cleared via glomerular filtration in the kidneys) (Gaumet et al., 2008). These nanosystems can also minimize side effects and sustain drug release over a prolonged period. Nanovectors are generally suitable for parental, oral, mucosal, pulmonary, ocular, dermal, and transdermal delivery. Opportunely designed, they are able to cross physiological barriers; that is, the blood–brain, blood–cerebrospinal fluid, and blood–retinal barriers, and the other barriers of the eye (corneal, tear film, and aqueous). A nanotechnology approach could be attractive for modulation of both drug pharmacokinetics and biodistribution, thus decreasing potential side effects by leaving the normal sensitive cells unharmed (Bilia et al., 2017).

Different types of nanomaterials can be used to prepare nanocarriers capable of being loaded with hydrophobic and hydrophilic drugs. Since the human body contains mostly water, the ability to successfully deliver hydrophobic drugs is a major therapeutic benefit of nanocarriers. Some of these, namely, vesicles, are able to contain either hydrophilic or hydrophobic drugs depending on the orientation of the amphiphilic molecules. Generally, nanocarriers are classified as polymer-based and lipid-based systems (Figure 12.2). Polymeric nanocarriers are made of natural (proteins and polysaccharides), semisynthetic, or synthetic polymers, generally

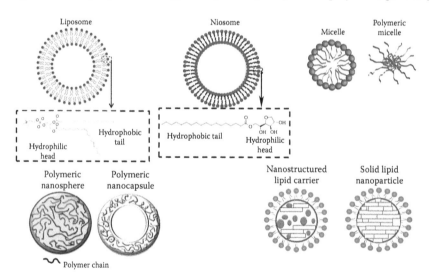

Figure 12.2 Structures of the diverse nanocarriers. (Reproduced from Bilia A.R. et al., *Planta Med*, 83(05), 366–381, 2017. With the permission of Thieme.)

represented by nanospheres or nanocapsules. Nanospheres are matrix systems, in which the drug is physically and uniformly dispersed, while nanocapsules are the system in which the drug is confined to a cavity surrounded by a polymeric membrane. Lipid-based nanocarriers include nanometric-scaled emulsions, namely, micro- and nanoemulsions, vesicles, micelles, and nanoparticles divided roughly into solid lipid nanoparticles (SLNs) and nanostructured lipid carriers (NLCs). They present a lipid core, which makes these carriers suitable for entrapment of lipophilic compounds. Lipids and surfactant agents are generally selected from a plethora of edible constituents and "Generally Recognized as Safe" compounds (Bilia et al., 2014, 2017). Nowadays, promising novel nanoformulations loaded with artemisinins, such as liposomes, micelles and nanoparticles, offer significant promise in improving half-life, controlled release, better permeability, resistance to metabolic processes, and highly specific site-targeted delivery of these therapeutic compounds, thus growing their medicinal value.

12.3.1 Vesicles

Vesicles are aggregates of amphiphilic lipids that organize themselves spontaneously. Vesicles are bilayer structures that are suitable for loading both hydrophilic and hydrophobic compounds (Figure 12.2). Liposomes, the most widely used vesicles, consist of one or more concentric lipid bilayers, separated by an aqueous medium. Lipids are phospholipids (synthetic or natural) plus cholesterol. They are versatile carriers designed to carry both hydrophilic substances (encapsulated in the aqueous compartment) and lipophilic molecules (inserted into the membrane). Liposomes were first proposed as carriers of biologically active substances in 1971, and have since been comprehensively studied. They can be classified as large multilamellar liposomes, small unilamellar vesicles, or large unilamellar vesicles, depending on their size and the number of lipid bilayers. They are very versatile, because water-soluble drugs can be included within the aqueous compartments and lipophilic or amphiphilic compounds can be associated with the lipid bilayers. Niosome is an acronym for "non-ionic surfactant-based" vesicle. Niosomes are structurally similar to liposomes, but are formed mostly by non-ionic surfactant and cholesterol as excipients that make them more stable; thus, they offer many advantages over liposomes (Bilia et al., 2014, 2017).

In 1996, the first approach to using liposomes as formulating agents for artemisinin and its derivatives was reported by Al-Angary et al., who evaluated some liposomes containing artheether (Al-Angary et al., 1996) *in vitro*. The purpose of the work was to study the physicochemical characteristics of different liposomal formulations and the impact of these characteristics on the stability of the formulations.

Formulations were based on the following phospholipids: dimyristoyl phosphatidylcholine (DMPC), dipalmitoyl phosphatidylcholine (DPPC), egg phosphatidylcholine (EPC), or dibenphenoyl phosphatidylcholine (DBPC), alone or in mixtures. Cholesterol was also added to some formulations. The type of phospholipid and the incorporation of cholesterol in the liposomal bilayer were found to influence the artheether-trapping efficiency, liposomal particle size, and drug release rate from the liposomes. The trapping of arteether in liposomal vesicles was increased by increasing the acyl chain length of the phospholipid and by the addition of cholesterol. EPC liposomes exhibited relatively low trapping efficiency, due to high drug adsorption. The trapping efficiency value was 48.45% for DMPC liposomes and 56.22% for liposomes prepared with DPPC, which contains a longer acyl chain. Using the phospholipid EPC, with its higher molecular weight and unsaturated phospholipids, the trapping efficiency was only 46.17%, perhaps due to the greater ability of the arteether molecules to be loosely adsorbed at the water/phospholipid bilayer interface. The release of arteether from the liposomal system was characterized by a fast phase for 2 days, followed by a slower phase. The fast phase was the highest with EPC liposomes, indicating the release of the adsorbed drug. Generally, the increase of the acyl chain length as well as the addition of cholesterol caused a decrease in the arteether release rate.

Based on these previous studies, the same group of researchers (Bayomi et al., 1998) tested *in vivo* a selected liposomal formulation composed of DPPC, DBPC, cholesterol, and arteether in the molar ratio of 1:1:2:1. This composition was found to give stable liposomes compared with other formulations and it gave 67.56% trapping efficiency and a particle size of 3.21 ± 0.76 µm. The liposomes were administered orally and intravenously to New Zealand rabbits at a dose of 50 mg/kg. Pharmacokinetic parameters after the oral administration of liposomes were compared with those after an oral administration of an aqueous suspension of micronized arteether. High bioavailability of arteether was evident in the case of the oral liposomes, where a faster rate and better absorption of arteether were observed compared with the aqueous suspension. Oral liposomes gave higher C_{max} and shorter T_{max}, as well as a higher value for AUC. Almost complete arteether absorption was observed for oral liposomes, where relative bioavailability was 97.91%, compared with 31.83% for the oral suspension.

Intersubject variations were found to be relatively high for oral liposomes. The obtained values for mean residence time (MRT) and mean absorption time (MAT) indicated that arteether remained in the gastrointestinal tract longer, and was absorbed over a longer period, in the case of the suspension compared with the liposomal formulation. In addition, arteether was successfully administered intravenously in liposomal formulations and showed a longer elimination half-life with respect to other artemisinin derivatives. Thus, an optimum oral liposomal formulation for arteether can be developed for fast and complete absorption of the drug from the gastrointestinal tract. Furthermore, liposomal formulation of arteether could allow for intravenous administration of the drug in high-risk malaria patients for a long duration of effect.

In conclusion, arteether liposomes were successfully prepared for oral and intravenous administration. Oral administration of liposomes showed high bioavailability of arteether compared with poor bioavailability of the aqueous suspension, as demonstrated by pharmacokinetic parameters. These liposomal formulations can allow the effective use of the potent antimalarial arteether in high-risk malarial patients.

Another *in vivo* study was reported by Chimanuka et al. (2002). β-Artemether liposomes based on egg phosphatidylcholine-cholesterol liposome formulations were prepared, analyzed for their encapsulating capacity, chemical stability, leakage, *in vitro* release, and their therapeutic efficiency against *Plasmodium chabaudi* infection in mice. This liposomal formulation containing β-artemether has also been evaluated for its therapeutic efficacy against the recrudescence of the virulent rodent malaria parasite strain *P. chabaudi chabaudi* (*P. c. chabaudi*) IP-PC1 in mice. The study demonstrated that liposomes with multilamellar vesicles composed of egg phosphatidylcholine/cholesterol in a 4:3 molar ratio could incorporate as much as 1.5 mg of β-artemether without any crystal formation. Good stability of these liposomes (3 months' storage) and a trapping efficiency of nearly 100% were demonstrated, and this formulation was successfully used to circumvent the recrudescent parasitemia in mice infected with *P. c. chabaudi*.

Other researchers (Gabriëls and Plaizier-Vercammen, 2003) investigated the physical and chemical evaluation of liposomes containing artesunate. Artesunate has a rapid onset of therapeutic effect, but suffers from very quick elimination, and as a consequence, frequent administration is required. Therefore, slow-release preparations, as can be developed with liposomal suspensions, seemed to be a logical approach in artesunate monotherapies, especially for parenteral administration. The aim of this study was to develop sterile liposomes and evaluate their chemical/physical stability, including chemical degradation and crystallization of artesunate, and release capacities. The maximal encapsulation degree of artesunate without crystals was 1.5 mg in 300 mg lipids per milliliter of suspension, containing egg phosphatidylcholine/cholesterol in a molar ratio of 4:3. The highest stability of artesunate was obtained with a phosphate buffer of pH 5, which could be expected, as artesunate is almost totally encapsulated. However, due to its instability in water, the suspension containing artesunate 1 mg/mL was preferred, as the encapsulation efficiency was 100%, and the *in vitro* release test proved that artesunate was reversibly encapsulated in liposomes.

Liposomes based on artesunate were also evaluated for their cytotoxic activity against HepG-2 cells. IC_{50} values of artesunate liposomes and artesunate were approximately 16 and 20 μg/mL, respectively. Additionally, IC_{50} values of the same drugs against L-O2 normal human liver cells were approximately 100 and 106 μg/mL, respectively. The tumor growth inhibitory effect of artesunate nanoliposomes was 32.7%, while that of artesunate was 20.5%. HepG-2 cells treated with artesunate liposomes showed dose-dependent apoptosis. The antitumor effect of artesunate liposomes on human hepatoma HepG-2 cells was stronger than that of artesunate at the same concentration (Jin et al., 2013).

Artemisinin-loaded conventional liposomes and polyethylene glycol (PEGylated) liposomes were also developed. Both liposomal formulations showed more than 70% encapsulation efficacy with a mean diameter of approximately 130–140 nm. The polydispersity index of the formulations ranged from 0.2 to 0.3, and they were therefore suitable for intraperitoneal administration. The pharmacokinetic profiles and main pharmacokinetic parameters of the liposomes were evaluated in healthy mice. Free artemisinin was rapidly cleared from plasma and hardly detectable 1 h after administration. Conversely, both liposomal formulations showed much longer blood circulation time than free artemisinin; artemisinin was still detectable after 3 and 24 h of administration for conventional and PEGylated liposomes, respectively. The AUC (0–24 h) values were increased approximately six-fold in both liposomal formulations, in comparison with free artemisinin. A strong effect of formulation on the half-life of artemisinin was enhanced more than fivefold by the incorporation of PEG into liposomes. Liposomes loaded with artemisinin, especially the long-circulating vesicles, could represent a genuinely new strategy for developing smart, well-tolerated, and efficacious therapeutic nanocarriers to treat tumors, but could also be very useful to treat parasitic disease (Isacchi et al., 2011).

The antimalarial efficacies of the developed novel liposomal delivery systems based on artemisinin or artemisinin-based combination therapy with curcumin have been investigated and reported in a further study. The *in vivo* activity was tested in *P. berghei* NK-65 infected mice, a suitable model for studying malaria because the infection presents structural, physiological and life cycle analogies with the human disease. Artemisinin, alone or in combination with curcumin, was encapsulated in conventional and PEGylated liposomes and its *in vivo* performance was assessed by comparison with the free drug. Mice were treated with artemisinin at the dosage of 50 mg/kg/day alone or with curcumin as a partner drug, administered at the dosage of 100 mg/kg/day. Non-formulated artemisinin began to decrease parasitemia levels only 7 days after the start of the treatment, and it appeared to have a fluctuant trend in blood concentration that is reflected in its antimalarial effectiveness. In contrast, treatments with artemisinin-loaded conventional liposomes (A-CL), artemisinin-curcumin-loaded conventional liposomes (AC-CL), artemisinin-loaded PEGylated liposomes (A-PL), and artemisinin-curcumin-loaded PEGylated liposomes (AC-PL) appeared to have an immediate antimalarial effect. Both nanoencapsulated artemisinin and artemisinin plus curcumin formulations cured all malaria-infected mice within the same post-inoculation period of time (Figure 12.3). Additionally, all formulations showed less variability in artemisinin plasma concentrations, which suggested that A-CL, AC-CL, A-PL, and AC-PL give a modified release of drug(s) and, as a consequence, a constant antimalarial effect over time. In particular, A-PL seems

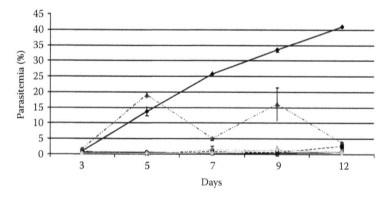

Figure 12.3 Parasitemia progression for mice groups infected with *P. berghei*. (Reproduced from Isacchi B. et al., *Eur J Pharm Biopharm.*, 80(3), 528–534, 2012. With the permission of Elsevier.)

to give the most pronounced and statistically significant therapeutic effect in this murine model of malaria. The enhanced retention in blood of A-PL suggests the use of these nanosystems as suitable passive targeted carriers for parasitic infections; this strong effect of formulation is added to the mechanism of action of artemisinin, which acts in the erythrocyte cycle stage of the human host as a blood schizonticide (Isacchi et al., 2012).

Conventional and PEGylated liposomes loaded with DHA have been also developed and tested for their cytotoxic effects against the MCF-7 cell line. Both developed formulations show desirable physical characteristics as drug carriers for parental administration, and good values of encapsulation efficiency (71% for conventional liposomes and 69% for PEGylated liposomes). Their physical and chemical stabilities were evaluated under storage conditions and in the presence of albumin. The cellular uptake efficiency of liposomes was determined by flow cytometry. Higher internalization occurred in the conventional liposomes than in the stealth liposomes, suggesting that the hydrophilic steric barrier of PEG molecules can reduce cellular uptake. Flow cytometry analysis was also used as an alternative technique for rapid size determination of liposomes. Cytotoxicity studies in the MCF-7 cell line confirmed the absence of toxicity in blank formulations, suggesting liposomes may be a suitable carrier for the delivery of DHA, avoiding the use of organic solvents. The cytotoxicity of DHA and of both liposomal formulations was evaluated in the same cell line, confirming a modified release of DHA from vesicles after cellular uptake (Righeschi et al., 2014).

The same group of researchers investigated the antitumor properties of artemisinin using a novel actively targeted nanocarrier, a PEGylated nanoliposome decorated with transferrin. Transferrin receptors are largely expressed in cancer cells, where the iron content is higher than in normal cells. Artemisinin-loaded liposomes were investigated for their cellular uptake and cytotoxicity properties (Figure 12.4) using the HCT-8 cell line, which was selected from several cell lines because of its transferrin receptor overexpression. The results confirmed the enhanced delivery of artemisinin loaded in liposomes actively targeted with transferrin in comparison with the delivery of other artemisinin-loaded liposomes, as well as improved cytotoxicity (Leto et al., 2016).

Recently, paclitaxel plus artemether liposomes functionalized with a mannose–vitamin E derivative conjugate (MAN–TPGS1000) and a dequalinium–lipid derivative conjugate (DQA–PEG2000–DSPE) were developed and tested on brain glioma cells *in vitro* and in brain glioma-bearing rats. The functional targeting was intended to transport drugs across the blood–brain barrier, destroying vasculogenic mimicry (VM) channels, and eliminating cancer stem cells and cancer cells in the brain. The transport mechanism across the blood–brain barrier was associated with receptor-mediated endocytosis of the MAN–TPGS1000 conjugate via glucose transporters and adsorptive-mediated endocytosis of the DQA–PEG2000–DSPE conjugate via electric charge-based interactions. The efficacy was related to the destruction of VM channels by regulating VM indicators, as well as the induction of apoptosis in brain cancer cells and cancer stem cells by activating apoptotic enzymes and pro-apoptotic proteins and inhibiting anti-apoptotic proteins (Li et al., 2014).

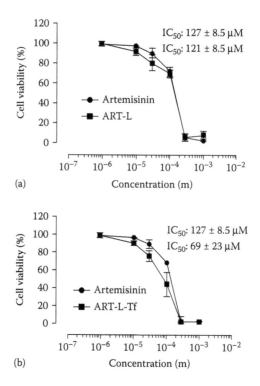

Figure 12.4 *In vitro* cytotoxicity of free ART, ART-L, and ART-L-Tf on HCT-8 colon cancer cells as a function of drug concentration. (Reproduced from Leto I. et al., *Chem. Med. Chem.*, 11, 1745–1751, 2016. With the permission of Wiley.)

Recently, artemisinin and transferrin-loaded magnetic liposomes in thermosensitive and non-thermosensitive forms have been developed and evaluated for their antiproliferative activity against MCF-7 and MDA-MB-231 cells for better tumor-targeted therapy. The entrapment efficiencies of artemisinin, transferrin, and magnetic iron oxide in the non-thermosensitive liposomes were approximately 89%, 85%, and 78%, respectively. Moreover, the thermosensitive formulation showed a suitable condition for thermal drug release at 42°C and exhibited high antiproliferative activity against MCF-7 and MDA-MB-231 cells in the presence of a magnetic field (Gharib et al., 2014).

Artemisone (10-amino-artemisinin derivative) was encapsulated in niosomes prepared with sorbitan monostearate: cholesterol (3:1 ratio), and their effects against human melanoma A-375 cells and human HaCaT keratinocytes were evaluated. The encapsulation efficiency was around 67%, with average particle sizes of approximately 211 nm, and a zeta potential of −38. After 7 h, 85% of the artemisone had been released from the niosomes. The MTT assay indicated that the formulation significantly suppressed melanoma cells ($P \leq 0.05$) in a dose-dependent manner. At 0.06 mg/mL, the free artemisone suppressed almost 50% of the melanoma cell growth, whereas the niosomes almost completely suppressed melanoma cell growth at this concentration. Highly selective cytotoxicity toward the melanoma cells, with negligible toxicity toward the normal skin cells was found. In the specific case of melanoma, as the nanovesicles enhance skin permeation, artemisone-loaded vesicles should be examined as a topical therapy for melanoma (Dwivedi et al., 2015).

12.3.2 Micelles

A micelle is an aggregation of amphiphilic molecules containing a polar "head" and a nonpolar hydrocarbon chain "tail" in a liquid medium, generally represented by water. When these molecules are added to water, they are dispersed in the medium as monomers, but if their concentration is increased, reaching a critical value called the *critical micellar concentration* (CMC), they arrange themselves so as to mini-mize the interaction of the hydrophobic tail with the water. In order to accomplish this, the hydrophobic tails align themselves alongside each other; in some cases, leading to the formation of a spherical structure called a *micelle*, leaving only the water-soluble ionic heads exposed to the solution. At higher concentrations, surfac-tants can also form elongated columns that pack into hexagonal arrays. The columns have hydrophobic cores and hydrophilic surfaces and are separated from one another by water. Micelles contain a variable number of monomers (generally 50–100 units) and some of them, especially the polymeric micelles, are currently used as pharma-ceutical nanocarriers for the delivery of poorly water-soluble drugs, which can be solubilized within the hydrophobic inner core. As a result, micelles can substantially improve the solubility and bioavailability of various hydrophobic drugs. The small size (10–100 nm) of micelles allows for efficient micelle accumulation in pathologi-cal tissues with permeabilized ("leaky") vasculature, such as tumors and infarcts, via the enhanced permeability and retention effect (Bilia et al., 2014, 2017).

Concerning studies with micelles, in 2002, Bilia and coworkers (2002) reported a diffusion-ordered nuclear magnetic resonance spectroscopy study of the solubi-lization of artemisinin by octanoyl-6-*O*-ascorbic acid (ASC8), a relatively novel surfactant that combines surface activity with powerful performance as a radical scavenger. The experimental solubility was in good agreement with a simple model in which a single artemisinin molecule was dissolved per 40-monomer ASC8 micelle, with no substantial perturbation of the self-assembled aggregate. Figure 12.4 shows the solubility of artemisinin in D_2O as a function of ASC8 concentration. There is clear evidence for the solubilization of artemisinin rising approximately linearly with ASC8 concentration above the CMC. Least-squares fitting of the experimen-tal data yielded a solubility of artemisinin in D_2O of 0.21 ± 0.02 mM, and a molar ratio of artemisinin to micellar ASC8 of 0.014 ± 0.002. This ratio corresponds to a partition coefficient $P = X_{micellar}/X_{free}$ for artemisinin of 4000, or a binding constant for the binding of artemisinin to micellar ASC8 of 0.07 mM–1. Artemisinin can be efficiently solubilized by ASC8 micelles, with no significant perturbation of the micellization.

The same authors have investigated the solubilization of artemisinin and cur-cumin, individually and in combination, in micelles of sodium dodecyl sulfate (SDS). The aqueous solubility of artemisinin was enhanced approximately 25-fold by 40 mM SDS, and 50-fold by 81 mM SDS, while that of curcumin was increased to 2 mM by 81 mM SDS. In addition, model studies on the use of the surface-active radical scavenger ASC8 to combine solubilization with protection against oxidation for the chemically labile artemisinin were investigated. A 16-fold enhancement of artemisinin solubility was measured in a solution containing 40 mM SDS and 60 mM ASC8. Even after treatment with 60 mM hydrogen peroxide, more than a 30-fold excess, almost half the artemisinin remained, suggesting a potentially useful combination of the surface activity and antioxidant properties of the novel binary SDS:ASC8 system (Lapenna et al., 2009).

A new polymeric amphiphilic micellar system was developed for the solubilization and controlled delivery of artemether. Methoxy polyethylene glycol (MPEG) 2000 and 5000 were used as a hydrophilic terminal, which was linked to the hydrophobic di-fluorene methoxycarbonyl-l-lysine and to the two amino groups of l-lysine by consecutive peptide linkages and deprotection for up to 2.5 generations. The half-generation (0.5 G, 1.5 G, and 2.5 G) dendritic micelles of MPEG 2000 and 5000 were used to solubilize artemether. A considerable enhancement of artemether's solubility, up to three to fifteen times, depending on the concentration, generation, and type of dendritic micelles used, was found. The dendritic carriers were found to form stable micelles at 10–30 μg/mL depending on the generation and type of MPEG used. The formulations increased the stability of the drug and also prolonged the release of artemether up to 1–2 days *in vitro* (Bhadra 2005).

12.3.3 Solid Lipid Nanoparticles and Nanostructures Lipid Carriers

SLNs are highly stable nanocarriers (50–1000 nm) that provide high protection against degradation to labile drugs and are easy to produce on a large scale. They contain highly purified triglycerides, composed mainly of lipids that are solid at room temperature, which are stabilized by surfactants. Due to their small size and biocompatibility, SLNs may be used in the pharmaceutical field for various routes of administration, such as oral, parenteral, and percutaneous. NLCs represent a secondary generation of SLNs, overcoming some disadvantages such as an unpredictable gelation tendency, polymorphic transition, and low incorporation due to the crystalline structure of solid lipids. NLCs contain a mixture of lipid and solid phases that forms a disorganized liquid lipid matrix, which accommodates active substances, thus increasing their stability and controlling drug extrusion (Bilia et al., 2014, 2017).

SLNs made of Tween 80, Pluronic F68, and soya lecithin were loaded with artheether to obtain homogeneous particle sizes of around 100 nm and an entrapment efficiency of about 69%. Pharmacokinetics studies in rats indicated that their absorption had been significantly enhanced in comparison to that of artheether in aqueous suspension and artheether in ground nut oil. The relative bioavailability of the SLNs loaded with artheether to that of artheether in ground nut oil and artheether in aqueous suspension in rats was approximately 170% and 7461%, respectively, which was found to be significantly high in both cases (Dwivedi et al., 2014).

The same authors reported a further study on SLNs prepared with a solid lipid and an emulsifier (2:1 monostearate:lecithin) and loaded with artemisone

(10-amino-artemisinin derivative). The effects of SLNs against human melanoma A-375 cells and human HaCaT keratinocytes were evaluated. The encapsulation efficiencies were around 80%, with average particle sizes of approximately 295 nm, and zeta potentials of −12 mV. The extent of drug release after 7 h was 85%. The MTT assay indicated that the formulation significantly suppressed melanoma cells ($p \leq 0.05$) in a dose-dependent manner. At 0.06 mg/mL, free artemisone suppresses almost 50% of melanoma cell growth, whereas the niosomes almost completely suppressed melanoma cell growth at this concentration (Dwivedi et al., 2015).

NLCs loaded with artemether (Nanoject) were formulated by employing a microemulsion template technique. The average particle size of the Nanoject was approximately 63 nm and the encapsulation efficiency was found to be about 30%, with a sustained release. *In vitro* hemolytic studies showed that Nanoject had lower hemolytic potential (at a rate of approximately 13%) as compared with all the components when studied individually. Nanoject showed significantly higher ($P < 0.005$) antimalarial activity as compared with the marketed injectable oily intramuscular formulation. Nanoject maintained its antimalarial activity for a longer duration (more than 20 days), indicating that Nanoject may be long-circulating *in vivo*. There was a significantly higher survival rate (60%) for Nanoject, even after 31 days, as compared with a marketed formulation that showed 0% survival (100% mortality) (Joshi et al., 2008).

The pharmacokinetics and tissue distribution after intravenous administration of DHA in NLCs and in solution were compared. Glycerol monostearate and Miglyol® 812 were used as solid lipid and liquid lipid materials, respectively. The surfactants used were Tween 80 (1%) and Poloxamer 188 (1%). Each preparation was injected through the tail vein at the DHA dose of 10 mg/kg. Following intravenous administration of the DHA solution, the mean measured peak plasma concentration achieved was 917.51 ng/mL, and for the drug-loaded NLCs it was 289.28 ng/mL. After 2 h, the plasma concentration was lower for the DHA solution than for the drug-loaded NLCs because of its solubility in plasma and ensuing rapid distribution and elimination, and the slower release of DHA from NLCs, which led to lower clearance. The AUC ($0 \sim \infty$) increased from 633.97 ng/mL/h for the drug solution to 1382.45 ng/mL/h for the drug-loaded NLCs. However, the clearance decreased from 15.77 to 7.23 (mg/kg/h·(ng/mL)) accordingly.

The mean residence time of the drug-loaded NLCs (25.99 h) was higher than that of the drug solution (0.98 h). The distribution half-lives of both formulations were equal (0.06 h), while the volume of distribution of the drug-loaded NLCs was 14.91 ((mg/kg)/(ng/mL)), and this was considerably larger than that (5.21) of the drug solution. Furthermore, the clearance rate of DHA solution was higher than that of DHA loaded in NLCs, suggesting that NLCs represent an effective sustained-release drug delivery system.

In the organs tested, the AUC values of the formulated DHA were higher than those of the drug solution in the liver, spleen, lung, brain, and muscle, and lower than the drug solution in the heart and kidney (Zhang, 2010).

NLCs based on glyceryl trimyristate and soybean oil, loaded with 10% artemether and surface-tailored with a combination of non-ionic, cationic, or anionic surfactants were developed. Their mean particle size, zeta potential, and encapsulation efficiency were approximately 120 nm, −38 mV, and 97%, respectively. Their hemolytic activity was within the acceptable range (7%), revealing a low toxicity risk of NLCs for the parenteral delivery of artemether. Their biocompatibility was confirmed by hepato- and nephrotoxicity analyses. *In vivo* antimalarial studies revealed enhanced activity of a SLN formulation in comparison to a conventional plain drug solution and to a marketed formulation used to treat malaria patients (Aditya, 2010).

12.3.4 Polymeric Nanoparticles

Nanoparticles may consist of either a polymeric matrix (nanospheres) or a reservoir system in which an oily core is surrounded by a thin polymeric wall (nanocapsules). Polymers suitable for the preparation of nanoparticles include natural polymers such as chitosan and gelatin, or synthetic derivatives including poly(alkylcyanoacrylates), poly(methylidene malonate), and polyesters such as poly(lactic acid), poly(glycolic acid), poly(ε-caprolactone), and their copolymers. Biodegradable polymeric nanoparticles are highly preferred because they provide controlled/sustained release properties and biocompatibility with tissues and cells. Lipophilic drugs, which have some solubility in the polymer matrix or in the oily core of nanocapsules, are more readily incorporated than hydrophilic compounds, although the latter may be absorbed onto the particle surface. Methods for the preparation of nanoparticles can start from either a monomer or from a preformed polymer (Bilia et al., 2014, 2017).

A study has reported on the preparation and characterization of chitosan/lecithin nanoparticles (below 300 nm) loaded with artesunate and artesunate complexed with β-cyclodextrin to boost the antimalarial activity. The drug entrapment efficiency was found to be maximum (90%) for nanoparticles containing 100 mg of artesunate. Increased *in vivo* antimalarial activity, in terms of lower mean percentage parasitemia, was observed in infected *P. berghei* mice after the oral administration of all the prepared nanoparticle formulations (Chadha et al., 2012).

DHA was encapsulated in gelatin or hyaluronan nanoparticles (30–40 nm). The entrapment efficiencies for DHA were approximately 13% and 35% for the gelatin and hyaluronan nanoparticles, respectively. The proliferation of A549 cells was inhibited by both nanoparticles. Fluorescent annexin V-fluorescein isothiocyanate (FITC) and propidium iodide (PI) staining revealed low background staining with annexin V-FITC or PI on DHA-untreated cells. In contrast, annexin V-FITC and PI staining dramatically increased when the cells were incubated with gelatin and hyaluronan nanoparticles, with greater anticancer proliferation activities than DHA alone in A549 cells (Sun et al., 2014).

A further study developed artemisinin-loaded poly lactic co-glycolic acid nanoparticles of approximately 221 nm in diameter, with a polydispersity index, zeta potential, drug loading, and entrapment efficiency of 0.1, −9.07 mV, 28%, and 68%, respectively. Atomic force microscopy and transmission electron microscopy studies indicated that the particles were spherical in shape. The drug release behavior, investigated using a dialysis method at pH 7.4 and 5.5, exhibited a biphasic pattern characterized by an initial burst release during the first 24 h, followed by a sustained release up to 100 h. The nanoparticles were stable when stored for a period of about 1 month at 4°C, showing that there was no significant change (P > 0.05) in the mean particle size, PDI, zeta potential, and drug loading of the nanoparticles. Investigation of the effects of the nanoparticles on murine macrophages revealed no significant toxicity, while native artemisinin exhibited significant toxicity at 200 µg/mL, with a drop in cell viability to 40%. The pentamidine that served as a standard anti-leishmanial drug also produced signs of toxicity in the murine macrophages. The nanoparticles significantly inhibited the growth of intracellular amastigotes compared with free artemisinin, while empty nanoparticles did not exhibit any anti-leishmanial activity. The IC_{50} value of nanoparticles for intracellular amastigotes, calculated by linear regression analysis, was found to be 2.9-fold lower than of free artemisinin (11.9 vs. 3.93 µg/mL). Treatment of amastigote-infested macrophages with artemisinin-loaded poly lactic co-glycolic acid nanoparticles also resulted in a significant reduction in the percentage of infected macrophages, yielding a 3.6-fold lower IC_{50} value compared with that of free artemisinin (14.86 vs. 4.16 µg/ml) (Want et al., 2014).

Albumin-based nanoparticles were developed as carriers of artemisinin. Their mean diameter was 339 nm, while the zeta potential was −43.8 mV. When artemisinin was loaded to the nanoparticles at a ratio of 1:10 with respect to albumin, rods were formed, as revealed by transmission electron microscopy. These freeze-dried nanoparticles had a mean diameter of 612 nm, with an artemisinin entrapment efficiency of 97.5%. When reconstituted, these nanoparticles showed satisfactory physical stability when stored at 4°C for 4 days. The increase in the mean diameter was 5.8%, with good homogeneity (PI < 0.25). The high zeta potential (−43.8 mV) may account for this stability by providing sufficient electrostatic repulsion, thus preventing particle aggregation. Artemisinin showed good chemical stability within nanoparticles both in its powdered (lyophilized) and reconstituted aqueous forms. After storage of lyophilized nanoparticles for 1 month at 4°C, the amount of artemisinin remaining in nanoparticles was 98.4%. The antiplasmodial activity of the developed formulation was investigated in an *in vitro* model of a chloroquine-resistant strain of *P. falciparum* (FcB1). This revealed improved activity compared with unformulated artemisinin (IC_{50} < 3.5 vs. 11.4 nM). The *in vivo* antimalarial activity of the developed formulation was assessed by the intravenous route in humanized mice infected with the human parasite *P. falciparum* (3D7 strain). A 4-day treatment with 10 mg/kg of nanoparticles achieved a sharp reduction in parasitemia (96%) measured the day after the end of the treatment. Moreover, mice survived for more than 18 days with no recrudescence until the end of the experiment (Ibrahim et al., 2015).

Artemisinin-loaded poly lactic co-glycolic acid nanoparticles were prepared with a particle size of approximately 220 nm, 29% drug loading, and 69% encapsulation efficiency. When administered at doses of 10 and 20 mg/kg body weight, they showed superior antileishmanial efficacy compared with free artemisinin in a BALB/c murine model of visceral leishmania. There was a significant reduction in hepatosplenomegaly, as well as in parasite load in the liver (ca. 85.0%) and spleen (ca. 82.0%) with nanoparticle treatment at 20 mg/kg body weight compared with free artemisinin (ca. 70% in liver and 63% in spleen). In addition, nanoparticle treatment restored the defective host immune response in mice with established visceral leishmania infection. The protection was associated with a Th1-biased immune response, as evident from a positive delayed-type hypersensitivity reaction, escalated IgG2a levels, augmented lymphoproliferation, and enhancement in pro-inflammatory cytokines (IFN-γ and IL-2), with significant suppression of Th2 cytokines (IL-10 and IL-4) after *in vitro* recall, compared with infected controls and free artemisinin treatment (Want 2015).

12.4 CONCLUDING REMARKS

Clearly, the discovery and development of artemisinin is the most exciting and successful breakthrough in medicinal plant drug development for the control of malaria. The rapid increase of multidrug-resistant parasites has rendered previous malarial control methods unpractical, risky, and no longer cost-effective. A characteristic of artemisinin and its derivatives is the rapid onset of action with clearance of parasites from the blood within 48 h in most cases, and they offer a greater log kill per cycle than other classes of antimalarials and the broadest stage specificity of drug action. The challenge of low bioavailability is being dealt with through pharmaceutical technology in order to obtain formulations with improved biopharmaceutical characteristics.

The application of nanotechnology to artemisinin will, in the future, have a significant impact on developing appropriate therapeutic forms of artemisinin. So far, the results obtained from nanoencapsulated natural products are very encouraging, as they generally have better stability, sustained release, and improved bioavailability at much lower doses, and offer increased long-term safety.

REFERENCES

Aditya N.P., Patankar S., Madhusudhan B., Murthy R.S., and Souto E.B. 2010. Arthemeter-loaded lipid nanoparticles produced by modified thin-film hydration: Pharmacokinetics, toxicological and *in vivo* anti-malarial activity. *European Journal of Pharmaceutical Sciences.* 40(5): 448–455.

Al-Angary A.A., Al-Meshal M.A., Bayomi M.A., and Khidr S.H. 1996. Evaluation of liposomal formulations containing the antimalarial agent, arteether. *International Journal of Pharmaceutics* 128(1–2): 163–168.

Ashton M., Nguyen D.S., Nguyen V.H., Gordi T., Trinh N.H., Dinh X.H., Le D.C., et al. 1998. Artemisinin kinetics and dynamics during oral and rectal treatment of uncomplicated malaria. *Clinical Pharmacology and Therapeutics* 63(4): 482–493.

Balint G.A. 2001. Artemisinin and its derivatives. An important new class of antimalarial agents. *Pharmacology and Therapeutics* 90(2): 261–265.

Bayomi M.A., Al-Angary A.A., Al-Meshal M.A., and Al-Dardiri M.M. 1998. *In vivo* evaluation of arteether liposomes. *International Journal of Pharmaceutics* 175(1): 1–7.

Bhadra D., Bhadra S., and Jain N.K. 2005. Pegylated lysine based copolymeric dendritic micelles for solubilization and delivery of artemether. *Journal of Pharmacy and Pharmaceutical Sciences* 8(3): 467–482.

Bilia A.R., Bergonzi M.C., Morris G.A., Lo Nostro P., and Vincieri F.F. 2002. A diffusion-ordered NMR spectroscopy study of the solubilization of artemisinin by octanoyl-6-*O*-ascorbic acid. *Journal of Pharmaceutical Sciences* 91(10): 2265–2272.

Bilia A.R., Guccione C., Isacchi B., Righeschi C., Firenzuoli F., and Bergonzi M.C. 2014. Essential oils loaded in nanosystems: A developing strategy for a successful therapeutic approach. *Evidence-Based Complementary and Alternative Medicine* 2014: 651593.

Bilia A.R., Piazzini V., Guccione C., Risaliti L., Asprea M., Capecchi G., and Bergonzi M.C. 2017. Improving on nature: The role of nanomedicine in the development of clinical natural drugs. *Planta Medica* 83(05): 366–381.

Chadha R., Gupta S., and Pathak N. 2012. Artesunate-loaded chitosan/lecithin nanoparticles: Preparation, characterization, and *in vivo* studies. *Drug Development and Industrial Pharmacy* 38(12): 1538–1546.

Chimanuka B., Gabriëls M., Detaevernier M.-R., and Plaizier-Vercammen J.A. 2002. Preparation of β-artemether liposomes, their HPLC–UV evaluation and relevance for clearing recrudescent parasitaemia in *Plasmodium chabaudi* malaria-infected mice. *Journal of Pharmaceutical and Biomedical Analysis* 28(1): 13–22.

Committee for Proprietary Medicinal Products (CPMP). 2010. Note for guidance on the investigation of bioavailability and bioequivalence. European Medicines Agency. http://www.ema.europa.eu/docs/en_GB/document_library/Scientific_guideline/2010/01/WC500070039.pdf, retrieved on 3 September 2016.

de Vries P.J. and Dien T.K. 1996. Clinical pharmacology and therapeutic potential of artemisinin and its derivatives in the treatment of malaria. *Drugs* 52(6): 818–836.

Dwivedi A., Mazumder A., du Plessis L., du Preez J.L., Haynes R.K., and du Plessis J. 2015. *In vitro* anti-cancer effects of artemisone nano-vesicular formulations on melanoma cells. *Nanomedicine* 11(8): 2041–2050.

Dwivedi P., Khatik R., Khandelwal K., Taneja I., Raju K.S., Wahajuddin, Mishra P.R., et al. 2014. Pharmacokinetics study of arteether loaded solid lipid nanoparticles: an improved oral bioavailability in rats. *International Journal of Pharmaceutics* 466(1–2): 321–327.

Gabriëls M. and Plaizier-Vercammen J. 2003. Physical and chemical evaluation of liposomes, containing artesunate. *Journal of Pharmaceutical and Biomedical Analysis* 31(4): 655–667.

Gaumet M., Vargas A., Gurny R., and Delie F. 2008. Nanoparticles for drug delivery: The need for precision in reporting particle size parameters. *European Journal of Pharmaceutics and Biopharmaceutics* 69(1): 1–9.

Gharib A., Faeizadeh Z., Mesbah-Namin S.A., and Saravani R. 2014 May 28. Preparation, characterization and *in vitro* efficacy of magnetic nanoliposomes containing the artemisinin and transferrin. *DARU Journal of Pharmaceutical Sciences* 22(1): 44.

Gu H.M., Lu B.F., and Qu Z.X. 1980. Antimalarial activity of 25 derivatives of artemisinin against cloroquine-resistant *Plasmodium berghei*. *Acta Pharmacologica Sinica* 1(1): 48–50.

Haynes R.K. and Krishna S. 2004. Artemisinins: Activities and actions. *Microbes and Infection* 6(14): 1339–1346.

Hien T.T. and White N.J. 1993. Qinghaosu. *The Lancet* 341(8845): 603–608.

Ibrahim N., Ibrahim H., Sabater A.M., Mazier D., Valentin A., and Nepveu F. 2015. Artemisinin nanoformulation suitable for intravenous injection: Preparation, characterization and antimalarial activities. *International Journal of Pharmaceutics* 495(2): 671–679.

Isacchi B., Arrigucci S., la Marca G., Bergonzi M.C., Vannucchi M.G., Novelli A., and Bilia A.R. 2011. Conventional and long-circulating liposomes of artemisinin: Preparation, characterization, and pharmacokinetic profile in mice. *Journal of Liposome Research* 21(3): 237–44.

Isacchi B., Bergonzi M.C., Grazioso M., Righeschi C., Pietretti A., Severini C., and Bilia A.R. 2012. Artemisinin and artemisinin plus curcumin liposomal formulations: Enhanced antimalarial efficacy against *Plasmodium berghei*-infected mice. *European Journal of Pharmaceutics and Biopharmaceutics* 80(3): 528–534.

Jin M., Shen X., Zhao C., Qin X., Liu H., Huang L., Qiu Z., and Liu Y. 2013. *In vivo* study of effects of artesunate nanoliposomes on human hepatocellular carcinoma xenografts in nude mice. *Drug Delivery* 20 (3–4): 127–133.

Joshi M., Pathak S., Sharma S., and Patravale V. 2008. Design and *in vivo* pharmacodynamic evaluation of nanostructured lipid carriers for parenteral delivery of artemether: Nanoject. *International Journal of Pharmaceutics* 364(1): 119–126.

Jung M., Lee K., Kim H., and Park M. 2004. Recent advances in artemisinin and its derivatives as antimalarial and antitumor agents. *Current Medicinal Chemistry* 11(10): 1265–1284.

Lapenna S., Bilia A.R., Morris G.A., and Nilsson M. 2009. Novel artemisinin and curcumin micellar formulations: Drug solubility studies by NMR spectroscopy. *Journal of Pharmaceutical Sciences* 98(10): 3666–3675.

Leto I., Coronnello M., Righeschi C., Bergonzi M.C., Mini E., and Bilia A.R. 2016. Enhanced efficacy of artemisinin loaded in transferrin-conjugated liposomes versus stealth liposomes against HCT-8 colon cancer cells. *Chem. Med. Chem.* 11(16): 1745–1751.

Li X.Y., Zhao Y., Sun M.G., Shi J.F., Ju R.J., Zhang C.X., Lu W.L., et al. 2014. Multifunctional liposomes loaded with paclitaxel and artemether for treatment of invasive brain glioma. *Biomaterials* 35(21): 5591–5604.

Li, Y., Wu, Y-L., 1998. How Chinese Scientists discovered Qinghaosu (artemisinin) and developed its derivatives? What are the future perspectives? *Med. Trop.* 58(3): 9–12.

Li Y., Yu P.L., and Chen Y.X. 1981. Studies on analogs of artemisinin. The synthesis of ethers, carboxylic esters, carbonates of dihydroartemisinin. *Acta Pharmacologica Sinica* 16: 429–439.

Liu, X., 1980. Study on qingausu derivatives. *Chinese Pharmaceutical Bulletin.* 15: 183.

Luo X.-D. and Shen C.C. 1987. The chemistry and clinical applications of qinghaosu and its derivatives. *Medicinal Research Reviews* 7(1): 29–52.

Myint P.T., Shwe T., Soe L., Htut Y., and Myint W. 1989. Clinical study of the treatment of cerebral malaria with arthemeter. *Transactions of the Royal Society of Tropical Medicine and Hygiene* 83(1): 72–76.

Navaratnam V., Mansor S.M., Sit N.W., Grace J., Li Q., and Olliaro P. 2000. Pharmacokinetics of artemisinin-type compounds. *Clinincal Pharmacokinetics* 39(4): 255–270.

Ngo T.H., Michoel A., and Kinget R. 1996. Dissolution testing of artemisinin solid oral dosage forms. *International Journal of Pharmaceutics* 138(2): 185–190.

Ngo T.H., Vertommen J., and Kinget R. 1997a. Formulation of artemisinin tablets. *International Journal of Pharmaceutics* 146(2): 271–274.

Ngo T.H., Quintens I., Roets E., Declerck P.J., and Hoogmartens A. 1997b. Bioavailability of different artemisinin tablet formulations in rabbit plasma—Correlation with results obtained by an *in vitro* dissolution method. *Journal of Pharmaceutical and Biomedical Analysis* 16(2): 185–189.

Niu X., Ho L., Ren Z., and Song Z. 1985. Metabolic fate of qinghaosu in rats: A new TLC-densitometric method for its determination in biological material. *European Journal of Drug Metabolism and Pharmacokinetics* 10(1): 55–59.

Qinghaosu Antimalaria Coordinating Research Group, 1979. Antimalaria studies on qing-haosu. *Chinese Medical Journal* 92(12): 811–816.

Righeschi C., Coronnello M., Mastrantoni A., Isacchi B., Bergonzi M.C., Mini E., and Bilia A.R. 2014. Strategy to provide a useful solution to effective delivery of dihydroarte-misinin: Development, characterization and *in vitro* studies of liposomal formulations. *Colloids and Surfaces B: Biointerfaces* 116: 121–127.

Sriram D., Rao V.S., Chandrasekar K.V.G., Yogeeswari P., Kannan R., Kumar K., and Sahal D. 2004. Progress in the research of artemisinin and its analogues as antimalarials: An update. *Natural Product Research* 18(6): 503–527.

Sun Q., Teong B., Chen I.F., Chang S.J., Gao J., and Kuo S.M. 2014. Enhanced apoptotic effects of dihydroartemisinin-aggregated gelatin and hyaluronan nanoparticles on human lung cancer cells. *Journal of Biomedical Materials Research Part B: Applied Biomaterials* 102(3): 455–462.

Titulaer H.A.C., Zuidema I., Kager P.A., Wetsteyn J.C.F.M., Lugt Ch. B., and Merkus F.W.H.M. 1990. The pharmacokinetics of artemisinin after oral, intramuscular and rectal administration to volunteers. *Journal of Pharmacy and Pharmacology* 42(11): 810–813.

van Agtmael M.A., Eggelte T.A., and van Boxtel C.J. 1999. Artemisinin drugs in the treatment of malaria: From medicinal herb to registered medication. *Trends in Pharmacological Sciences* 20(5): 199–204.

van Nijlen T., Brennan K., Van den Mooter G., Blaton N., Kinget R., and Augustijns P. 2003. Improvement of the dissolution rate of artemisinin by means of supercritical fluid tech-nology and solid dispersions. *International Journal of Pharmaceutics* 254(2): 173–181.

Want M.Y., Islamuddin M., Chouhan G., Dasgupta A.K., Chattopadhyay A.P., and Afrin F. 2014. A new approach for the delivery of artemisinin: Formulation, characterization, and *ex-vivo* antileishmanial studies. *Journal of Colloid and Interface Science* 432: 258–269.

Want M.Y., Islamuddin M., Chouhan G., Ozbak H.A., Hemeg H.A., Dasgupta A.K., Afrin F., et al. 2015. Therapeutic efficacy of artemisinin-loaded nanoparticles in experimental visceral leishmaniasis. *Colloids and Surfaces B: Biointerfaces* 130: 215–221.

Zhang X., Qiao H., Liu J., Dong H., Shen C., Ni J., Xu Y. et al. 2010. Dihydroartemisinin loaded nanostructured lipid carriers (DHA-NLC): Evaluation of pharmacokinetics and tissue distribution after intravenous administration to rats. *Die Pharmazie—an International Journal of Pharmaceutical Sciences* 65(9): 670–678.

Zhao K.C., Chen Q.M., and Song Z.Y. 1986. Studies on the pharmacokinetics of qinghaosu and two of its active derivatives in dogs. *Acta Pharmacologica Sinica* 21: 736–739.

Zhao S. 1987. High-performance liquid chromatographic determination of artemisinin in human plasma and saliva. *Analyst* 112(5): 661–664.

CHAPTER 13

In Situ Cultivation of *Artemisia annua*

Salisu Muhammad Tahir

CONTENTS

13.1 INTRODUCTION

Artemisia annua L (Wormwood) belongs to the family Asteraceae (Ferreira and Janick, 2002), which consists of about 400 species (Bailey and Bailey, 1976; Bennett et al., 1982; McVaugh, 1984; El-haq et al., 1991; Klayman, 1993; Jaime and Da Silver, 2003). The family is characterized by extreme bitterness in all parts of the plant (Tripathi et al., 2000, 2001; Ferreira and Janick, 2009).

Artemisia is an herb of Asiatic (probably Chinese) and Eastern European origin, and it is the source of the traditional Chinese herbal medicine that has been used for over 2000 years to alleviate fevers (Antony, 2010). More than 400 species have been identified, and it is widely dispersed throughout the temperate regions of the world and in high-altitude regions with a pronounced cool period. Other *Artemisia* species are aromatic perennials and are also used for medicinal purposes (Klayman et al., 1984).

The plant has become naturalized and is cultivated in many countries including Argentina, Spain, the United States, and Africa—mainly Kenya, Tanzania, and Nigeria—to support the production of the antimalarial artemisinin (Bailey and Bailey, 1976; Klayman, 1993; Ferreira et al., 1997). Cultivation of *A. annua* in Africa was a response to the call by the World Health Organization (WHO) for the adoption of artemisinin combination therapies (ACTs) as a replacement for artemisinin monotherapy in the fight against multidrug-resistant *Plasmodium falciparum*. Large-scale and commercial propagation of *Artemisia* in Africa is limited to Kenya and Tanzania. However, current demand for artemisinin has led to a significant increase in the large-scale propagation, both in established and new areas (Ferreira et al., 1997).

In situ production of *A. annua* is presently the most commercially feasible approach to produce artemisinin and related compounds.

The cultivation of *A. annua* has expanded from its center of origin (China) to other parts of the globe; mainly Kenya, Tanzania, and Nigeria, in response to the call by the WHO for the use of ACTs for treating malaria fever. Its effectiveness has been demonstrated in the treatment of skin diseases and it has also been shown to be as effective a non-selective herbicide as glyphosate (Duke et al., 1987; Paniego and Giulietti, 1994).

Members of some plant families exhibit erratic germination due to seed dormancy (Bewly and Black, 1994). They readily germinate within the native environment, but fail to show good germination under alien conditions. Depending on the plant species and type of dormancy, several methods may be used to break dormancy in order to induce germination (Zare et al., 2011).

Artemisia seeds were observed to undergo chemical dormancy due to the presence of some chemical compounds such as phenolics on the surface. This was linked with seed germination inhibition and dormancy of the plant. Accumulation of phenolics played a protective role in strengthening the plant cell walls during growth by polymerizing into lignin (Farouk et al., 2008). Productive *Artemisia* seed is expensive and not readily available. Successful and quantitative production of biomass is an important step toward maximizing artemisinin content in *Artemisia* for the treatment of malarial fever. However, this has been hindered by several biotic and abiotic factors such as pests, diseases, and climatic constraints. Consequently, this has led to poor performance, and hence, decrease in yields. Similarly, there is inadequate agro-technological information regarding the ideal planting dates, seed density, harvesting system, post-harvesting, and optimum fertilizer application rates required for higher yields. There is therefore, the need to determine the most effective treatment for the germination of *Artemisia* seeds as a commercially viable means for the production of artemisinin.

13.2 SIGNIFICANCE OF *A. ANNUA*

Besides its antimalarial properties, *A. annua* has been shown to be effective in the treatment of skin diseases and also to be an effective non-selective herbicide (Duke et al., 1987; Paniego and Giulietti, 1994). The antibacterial and antioxidant properties of the plant's extracts and essential oil are a welcome development for the livestock industry (Xiao and Catto, 1989; Allen et al., 1997; Xiao et al., 2000; Lescano et al., 2004; Arab et al., 2006; Brisibe et al., 2008). The plant has also been used to impair the growth and development of insects that attack stored grains (Tripathi et al., 2000; Tripathi, 2001). Three common derivatives found in *Artemisia* are artesunate, artemether, and artemisinin (Duke et al., 1987; Kindscher, 1992). Artemisinin contains two oxygen atoms, linked together in an endoperoxide bridge, which react with iron atoms. When it reacts with the high iron content of the parasites, it generates free radicals, which damage the parasite (Luo and Shen, 1987). By the same mechanism, artemisinin becomes toxic to cancer cells, which sequester relatively large amounts of iron compared with normal, healthy human cells. Tests have revealed that artemisinin causes rapid and extensive damage and death in cancer cells, yet it has relatively low toxicity to normal cells (Luo and Shen, 1987).

Artemisinin has also been shown to regulate plant growth by inhibiting lateral root growth (Duke et al., 1987). Artemisinin has been more effective than glyphosate, when tested as a herbicide in the mung bean (*Vigna radiata*) (Chen et al., 1991).

13.3 USE OF *A. ANNUA* IN THE TREATMENT
OF MALARIA FEVER

A. annua has received considerable attention because of its antimalarial properties. This activity has been established to be due to artemisinin (qinghaosu), a cadinane-type sesquiterpene lactone endoperoxide, present in the aerial parts. Its semisynthetic derivatives, artemether, arteether, and artesunate are effective against multidrug-resistant malaria caused by *P. falciparum* and against the life-threatening complication, cerebral malaria (Klayman et al., 1984).

The WHO estimates that, in 2010, malaria caused 219 million clinical episodes, and 660,000 deaths. An estimated 91% of deaths in 2010 were in the African Region, followed by the South-East Asian Region (6%), and the Eastern Mediterranean Region (3%). About 86% of deaths globally were in children (Nigeria Malaria Fact Sheet, United States Embassy in Nigeria, 2011).

Currently, efforts are being made to make these drugs available worldwide, and in order to meet the industrial demand, many research programs have started focusing on the selection and cloning of high-artemisinin-yielding chemotypes of *A. annua*. A review of recent agricultural techniques to improve yields from cultivated plants is also available (Laughlin, 1994). At present, the total synthesis of artemisinin is not economical for large-scale production, and so alternative approaches to improve the economics of drug production are being studied through the extraction of artemisinin-related sesquiterpenes from *A. annua* plants and their subsequent conversion into artemisinin semisynthetically. Among these, artemisinic acid is the most promising (Laughlin, 1994).

Use of artemisinin as a monotherapy is explicitly discouraged by the WHO, as there have been signs that malarial parasites are developing resistance to the drug. Therapies that combine artemisinin with some other antimalarial drug are preferred for the treatment of malaria and are both effective and well tolerated in patients. The drug is also increasingly being used in *P. vivax* malaria as well as being a topic of research in cancer treatment (WHO, 2002).

13.4 USE OF *A. ANNUA* IN THE LIVESTOCK INDUSTRY

The uses of *A. annua* and artemisinin in the livestock industry are currently expanding based on current reports of the antiprotozoal, antibacterial, and antioxidant activities of its extracts and essential oil on *Babesia eimeria* and coccidiosis (Allen et al., 1997; Arab et al., 2006; Brisibe et al., 2008), and others including the trematodal blood fluke, *Schistosoma* (Xiao and Catto, 1989; Xiao et al., 2000; Lescano et al., 2004). Recently, different tissues of the plant have been analyzed for their potential use in animal feed and scored high values for antioxidant capacity and as a source of amino acids, with negligible amounts of anti-nutritive components such as phytates and oxalates (Lescano et al., 2004).

13.5 USE OF *A. ANNUA* IN THE TREATMENT OF CANCER

Cancer is a general term used to describe the uncontrolled proliferation of somatic cells, which can develop into a tumor or spread around the body affecting one or more vital systems. In the United States alone, cancer kills over 500,000 people a year, being the second-biggest killer after heart disease (Brown, 1990). Among the types of cancer, lung cancer kills the most, irrespective of gender, followed by prostate cancer in men and by breast cancer in women. Cancer is an economic burden to the United States, costing over $585 million in healthcare alone, not to mention disability and productivity losses. Another possible use of artemisinin is in the treatment of this deadly menace. The antimalarial activity of artemisinin is thought to be due to its interaction with iron, present in very high concentrations in the malaria parasite. Since some cancer cells, particularly leukemia cells, also have high iron concentrations, they may also be killed by artemisinin, as has been demonstrated in some initial studies with cancer cells in tissue culture (Luo and Shen, 1987). The mechanism of action of artemisinin exerted on *Plasmodium* was also observed on cancerous cells, which sequester relatively large amounts of iron compared to normal, healthy human cells. Tests have shown that artemisinin causes rapid and extensive damage and death in cancer cells, yet it has relatively low toxicity to normal cells (Luo and Shen, 1987). Traditional treatment includes radiation therapy or chemotherapy; the latter uses synthetic or semisynthetic drugs that kill cancer cells as well as healthy cells. Chemotherapy also causes undesirable side effects, such as loss of appetite, nausea, and depression (Luo and Shen, 1987). The potential of artemisinin and its derivatives as cancer chemotherapeutic agents is being actively investigated in a variety of anticancer screens. The combination of high demand for artemisinin-based antimalarials and limited commercial-scale production of *A. annua* (in only a few locales in China and Vietnam) has left artemisinin-based therapies in short supply (Chen et al., 1991). The WHO has, therefore, stepped in to develop a plan to bolster the production of *A. annua* (Duke et al., 1987; Chen et al., 1991).

13.6 CURRENT STATUS OF *A. ANNUA* PRODUCTION

The areas of most concentrated *Artemisia* production are Asia, Europe, the United States, and, recently, East Africa (Klayman, 1993). However, global estimates for *A. annua* revealed a gradual increase in production area of 2,000 ha in 2003, 3,000 ha in 2004, 9,500 ha in 2005, 26,000 ha in 2006, 14,500 ha in 2007, 4,500–5,000 ha in 2008, 6,000 ha in 2009, 15,000 ha in 2010, 17,500 ha in 2011, and 16,500 ha in 2012 (Artepal, 2008; Pilloy, 2008; Roll Back Malaria, 2009; Hiey, 2010; Ut, 2011). The related biomass volumes were 3,000, 10,000, and 14,000 tons, respectively (Tan, 2006). Production in 2013 was very uncertain due to labor shortages, competition from other crops, and currency devaluation, but above all due to the reductions in artemisinin prices, market uncertainties, and the introduction of semisynthetic artemisinin (Pilloy, 2010; Roll Back Malaria, 2009; Antony, 2010). East African production was placed at about 4000 ha in 2006 and 2007, with an average biomass yield of 2.5 t/ha (EABL, 2005). Similar estimates were not available for other countries and areas (Cutler, 2008). In China, production rates of 2000 ha in 2004, 6000 ha in 2005, and 9000 ha in 2006 were reported (Tan, 2006). Vietnam is also a major producer, but its production was reported to have dropped from 10,000 ha in 2006 to 3,000 ha in 2007, 1,000 ha in 2008, 700 ha in 2009, and 500–700 ha in 2010, before then increasing to 1,500 ha in 2011 (Artepal, 2008; Hiey, 2010; Ut, 2011). India has the potential for *Artemisia* cultivation but no estimate of production is available

(Sharma, 2006; Cutler, 2008). In 2005, Kenya, Uganda, and Tanzania recorded an annual production of 1650 ha. The yield rose in 2007 to between 3500 and 4000 ha (Advanced Bio-Extracts, 2007).

In Nigeria, propagation of *A. annua* started in 2003, while artemisinin extraction commenced in 2005. As of August 2007, over 1500 ha were prepared for planting with seeds from Brazilian, Chinese, and local breeds. No further information was received (Anonymous, 2007; NA, 2007). Similarly, Tahir et al. (2015) reported field propagation of *A. annua* using seeds of the Chiyong variety. *In vitro* work on *Artemisia* was also reported by Jamaleddine et al. (2011) and Tahir et al. (2013, 2016).

13.7 TAXONOMY OF *A. ANNUA*

Artemisia belongs to the tribe Anthemideae of the Asteroideae, a subfamily of the Asteraceae. The genus *Artemisia* was sub-divided into various subgeneric sections; *A. annua* has been considered to belong to the subsection *Absinthium* or to a combined subsection of *Absinthium* and *Abrotanum*. The greatest number of species has been reported in Asia, with 150 accessions in China, 174 in the former U.S.S.R., and about 50 in Japan; 57 species were identified in the European region, and 30 species were documented in the New World (Colin, 2002).

The first natural classification of the genus was given by dividing the genus into four sections based on fundamental differences in the floral structure. *Abrotanum* and *Absinthium* are considered the most primitive sections, while *Dracunculus* and *Seriphidium* are the most advanced (Colin, 2002).

Some authorities have elevated Besser's sections to the subgenus level and reduced the number of subgenera to three, by combining *Abrotanum* with *Absinthium* to form the subgenus *Artemisia*, which was further divided into three sections: *Artemisia, Abrotanum,* and *Absinthium.* The shrubby members of the subgenus *Seriphidium*, endemic to North America, have been recognized as being closely related, distinct from the Old World *Seriphidium*, and grouped together in the section *Tridentatae* of the subgenus *Seriphidium.* Many authors have not accepted this assignment and have argued that the New World *Tridentatae* and the rest of the New World *Seriphidium* species should be considered as separate taxonomic entities. A similar close relationship among different species of the *Abrotanum* was recognized, leading them to be grouped in the subsection *Vulgares* (Colin, 2002).

13.8 MORPHOLOGY OF *A. ANNUA*

A. annua, is an annual, short-day plant with a critical photoperiod of 13.5 h (Ferreira et al., 1995). The chromosome number of *A. annua* is $2n=36$ (Bennett et al., 1982). The plant usually reaches about 2 m in height with alternate branches and alternate, deeply dissected, aromatic leaves ranging from 2.5 to 5.0 cm in length. Tiny yellow flowers (capitula), only 2 or 3 mm across, are displayed in lose panicles containing numerous, greenish or yellowish, bisexual central (disc) florets containing little nectar and pistillate marginal (ray) florets. The central flowers are perfect and can be either fertile or sterile (Ferreira et al., 1995). The ovaries are inferior and unilocular, and each generates one achene, 1 mm in length and faintly nerved. The pistillate marginal florets in the capitulum produce numerous achenes without a pappus. The pollen is tricolpate and smooth, typical of anemophilous species, and has vestigial or no spines (Stix, 1960). It has an internal, complex, columellae tecta configuration in the exine, which is common to all taxa of the tribe Anthemideae and varies from two to three layers in *A. annua*. The plant is naturally cross-pollinated by insects and wind action, which is unusual in the Asteraceae. The leaves alternate, and are dark green or brownish-green, rolled, and crumpled. Their odor is characteristically aromatic and they taste bitter (Ferreira et al., 1995) (Figure 13.1).

Non-glandular T-shaped trichomes and ten-celled biseriate glandular trichomes occur on the leaves, stems, and inflorescences. The morphology and origin of the glandular trichomes have been described for the leaves (Duke and Poul, 1993) and capitula (Ferreira and Janick, 1995) using light and/or scanning electron microscopy. The fruit of *A. annua* is an achene with a single seed inside. The seeds are approximately 1 mm in length, oblong, and yellow-brownish with a lustrous surface marked by vertical furrows; the seed endosperm is creamy white in color and fatty in content. The seeds weigh approximately 0.03 g per 1000 (WHO, 2006).

13.9 SOIL AND CLIMATE

Artemisia grows very well in slightly alkaline loamy and well-drained neutral soils. The plant grows well in places exposed to good sunlight throughout the day. *Artemisia* is quite tolerant of drought and low-moisture conditions (Ferreira and Janick, 2009). In fact, plants grown in poor and dry soils do produce a high aromatic quality. However, *Artemisia* tolerates temperatures that are as low as −5°C (Bui et al., 2011).

Figure 13.1 *A. annua* in the field. (Source: Eric Sawyer, 2012.)

Another notable feature of all the plants included in this genus is their resistance to the honey fungus—a common fungus that affects many plants. At the same time, plants in this genus are rarely troubled by deer and other browsing animals.

Artemisia is usually propagated using the seeds. These seeds are sown on the soil surface in late winter and sometime in the early summer when they are grown in a greenhouse. When the seeds germinate and seedlings become large enough to handle, they are pricked and sorted out into individual pots and then planted out in the summer months. Plants are also grown from cuttings of the half-ripe wood; these are placed in a frame sometime in July or August. They are then divided in the spring or autumn.

A. annua is a determinate short-day plant that is very responsive to photoperiodic stimuli. The critical photoperiod is about 13.5 h, but there are likely to be photoperiod×temperature interactions. It thrives in many temperate to subtropical ecologies, but the plant is not adapted to the tropics because flowering will be induced when the plants are very small, with the possible exceptions of high-altitude plateaus or regions with a pronounced cool period (Klayman, 1993)

13.10 DORMANCY IN *A. ANNUA*

Artemisia seeds were observed to undergo chemical dormancy due to the presence of some chemical compounds such as phenolics. This was linked with seed germination inhibition and dormancy of the plant. Their concentrations in plants were subject to seasonal alterations. The accumulation of phenolics plays a protective role in strengthening the plant cell walls during growth by polymerization into lignins. These chemicals may be leached out of the tissues by washing or soaking the seed either in cold or warm water, or they may be deactivated by other means. Other chemicals that prevent germination are washed out of the seeds by rainwater or snow melt. Other possible methods of breaking dormancy include heating, soaking in acid (e.g., sulfuric acid), and scarification (e.g., rubbing over sand) (Farouk et al., 2008).

13.11 PROPAGATION CONSTRAINTS

Some of the challenges faced by farmers that hinder normal propagation of *A. annua* through the seeds include access to good-quality seeds, seed shortages, and the presence of phenolic secretions on the surface of the seeds, which has been linked to germination inhibition and dormancy of the plant (Farouk et al., 2008). The devastating decline in natural habitats due to destruction of plant resources as a result of man's economic and social activities, lack of sufficient training, difficulty in meeting quality standards for processing and extraction, lack of a steady market and profitable price to attract farmers, and the need for weed control and specialized harvesting and drying techniques all present challenges to *A. annua* production (Advanced Bio-Extracts, 2006). Artemisinin content and production is influenced by climatic conditions together with the time of planting and method of harvesting of the plant

(Ram et al., 1997; Wallaart et al., 2000). The cost of production of metabolites from cultured plant cells is another barrier that must be overcome to make their commercial regeneration a reality (Whitaker and Hashimoto, 1986; Giles and Songstad, 1990; Razdan, 2002).

Given these challenges, there is a need to enhance the cultivation of the plant by the development of effective *in situ* culture techniques and to improve the artemisinin extraction methods (Trigiano and Gray, 2000). This may be achieved through the use of efficient *in situ* techniques for year-round plant propagation. This is expected to bring about increased production of *Artemisia* and artemisinin, create job opportunities, and improve the income of small-scale growers of *Artemisia*. It will also supplement the efforts of several studies and will help toward improving health services across the globe.

13.12 *IN SITU* PROPAGATION OF *A. ANNUA*

Field production of *A. annua* is presently the only commercially viable method to produce artemisinin because the synthesis of the complex molecule is uneconomic. The currently used selections reach their peak artemisinin content before flowering and at the end of vegetative growth, allowing maximal biomass accumulation of artemisinin before harvest. The most important management problems involve planting, the achievement of uniform stands, weed control, and post-harvest drying of the crop. The plant is extremely vigorous, and essentially disease- and pest-free. Most researchers transplant seedlings, but direct seeding and mechanical transplanters have been used in commercial production. Traditional medicine using plant extracts continues to provide health coverage for over 80% of the world's population, especially in the developing world (WHO, 2002). Natural products have always played an important role in medicine. It is well known that secondary metabolites are a source of biologically active natural products with various functions, including antibacterial, antifungal, antiviral, antineoplastic, and anticancer activities, and also act as inhibitors and plant growth promoters (Syed, 2001).

13.13 PROPAGATION FROM SEEDS

The propagation of *Artemisia* from seeds seems to be a very difficult exercise. Therefore, there is the need to use a variety of seed dormancy breaking techniques for successful and profitable generation of *Artemisia* from the seeds. For a high germination rate, a farmer is expected to select fresh and healthy seeds for planting. Old seeds tend to be dormant, with poor germination and vigor. The seeds germinate fast at a relatively warm temperature. *Artemisia* seeds can be obtained from a local garden, research institute, or certified seed company. *Artemisia* seeds can be sown on the surface of fertile, moist, and aerated soil with good drainage. (www.SeedsNow. com; Ferreira and Janick, 2009 and Bui et al., 2011)

13.13.1 Seed Treatments

For rapid germination, the seeds of *A. annua* may be subjected to any of the following treatments:

13.13.1.1 Chemical Treatment

Seeds are soaked in 10%, 20%, and 30% sulfuric acid (H_2SO_4) for 1, 2, and 3 min, respectively, with a control (0%).

13.13.1.2 Hormone Treatment

The seeds are soaked in different plant growth hormones at varying concentrations and timing. For example, different GA_3 concentrations of 100, 300, 500 ppm with a control (0%) for 6, 12, and 24 h, respectively.

13.13.1.3 Hot Water Treatment

Artemisia seeds can be soaked in a water bath and incubated at different temperatures such as 20°C, 40°C, and 60°C for 1, 3, and 5 min, respectively.

13.13.1.4 Cold Water Treatment

The seeds may be washed and soaked in cold water for 2, 4, and 6 h before planting.

13.13.2 Ridge Preparation

The ridges are prepared after ploughing and harrowing the sandy loam soil using a tractor. This provides good drainage and moist but damp soil conditions. Plots, each measuring 5.0×4.5 m, are marked out before transplanting (Ferreira et al., 1997).

13.13.3 Sowing

A bright, incandescent light on the seeds may provide both the heat and light they need for germination. Try to maintain temperature levels of about 20°C to promote healthy germination of the seeds.

In order to minimize seed waste and for maximum seedlings supply, *Artemisia* seeds may be sown in plastic containers with sterilized river soil (Ferreira et al., 1997) and monitored for germination. Transparent polythene material is used to cover the seeds.

This is to help maintain adequate moisture, temperature, and humidity levels in the soil, which are essential for *Artemisia* seed germination.

Even under the best of conditions, some seeds will not germinate, so it should be ensured that more seeds are available than are desired for planting. To prevent overcrowding, the seedlings usually require thinning. To minimize disturbance to a seedling that is retained, the soil is pressed around it after thinning the adjacent seedlings. The newly establishing seedlings are watered frequently until the roots have developed (www.SeedsNow.com; www.raganandmassey.com/riata.html).

Two weeks after germination, the tender seedlings are thinned by separating them individually into different containers or spaces with a mixture of 50% each of river sand and compost to facilitate drainage.

Depending on the viability of the seeds and environmental conditions, germination may occur in 3–7 days or 14–30 days. Days to germination are determined by counting the number of days to germinate per seed. However, germination percentage is calculated according to Wiese and Binning (1987):

$$Gr = (\text{number germinating since } n-1)/n$$

where:

Gr is the germination rate
n is the number of days of incubation

13.13.4 Watering

While waiting for the *Artemisia* seeds to germinate, depending upon the humidity in the area, the soil should be kept moist and damp by regular watering morning and evening.

13.13.5 Transplanting

Sixty days after germination, the seedlings are hardened gradually by exposing them to sunlight while reducing watering, prior to transplanting. Twenty seedlings are transplanted per plot of 5.0×4.5 m. In this case, there are four rows per plot and inter- and intra-row spacing of 1.5 and 1.0 m, respectively. The ridges are labeled and irrigated with the same volume of water or rain fed up to harvesting.

13.13.6 Weeding

Weeds are controlled mechanically by hoeing at two-week intervals or chemically through the use of herbicides depending on the type of the weeds.

13.13.7 Data Collection

Depending on the scope of the research, as well as the species type, the following data are collected from *in situ* experiments at two-week intervals up to the flowering stage:

1. Seedling vigor is determined based on morphological appearance and the presence of 5–6 leaves, per Gibson (1980). A scale of 1–5 is used, where 1 = very high vigor and 5 = very low vigor.

2. Plant height is determined using measuring tape from the level of the soil to the top of the shoot apex.
3. The plant canopy spread is determined by stretching measuring tape horizontally across the plant.
4. Stem girth is determined 5 cm above the ground using a thread.
5. Fresh leaf and root weight are obtained by using an electronic weighing scale.
6. Dry leaf weight is determined using an electronic weighing balance after air drying under the shade for three weeks.

13.13.8 Germination and Seedling Growth

Germination is a vital phenomenon during the life cycle of a plant (Geraldine and Lisa, 1999). The germinating seeds of *A. annua* exhibit a hypogeal type of germination by having the cotyledon remain below the soil surface. A seed is considered germinated when the tip of the radicle has grown free of the seed coat, emerging through the outer covering (Wiese and Binning, 1987; Auld et al., 1988). A hypogeal type of germination is observed to occur 2–3 days after sowing with the leaves being aromatic, deeply dissected, and ranging from 3.0 to 8.5 cm in length. This is similar to the findings of Ferreira and Janick (2009). Exposure of the shoot tip to light enables it to photosynthesize, thereby straightening the epicotyls (Moore, 1979; Osborne, et al, 1985).

Early germination has been observed after covering *Artemisia* seeds with polythene material. This might help in maintaining adequate moisture, warmth, and humidity levels in the soil, which are essential for their germination. Dormancy of some seeds is reported to be inhibited when soil temperatures are too warm. They therefore germinate only at high temperatures (Nicolas, 2003). Similarly, germination depends on weakening the seed coat by heating, thus providing the optimum temperature for influencing the rate of enzyme-controlled reactions (Taylor, 1997; Style, 2008). This physiology contributes to the maintenance of dormancy by impeding water and gas to and from the embryo; chemically, by inhibiting germination and mechanically, by restricting the growth of the embryo (Mott and Groves, 1981; Farouk et al., 2008; Style, 2008). This eventually serves as a barrier that restricts water uptake by the impermeable outer part of the epidermal layer of Malphigian cells, thus restraining expansion of the radicle and manifestation of germination (Bewley and Black, 1994).

After planting, the use of transparent polythene exposes the seeds to light, an important regulatory environmental signal that triggers germination (Baskin and Baskin, 1993), and produces a favorable response (Modares and Hashemi, 2003). This is contrary to the findings of Jamaleddine et al. (2011) who reported that by using *in vitro* techniques, after planting, seeds of *A. annua* germinated after exposure to dark.

In a field experiment reported by Tahir et al. (2015), germination of seeds of *A. annua*, commenced 2–3 days after sowing, and 96% of the seeds responded to treatment. This is contrary to the findings of Mannan et al. (2012) and other workers, who observed that seeds of *A. annua* and *A. absinthium*, as obtained, germinated in 6–7 days when grown under field conditions. An insignificant height may be observed with the young *Artemisia* seedlings. This can be attributed to a lack of ample reserved nutrients such as carbohydrate, lipid, and protein to enable the seedlings to achieve a critical size advantage (Michael, 1993).

13.14 VEGETATIVE PROPAGATION

In vitro propagation and the use of seed are common practices for the propagation of *A. annua* that are not always suitable under all conditions. As a matter of fact, the first method is expensive and time-consuming, whereas the second is characterized by high genetic variability and scarcely homogeneous production. Hence, propagation by cuttings remains an appropriate method that gives rapid propagation with reduced costs (Saranga and Cameron, 2006). Also, insufficient agro-technological information about ideal planting dates, seed density, harvesting system, and post-harvesting, coupled with harsh weather, are some of the challenges facing *Artemisia* producers using seed, especially in the tropical regions. Therefore, propagation by cuttings represents the most effective, economically viable, and rapid asexual method of propagating *Artemisia* to provide raw materials to the pharmaceutical industry for the extraction of artemisinin, which is eventually used for the synthesis of antimalarial drugs.

As with the seeds, *Artemisia* cuttings may also be treated by soaking them in different plant growth hormones such as NAA and GA_3 at varying concentrations (0.1%, 0.2%, and 0.3%), with a control (0%), for different durations.

Artemisia cuttings of varying lengths (for example, 8, 12, and 16 cm) are planted in polythene bags containing a mixture of sandy, loamy soil and cow dung at a ratio of 3:1 and monitored for regeneration.

The stem cuttings should be planted as soon as possible to prevent them drying out. If, for any reason, the transplanting is to be delayed, the cuttings should be kept moist in a sealed plastic bag under shade. After planting, the stem cuttings in the polythene bags are placed in shade to minimize the rate of evaporation from the stems, which in turn prevents them drying out. This process enhances the regeneration process of *A. annua* and is crucial to its asexual propagation.

The rooting capacity of cuttings is influenced by internal factors, such as genotype and nutritional status, and external factors, including temperature and light intensity (Hartmann et al., 1997; Agbo and Obi, 2008; Priadjati et al., 2001). Internal factors are strictly related to the amounts of plant growth regulators that are physiologically necessary for the rooting phase (Guo et al., 2009, Amri et al., 2010; Zobolo, 2010). Exogenous auxins (NAAs) are commonly used to improve natural rooting efficiency in root and stem cuttings, but it has been demonstrated in various plant species that relatively high NAA concentrations are required only during the induction phase, while these plant growth regulators become inhibitory during development (Hartmann et al., 1997). Root formation by cuttings is also affected by physical and chemical characteristics of the rooting substrate, such as bulk density, porosity, water-holding capacity, and pH, which can promote or inhibit root growth

(Hartmann et al., 1997). Moreover, adventitious rooting is often related to the season in which cuttings are collected, as the availability of internal NAA, as well as the nutrient content of plant tissues, may determine significant variations in the rooting capacity of cuttings (Rosier et al., 2004).

Tahir and Muhammad (2016) reported that all stem cuttings treated with NAA did not regenerate and dried out 16 days after planting. However, using hormone treatment, stems of 16 cm length exposed to 0.03 mg/L GA_3 regenerated in 6 days after planting; those treated with 0.02 mg/L regenerated in 9 days; and the control regenerated in 12 days. The 8 cm stem cuttings treated with GA_3 at concentrations of 0.01, 0.02, and 0.03 mg/L did not regenerate, and dried out in 10–14 days after planting (Tables 13.1 and 13.2).

Using a GA_3 concentration of 0.03 mg/L with 16 cm stem cuttings resulted in the highest percentage regeneration of 100%, as compared with the control's 33.3%. This was followed by 66.66% regeneration for 16 cm cuttings and 0.02 mg/L GA_3, and lastly by 33.33% for the 12 cm cuttings with 0.03 mg/L GA_3 (Tables 13.1 and 13.2). The 16 cm stem cuttings with a GA_3 concentration of 0.03 mg/L produced the greatest seedling height, followed by 16 cm cuttings with 0.02 mg/L, and lastly by the 12 cm cuttings with 0.03 mg/L GA_3 (Tables 13.1 and 13.2).

TABLE 13.1   The Effects of Varying Concentrations of GA_3 on Asexual Propagation of *A. annua*

Treatment	DR	PR (%)	PH (cm)	NL	VG
GA_3					
0.01 mg/L	0.00[b]	0.00[bb]	0.00[b]	0.00[bb]	0.00[c]
0.02 mg/L	2.97[ab]	33.33[a]	5.95[aa]	25.33[aa]	0.88[b]
0.03 mg/L	4.00[aa]	37.03[aa]	7.25[aa]	25.22[a]	1.55[a]
CONTROL	2.55[ab]	14.81[b]	3.55[ab]	8.22[b]	0.22[cc]
LSD	3.66	14.91	3.94	11.59	0.85

Means within a column followed by the same letter along columns are not significantly different ($p = 0.05$).
Key: DR, days to regeneration; PR, percentage regeneration; PH, plant height; NL, number of leaves; VG, vigor.

TABLE 13.2   The Effect of Length of Stem Cuttings on Asexual Propagation of *A. annua*

Stem Cutting Length	DR (days)	PR (%)	PH (cm)	NL	VG
16 cm	4.91[aa]	55.55[a]	10.25[a]	40.33[a]	1.83[a]
12 cm	2.25[ab]	8.33[bb]	2.18[bb]	3.75[bb]	0.17[bb]
8 cm	0.00[b]	0.00[b]	0.00[b]	0.00[b]	0.00[b]
LSD	3.26	12.91	3.33	10.03	0.41

Means within a column followed by the same letter along columns are not significantly different ($p = 0.05$).
Key: DR, days to regeneration; PR, percentage regeneration; PH, plant height; NL, number of leaves; VG, vigor.

The highest number of leaves (76) was produced from 16 cm stem cuttings treated with a concentration of 0.02 mg/L of GA_3, followed by 16 cm cuttings with 0.03 mg/L of GA_3 (41), compared with the 37 leaves grown by the control. The lowest number of leaves was obtained with 12 cm cuttings and 0.03mg/L (Tables 13.1 and 13.2). The best plant vigor of 2 was produced by the 0.03 and 0.02 mg/L concentrations of GA_3 and 16 cm stem cuttings. This was followed by the 0.03 mg/L concentration of GA_3 and 12 cm stem cuttings, which had a vigor of 4 (Tables 13.1 and 13.2).

According to Tahir and Muhammad (2016), the significant effect observed using different concentrations of GA_3 on the asexual regeneration and phenotypic characteristics of *A. annua* is similar to the findings of Jennifer (2010), who studied the effect of GA_3 on *A. annua* in a hydroponic system, and found that the cuttings treated with GA_3 showed the highest growth rate and best seedling characteristics.

Similarly, it has been reported that the higher the concentration of GA_3 the greater the improvement in regeneration rate and other seedling characteristics. This is contrary to the findings of Brian (1958), who described the effect of varying GA_3 concentrations on *Pisum sativum* and found that low concentrations led to greater increases in internode extension and other morphological characteristics.

The length of stem cuttings was also observed to be a significant factor, as higher regeneration percentages were obtained with 16 cm stem cuttings. This might be linked to their higher nutrient content compared with shorter cuttings. However, 8 cm long stem cuttings did not survive, and dried out within 10–14 days of planting.

All the stem cuttings treated with NAA did not survive, turned brown, and dried out within 2 weeks of planting. The poor performance of the NAA-treated cuttings compared with those treated with GA_3 indicated its inhibitory effect on shoot and root initiation of *A. annua*. This was also observed by Rao et al. (2000), who reported 86% rooting in *Tinospora cordifolia* without any hormone treatment. A very low concentration of IBA was sufficient to increase rooting. However, with increasing concentrations of hormones, rooting percentage was observed to decline. This indicated that these chemicals, which would otherwise act as stimulants of rooting, could also have an inhibitory effect on initiation of rooting.

Asexual propagation *A. annua* is a cost-effective method for propagating *A. annua*. This in turn provides the biomass materials for the extraction of artemisinin, which is a raw material for the manufacture of antimalarial drugs for treating malaria fever.

The application of manures and compost as soil additives is a well-established and ancient agricultural practice used by small- and large-scale farmers alike (Muchena, 1986). Since poultry manure is organic, it has a beneficial effect on the physical and chemical properties of the soil (Asafu-Agyei et al., 1997). The purpose of organic fertilization, besides the addition of nutrients to the soil, is to improve the soil organic matter content. It is also known to improve soil aeration, water-holding capacity, permeability, and resistance to erosion.

Miller and Turk (1991) reported that poultry manure promotes the growth of plants by providing microorganisms, micronutrients, organic matter, and regulating substances. According to Bandel et al. (1972), poultry manure contains appreciable amounts of N, P, and K at levels of 45%, 2.5%, and 2%, respectively. Ahn (1993) confirmed that poultry manure was rich in nitrogen and was the most potent farmyard manure to increase yield. Kallah and Adamu (1988) conducted a study involving the use of different organic manures. They reported that the relative efficiency of organic manure in improving soil fertility followed the order poultry manure > pig manure > farmyard manure. De Ridder (1990) reported significant yield responses from the application of between 2.5 and 5.0 t/ha of manure. The manure tonnage is high because the concentration of nitrogen in animal manure varies as much as fourfold, depending on the type of feed fed to the animals. This was confirmed by Ofori et al. (1997), who reported a variation from 0.5% to 2.0% nitrogen in manure. Liebhardt (1976) reported that the addition of 5 t/ha of poultry manure doubled the yield of *Artemisia* as compared with a control. Hileman (1971) also reported that poultry manure enhanced the rapid release of ammonia. Application of organic fertilizer increased the biomass yield and total essential oil yields of *Artemisia* (Parakasa, 1997). Poultry manure improves the physical and chemical properties of the soil by increasing its water infiltration rate, water-holding capacity, cation exchange capacity, and structural stability (Moore et al., 1995). Bationo and Mokwunye (1992) also confirmed that the addition and incorporation of organic materials either in the form of manures or crop residues had beneficial effects on the chemical, physical, and biological properties of soil.

According to Khalid and Shafei (2005), treating plants with different combinations of organic fertilizers and at different rates resulted in a significant increase in growth, yield characteristics, and artemisinin content. Marculescu et al. (2002) observed that soil enriched with macro- and micro-elements through the use of organic fertilizers plays an essential role in the growth and development of plants and in the biosynthesis of organic substances at all levels. The reason for the increase in crude extract yield may be the influence of poultry manure in promoting vegetative growth, which results in increased herbage production (Singh, 2001; Singh et al., 2004). A possible reason for increased shoot weight is the increase in shoot length when poultry manure is applied, as its effect on the photosynthetic efficiency of a crop results in the production of more leaves and stems or increased vegetative growth (Majbur Rahman Islam and Osman, 2003).

High fresh leaf weight and dry leaf weight were observed by Tahir et al. (2015) with the application of 6 t/ha of poultry manure. Similarly, Arul Navamani (2002) and Khalid and Shafei (2005) observed that application of poultry manure increased the nitrogen level of soil, which in turn increased leaf production, resulting in increased fresh herbage yield. The use of organic fertilizers plays an essential role in plant growth and development, biosynthesis of organic substances at all levels, and growth yield characteristics such as biomass yield, as reported by Marculescu et al. (2002).

Similarly, Dixit (1997) and Mathias (1997) reported that poultry manure increased plant shoots. This was attributed to probable low-nutrient status of soil by Yeboah (2010) Similar results were also reported by Agyenim-Boateng (1999), who was working on maize and found that the total weight of dry matter was higher in plants treated with poultry manure. However, 6 t/ha of poultry manure was reported by Tahir et al. (2015) to give the greatest plant height. This contradicts what was reported by Martinez and Staba (1988): that N fertilization stimulates the vegetative development of the plant, and the greater the application of N, the greater the height of the plant. This may also be attributed to an adequate supply of nutrients that influenced cell division and cell enlargement, resulting in better plant height, as suggested by Martinez and Staba (1988) and Gandhi (1996).

This suggests that application of poultry manure at a rate of 6 t/ha may be considered the optimum rate above which *Artemisia* does not respond. Application of 8 t/ha of poultry manure may be considered excessive, as it is not efficiently utilized by plants (Tahir et al., 2015). Consequently, high quantities of dissolved salt accumulate in the soil, making nutrients unavailable (Hileman, 1971).

Similarly, Ferreira et al. (1995) reported that organic amendments promoted vigorous growth and development, and strong root system and stem development of *Artemisia*. Poultry manure contains a considerable amount of nitrogen that results in the production of greater stem width and longer internodes (Bandel et al., 1972; Kallah and Adamu, 1988).

Inorganic manure is important for improving soil characteristics, including organic carbon and total nitrogen. If applied, organic matter improves soil structure, aeration, root penetration, water percolation, and efficiency in utilizing irrigation water. The addition of 20 t/ha of well-rotted manure has been recommended as both a short-term crop stimulant and a long-term measure to improve soil fertility (Rodriquez, 1986). Organic fertilizer, in addition to supplying nutrients, also improves the physicochemical condition of soils, enhancing nutrient cycling and building the soil's organic matter resources. Animal dung also has a favorable effect on artemisinin content. Land application of sewage sludge enhances microbial activity that may affect soil cycling, and therefore influences plant metabolism. Application of animal dung and other compost improves the soil's physical, biological, and chemical properties and their promotive effect on artemisinin content may be attributed to the release of nutrients, mainly nitrogen, which favors primary metabolism, growth, photosynthetic pigments, and nutrient status (Gandhi et al., 2000).

The kind of fertilizer applied is important not only for the economic returns the producer expects but also for environmental reasons. Therefore, soil analysis has been recommended for predicting fertilizer needs (Sallah, 1991). Plants usually respond positively to N application, but this is not always true, and depends on the agro-ecology and the cropping history of the field. Adequate and balanced nutrients are necessary to obtain high yields. Rodriquez (1986) and Agboola (1986), in their report on improving maize grain yield, suggested that responses were obtained with up to 100 kg N/ha in the Sudanese savannah, 150 kg N/ha in the Guinean savannah, and even higher doses for the forest region. Thus, optimum application of fertilizer is important in crop production but may vary from one agro-ecology to another. Dennis (1990) observed that the optimum application of fertilizer may be around 80 kg N/ha, but higher rates are needed on savannah soils. Simon et al. (1990) reported that both optimum essential oil of 85 kg/ha and fresh whole plant biomass yield of 35 t/ha were achieved at 67 kg N/ha at high plant density. However, the plants had a lower leaf-to-stem ratio. It was further observed that a fresh whole plant biomass of 270–750 g/plant was produced. Nitrogen deficiency was associated with a large decrease in artemisinin and leaf biomass yield (Fiqueira, 1996). Fiqueira (1996) observed that the omission of nitrogen or phosphorus drastically reduced plant growth and dry matter production in hydroponic studies in Brazil. Magalhaes et al. (1996) conducted a study with four levels of nitrogen (0, 32, 64, and 97 kg N/ha) applied as urea. The results revealed that the highest fresh whole plant biomass yield 3880 kg/ha and artemisinin yield of 40.4 kg/ha. In the same study, no significant differences were found for leaf biomass and artemisinin yield when ammonium sulfate and ammonium nitrate were compared as sources of nitrogen. To obtain high efficiency of applied nitrogen, Rodriquez (1986) recommended that the nitrogen be applied in band or side-dressed. Moll et al. (1982) defined nitrogen use efficiency of a cultivar as yield per unit of available N (soil + compound fertilizer N). Easst African Botancials Ltd (EABL) (2005) reported that a 50–50 split application of N, 50% at planting and 50% when the crop reached 50 cm tall, increased leaf biomass and artemisinin content.

According to Ferreira et al. (1995), flowering could be delayed by cutting the apical meristem and providing nitrogen fertilizer. This causes the plant to branch out and potentially increase leaf biomass. The WHO (1988) reported significant increases of total plant and leaf dry matter were obtained where a complete fertilizer mixture containing 100 kg N/ha, 100 kg P/ha, and 100 kg K/ha was applied. Dry leaf yields of *Artemisia* between 6 and 12 t/ha were obtained from a mixed fertilizer containing 60 kg N/ha, 60 kg P/ha, and 50 kg K/ha (Laughlin, 1994). Muchow (1988) reported that nitrogen supplementation to plants increased leaf area development, delayed leaf senescence, and increased the photosynthetic capacity of the leaf canopy. It was also noted that maximum yield was obtained by the application of 83 and 95 kg N/ha in the forest and transition zones, respectively. The response to P was non-significant. Wright (2002) noted that nitrogen is a very mobile element and easily leaches out of the root zone in tropical and subtropical conditions. He further observed that the yield of dry leaf varies from 1 to 40 t/ha.

Artemisia responds well to balanced fertilizers, and appears to be responsive to nitrogen, but few data are available on the accumulation of artemisinin relative to fertility. It has also been noted that nitrogen is the main constituent of protein and nucleic acid, which influence cell division and cell enlargement, causing proliferation of roots and increase in shoot length, and resulting in better plant height (Gandhi, 1996). Veldkamp (1992) reported that increases in shoot weight due to increased N levels may be attributed to the adequate supply of plant nutrients needed for protein, amino acid, and energy synthesis and improved metabolic activities. The number of branches increased substantially with treatments containing average levels of manure or compound fertilizer additions (Ming, 1994). The number of branches per plant was significantly enhanced by the greater availability of chemical or organic nutrients due to their positive effect on shoot biomass production (Rao et al., 1985). Magalhaes et al. (1996) reported that nitrogen application increased the dry leaf yield approximately twofold. Martinez and Staba (1988) also observed that N fertilization stimulates the vegetative development of plants: the greater the application of N, the greater the height of the plant. Plants with no manure or compound fertilizer additions were generally shorter and produced fewer nodes (Silva et al., 1971). Application of 50 kg/ha of NPK produced the highest number of branches in the report by Tahir et al. (2015). This is similar to the findings of Ming (1994), who observed a substantial increase in the number of branches with application of average levels of compound fertilizer. However, this is contrary to the findings of Arul Navamani (2002) and Rao et al. (1985) who recorded the highest number of branches with poultry manure.

Combined analysis of the effect of season, organic and inorganic manure on the growth of *A. annua* indicated significant differences between treatments ($P < 0.05$) on the plant traits studied, except on plant height, number of branches, fresh leaf weight, dry leaf weight, fresh whole weight before flowering, and dry leaf weight after flowering during the irrigation trial. The *Artemisia* seedlings established on the field, three weeks after transplanting. Likewise, 6 t/ha of poultry manure during the rainy season gave the highest plant height, followed by 4 t/ha of poultry manure during the irrigation trial (Taiz and Araya, 2007). Similarly, the highest number of branches was produced by 50 kg/ha of NPK in the rainy season trial, followed by the same quantity of NPK during the irrigation trial (Tables 13.1 and 13.2).

It was also reported that the best fresh leaf weight after flowering was obtained in the rainy season trial with 6 t/ha of poultry manure, followed by the irrigation trial with 4 t/ha of manure. The best dry leaf weight before flowering was observed with 25 kg/ha during the rainy season, followed by the same quantity of NPK during the irrigation trial. Similarly, the highest leaf fresh weight before flowering was found for 2 t/ha of poultry manure in the rainy season, followed by the same quantity during the irrigation trial. Also, the rainy season had the best result with respect to dry leaf weight after flowering with 8 t/ha of poultry manure, followed the irrigation trial with the same quantity (Tables 13.3 and 13.4).

No significant difference was observed between the two seasons with respect to canopy spread. However, the rainy season gave the best result with 100 kg/ha of NPK, followed by the irrigation trial with 25 kg/ha of NPK (Tables 13.3 and 13.4). Also, 25 kg/ha of NPK produced the highest fresh whole plant weight before flowering during the rainy season trial (Figures 13.2 and 13.3) followed by 6 t/ha of poultry manure. The lowest fresh whole plant weight was observed with 2 t/ha of poultry manure during the irrigation trial (Table 13.4).

TABLE 13.3 Effects of Organic and Inorganic Manure on the Growth of *A. annua* during Irrigation

| Treatment | PH (cm) | NB | CS (cm) | Growth Parameters | | | | | |
				LFWBF (g)	WFWBF (g)	LDWBF (g)	LDWAF (g)	LFWAF (g)	WFWAF (g)
				NPK					
25 kg/ha	42.26a	25.67a	39.62a	63.72a	208.97a	36.97a	15.13a	25.30bcd	265.92ab
50 kg/ha	49.05a	26.00a	32.24ab	42.67a	262.67a	26.87a	14.57a	16.38bcd	363.80a
75 kg/ha	42.31a	24.17a	25.61b	33.98a	187.55a	29.77a	15.07a	19.73bcd	199.65b
100 kg/ha	46.45a	23.83a	30.49b	39.65a	246.93a	32.75a	13.83a	31.02ab	162.90b
				Poultry Manure					
2 t/ha	47.23a	24.00a	28.25b	63.77a	250.70a	30.40a	13.72a	18.42bcd	256.77ab
4 t/ha	50.00a	24.00a	29.57b	46.13a	214.87a	25.50a	18.70a	44.68a	181.00b
6 t/ha	46.89a	25.67a	27.58b	47.87a	248.40a	26.10a	17.48a	12.30d	177.55b
8 t/ha	46.22a	24.83a	30.13b	53.08a	107.53a	18.37b	19.28a	26.88bc	234.63b

Means with same letters are not significantly different using LSD at 0.05%.
PH, plant height; NB, no. of branches; CS, canopy spread; LFWBF, fresh leaf weight before flowering; WFWBF, fresh whole weight before flowering; LDWBF, dry leaf weight before flowering; LDWAF, dry leaf weight after flowering; LFWAF, fresh whole leaf weight after flowering; WFWAF, fresh whole weight after flowering.

TABLE 13.4   Effects of Organic and Inorganic Manure on the Growth of *A. annua* during the Rainy Season

Treatment	PH (cm)	NB	CS (cm)	LFWBF (g)	WFWBF (g)	LDWBF (g)	LDWAF (g)	LFWAF (g)	WFWAF(g)
					NPK				
25 kg/ha	71.17c	39.33ab	40.00b	125.80a	758.20a	36.97a	16.32a	174.88ab	4455.00a
50 kg/ha	80.50ab	41.50a	43.00ab	122.46a	705.10a	26.87a	15.67a	179.93ab	3892.50a
75 kg/ha	77.00ac	35.50b	44.83ab	120.38a	606.70a	29.77a	16.20a	200.60ab	3384.20a
100 kg/ha	74.00c	38.167ab	46.33a	188.67a	571.30a	32.75a	14.75a	261.60a	4039.50a
					Poultry Manure				
2 t/ha	73.67c	39.178ab	44.50ab	179.87a	502.50a	30.40a	14.75a	231.83ab	4189.20a
4 t/ha	76.00c	27.33ab	44.83ab	149.54b	598.10a	25.50a	19.68a	209.35ab	4333.80a
6 t/ha	84.5a	35.00b	44.50ab	145.67b	672.10a	26.10a	17.78a	258.58a	4150.20a
8 t/ha	76.92c	36.83ab	45.17ab	147.55b	649.40a	18.37b	20.38a	152.77b	3619.40a
Control	97.167ab	38.500ab	46.00a	139.40b	673.30a	27.13a	25.55a	202.20ab	4457.60a

Means with same letters are not significantly different using LSD at 0.05%.
PH, plant height; NB, no. of branches; CS, canopy spread; LFWBF, fresh leaf weight before flowering; WFWBF, fresh whole weight before flowering; LDWBF, dry leaf weight before flowering; LDWAF, dry leaf weight after flowering; LFWAF, fresh whole leaf weight after flowering; WFWAF, fresh whole weight after flowering.

Figure 13.2   *A. annua* plants at the pre-flowering stage.

Figure 13.3   *In situ A. annua* plants at the flowering stage.

13.15 CONCLUSION

Covering *A. annua* seeds with a transparent polythene material under moist conditions is a simple and affordable technique, especially for local farmers. It gives good results in terms of germination and seedling production of the plant in comparison to leaving them uncovered. Planting in March can be tried in the tropics for *in situ* germination of *A. annua*. However, more work needs to be done to ascertain more suitable planting dates to facilitate timely transplanting of the seedlings to the field at the onset of rainfall. The use of poultry manure and NPK fertilizers for *in situ* biomass and artemisinin production from *A. annua* is highly recommended. Vegetative propagation using stem and root cuttings is also a viable means of large-scale propagation. More research is required on the economic exploitation of *A. annua* and its resources.

REFERENCES

Advanced Bio-Extracts 2006. The accelerated production of artemisinin in East Africa, Project Summary. Advanced Bio-Extracts Limited, Nairobi, Kenya.

Advanced Bio-Extracts 2007. Project status report. Advanced Bio-Extracts Limited, Nairobi, Kenya.

Agbo C.U. and Obi I.U. 2008. Patterns of vegetative propagation of stem-cuttings of three physiological ages of *Gongronema latifolia* Benth over two seasons in Nsukka. *Journal of Tropical Agriculture, Food, Environment and Extension* 7: 193–198.

Agyenim-Boateng K. 1999. Chicken and green manure as possible alternatives to compound fertilizer in semi-deciduous forest zones of Ghana. BSc. thesis, College of Agriculture and Natural Resources. Kwame Nkruma University of Science and Technology, Kumasi Ghana. 21

Amri E., Lyaruu H.V.M., Nyomora A.S., and Kanyeka Z.L. 2010. Vegetative propagation of African Blackwood (*Dalbergia melanoxylon* Guill. & Perr.): Effects of age of donor plant, IBA treatment and cutting position on rooting ability of stem cuttings. *New Forests* 39(2): 183–194.

Anonymous. 2007. Nigeria set to produce drugs for malaria. Daily Champion. Lagos, Nigeria. September 21.

Antony E. 2010. Cultivation of *Artemisia annua* in Africa and Asia. Research Information Ltd. www.pestoutlook.com

Arab H.A., Rahbari S., Rassouli A., Moslemi M. H., and Khosravirad, F. 2006. Determination of Artemisinin in Artemisia sieberi and anticoccidial effects of the plant extract in broiler Chicken. *Tropical Animal Health and Production.* 38: 497–503.

Artepal 2008. Artemisinin market situation in June 2008, Artepal Newsletter [info@artepal. org], Aug. 12 update.

Arul Navamani V.J. 2002. Integrated nutrient management of Ashwagandha (*Withania somnifera* Dunal.) for growth, yield and quality. M.Sc. (Hort.) thesis, Agricultural College and Research Institute, Tamil Nadu Agricultural University, Madurai.

Bailey L.H. and Bailey E.Z. 1976. *Hortus Third.* MacMillan, New York.

Bandel V.A., Shaffeur C.S., and McClare C.A. 1972. Chicken manure; a valuable fertilizer, University of Maryland Coop. Extension Service Fact Sheet, 39–40.

Bennett M.D., Smith J.B., and Heslop-Harrison J.S. 1982. Nuclear DNA amounts in angiosperms. *Proceedings of Royal Society London B* 216(1203): 179–199.

Bewley, J.D., and Black, M. 1994. Seeds: Physiology of development and germination. The language of science. 2nd edition. Plenum Press, New York, NY. 445–46.

Brian P.W. and Hemming H.G. 1958. Complementary action of gibberellic acid and auxin in pea internode extension. *Annals of Botany* 22(1): 1–17.

Brisibei E.R., Umoreu U.E., Owai P.U., and Brisibe F. 2008. Dietary inclusion of dried Artemisia annua leaves for management of coccidiosis and growth enhancement in Chickens. *African Journal of Biotechnology* 7: 4083–4092.

Brown, M.L. 1990. The national economic burden of cancer: an update. *Journal of the National Cancer Institute* 82: 1811–1814.

Bui T.T.T., Tran V.M, Boey P.L, and Chan L.K. 2011. Effects of environmental factors on growth and artemisinin content of *Artemisia annua* L. *Tropical Life Sciences Research* 22(2): 37–43.

Chen P.K., Leather G. and Polatnick M. 1991. Comparative study on artemisinin, 2,4-D, and glyphosate. *Journal of Agriculture and Food Chemistry* 39: 991–994.

Cutler M. 2008. *Artemisia*/artemisinin industry observations, unpublished paper. Artemisinin Forum. Guilin, China.

Dixit S.P. 1997. Response of onion to nitrogen and farmyard manure in dry temperature high hills of Himachal Pradesh. *Indian Journal of Agricultural Science* 67(5): 222–223.

Duke S. and Paul R. 1993. Development and fine structure of the glandular trichomes of *Artemisia annua* L. *International Journal of Plant Sciences* 154: 107–118.

El-Haq F., EI-Feraly J.S., and Hafez M. 1991. *In vitro* propagation of *Artemisia annua* L. *Journal of King Saud University. Agricultural Sciences* 3(2): 251–2259.

Eric, S. (2012). *Artemisia annua*: A Vital Partner in the Global Fight against Malaria. *Science* 336: 80.

Farouk A., Abeer I. and Fawzia M.R. A. 2008. Effect of Chlorogenic and Caffeic Acids on Activities and Isoenzymes of G6PDH and 6PGDH of Artemisia Herba Alba Seeds Germinated for One and Three Days in Light and Dark. *Jordan Journal of Biological Sciences.* 1(2): 85–88

Ferreira J.F.S and Janick J. 1995. Floral morphology of Artemisia annua with special reference to trichomes. *International Journal of Plant Sciences* 156: 807–815.

Ferreira, J.F., Simon, J.E., and Janick, J. 1997. Artemisia annua: botany, horticulture and pharmacology. J. Janick, (Ed). Horticultural reviews, John Wiley & Sons, Inc., New York, NY 319–371.

Ferreira J.F., Simon J.E., and Janick J. 1995. Developmental studies of Artemisia annua: Flowering and artemisinin production under greenhouse and field conditions. *Planta Medica* 61: 167–70. doi:10.1055/s-958040.

Ferreira J.F., Simon J.E., and Janick J. 1997. *Artemisia annua*: Botany, horticulture and pharmacology. In: J. Janick (ed) *Horticultural Reviews*. Wiley, New York, 319–371.

Ferreira J.F.S and Janick J. 2002. Production of artemisinin from *in vitro* cultures of *Artemisia annua* L. *Biotechnology in Agriculture and Forestry* 51: 1–12.

Gandhi K.P. 1996. Studies on the effect of plant density, nitrogen and *Azospirillum* on growth, herbage and essential oil of davana (*Artemisia pallens* wall). M.Sc. (Hort.) thesis, Tamil Nadu Agricultural University, Coimbatore.

Guo X., Fu X., Zang D., and Ma Y. 2009. Effect of auxin treatments, cuttings collection date and initial characteristics on *Paeonia*'Yang Fei Chu Yu' cutting propagation. *Scientia Horticulturae* 119(2): 177–181.

Hartmann H.T., Kester D.E., Davies F.T., and Geneve R.L. 1997. *Plant Propagation: Principles and Practices,* 7th edn. Prentice-Hall, Englewood Cliffs, NJ.

Hiey H.N. 2010. Artemisinin production in Vietnam. Artemisinin Conference, Madagascar, October [www.mmv.org/newsroom/events/pastevents/ past-artemisinin-events].

Hileman L.H. 1971. The effect of rate of chicken manure application on selected soil chemical properties. America Society of Agricultural Engineers, St. Jos, Michigan, 246–251.

Jamaleddine Z. O., Lyam P., Fajimi O., Giwa A., Aina A., Lawyer E. F., Okere A. U. et al. 2011. Invitro growth response of Artemisia annua seeds to different concentrations of plant growth regulators. *African Journal of Biotechnology.* 10 (77): 17841–17844. DOI: 10.5897.

Kallah P. and Adamu R. 1988. Annual manure intensive vegetable garden for profit and self sufficiency. Peace Corp Information Collection and Exchange. Washington, DC, 5–8.

Khalid H.A. and Shafei A.M. 2005. Production of dill (*Anethum graveolens* L.) as influenced by different organic manure rates and sources. *Journal of Agricultural Science* 13(3): 901–913.

Kindscher K. 1992. Medicinal wild plants of the prairie. An ethnobotanical guide. University Press of Kansas, KY. 340.

Klayman D.L. 1993. *Artemisia annua*: From weed to respectable anti-malarial plant. In A.D. Kinghorn and M.F. Balandrin (eds) *Human Medicinal Agents from Plants.* Symposium Series. American Chemical Society, Washington, DC, 242–255.

Klayman D.L., Lin A.J., Acton N., Scovill J.P., Hoch J.M., Milhous W.K. 1984. Isolation of artemisinin (qinghaosu) from Artemisia annua growing in the United States. *Journal of Nutritional Production* 47: 715–717.

Laughlin, J.C. 1994. Agricultural production of artemisinin-A review. *Transaction of the Royal Society of Tropical Medicine and Hygiene* 88(1): 21–22.

Lescano S.Z., Chieffi P.P., Canhassi R.R., Boulos M., and Neto V.A. 2004. Anti-schistosoma activity of Artemether in experimental Schistosomiasis Mansoni. *Revista de Saude Publica* 38: 71–75.

Luo X-D. and Shen C-C. 1987. The chemistry, pharmacology, and clinical applications of qinghaosu (artemisinin) and its derivatives. *Medicinal Research Reviews* 7: 29–52.

Majbur Rahman Islam A. and Osman K.T. 2003. Effect of N, P and K fertilizers on growth and herb and oil yield of *Mentha arvensis* in Bangladesh. *Journal of Medical and Applied Sciences* 25: 661–667.

Marculescu A., Sand C., Barbu C.H., Babit D., and Hanganu D. 2002. Possibilities of influencing the biosynthesis and accumulation of the active principals in *Chrysanthemum balsamita* L. species. *Romanian Biotechnological Letters* 7(1): 548–577.

Martinez B.C. and Staba E.J. 1988. The production of artemisinin in *Artemisia annua* L. tissue cultures. *Advances in Cell Culture* 6: 69–87.

Mathias K.A. 1997. Effect of chicken manure and inorganic fertilizer application on the growth and yield of cowpea. BSc. thesis, Kwame Nkrumah University of Science and Technology, Kumasi. p. 21.

McVaugh R. 1984. *Flora Novo-Galiciana: A Descriptive Account of the Vascular Plants of Western Mexico.* Vol. 12, Compositae. University of Michigan Press, Ann Arbor, MI, 1157.

Ming L.C. 1994. Influence of the organic manuring in the biomass production and tenor of essential oils of *Lippa Alba*. *Horticultura Brasileria* 12: 49–52.

National Academies (NA). 2007. Mobilizing Science-Based Enterprises for Energy, Water, and Medicines in Nigeria. National Research Council of the National Academies, National Academies Press, Washington, DC, 45–54.

Nigeria Malaria Fact Sheet/ United States Embassy in Nigeria. December 2011.

Paniego N. B. and Giulietti A. M. 1994. Artemisia annua L.: Dedifferentiated and differentiated cultures. *Plant Cell, Tissue, and Organ Culture* 36: 163–168.

Parija S.C. 2010. PCR for diagnosis of malaria. *Indian Medical Journal.* 3: 342–49

Pilloy J. National Academies (NA). 2007. Mobilizing Science-Based Enterprises for Energy, Water, and Medicines in Nigeria. National Research Council of the National Academies, National Academies Press, Washington, DC. 45–54.

Priadjati A., Smits W.T.M., and Tolhamp G.W. 2001. Vegetative propagation to assure a continuous supply of plant material for forest rehabilitation. In P.J.M. Hillegers and H.H. DeIongh (eds) *The Balance between Biodiversity Conservation and Sustainable Use of Tropical Rain Forests.* The Tropenbos Foundation, Wageningen, the Netherlands, 19–30.

Ram M., Gupta M. M., Dwivedi S. and Kumar S. 1997. Effect of plant density on the yields of artemisinin and essential oil in *Artemisia annua* cropped under low input cost management in north-central India. *Planta Medica* 63: 372–374.

Rao E.V.S.P., Narayana M.R., and Rao B.R.R. 1985. The effect of nitrogen and farm manure on yield nutrient uptake in davana (*Artemisia pallens* Wall. Ex DC.). *Journal of Herbs and Medicinal Plants* 5: 39–48.

Rao P.S., Vankajah K., Murali V., and Satyaranayana V.V. 2000. Macro-propagation of some important medicinal plant Andhra Pradesh. *Indian Forester* 126(12): 1265–1267

Roll Back Malaria 2009. Artemisinin requirements for the manufacture of ACTs. An update on the position paper of the Procurement and Supply Chain Working Management Group, Tech. Paper, Nov. 27, Roll Back Malaria Partnership, World Health Organization, Geneva.

Rosier C.L., Frampton J., Goldfarb B., Blazich F.A., and Wise F.C. 2004. Growth stage, auxin type, and concentration influence rooting of stem cuttings of Fraser fir. *HortScience*, 39(6): 1392–1396.

Saranga J. and Cameron R. 2006. Adventitious root formation in *Anacardium occidentale* L. in response to phytohormones and removal of roots. *Scientia Horticulturae* 111(2): 164–172.

Sharma V.P. 2006. Artemisinin drugs in the treatment of *Plasmodium falciparum* malaria in India. *Current Science (Bangalore)* 90: 1323–1324.

Stix E. 1960. Pollenmorphologische untersuchungen an Compositen. *Grana Palynologist* 2: 41–104.

Style, M.L.A. 2008. Plant Development. Encyclopedia Britanica, Ultimatte Reference Suit, Chicago, IL, 8–9.

Tahir S M, Usman I S, Katung M D, and Ishiyaku M. F. 2013. Micro Propagation Of Wormwood (*Artemisia Annua* L.) Using Leaf Primordia. *Science World Journal.* 8(1): 1–7

Tahir S.M., Muhammad A.S, Abdullahi A.Y., Abdulrahman M.D. Abdullahi F.A., Yakubu S., and Ibrahim H. 2016. Asexual propagation of worm wood (*Artemisia annua* L.) using stem cuttings. Unpublished project. Kaduna State University, Kaduna, Nigeria.

Tahir S.M., Usman I.S., Katung, M.D., and Ishiyaku M.F. 2015. In Vitro and in Situ Biomass and Artemisinin Production in Worm Wood (Artemisia Annua L. Var. Chiyong). Unpublished PhD Thesis. Department of Plant Science. Faculty Of Agriculture. Ahmadu Bello University, Zaria, Nigeria 166.

Tahir S.M., Usman I.S., Katung M.D., and Ishiyaku M.F. 2016. Effects Of Plant growth Regulators On Callus Initiation In Worm wood (*Artemisia annua.* L). *Bayero Journal Of Pure And Applied Sciences*, 9(1): 160–165.

Taiz L. and Eduardo Z. 1998 *Plant Physiology.* 2nd edn. Sinauer Associates, Sunderland, MA.

Tan N. 2006. Global situation of Artemisia annua and artemisinin production. Meeting on the production of artemisinin and artemisinin-based combination therapies, 6–7 June 2005, Arusha, United Republic of Tanzania, World Health Organization, Global Malaria Programme, Geneva.

Tripathi A.K., Projapati V., Aggarwal K.K., Khanuja S.P., and Kumar S. 2000. Repellency and toxicity of oil from *Artemesia annua* to certain stored product beetles. *Ecotoxicology* 93: 43–47.

Tripathi A.K., Projapati V., Aggarwal K.K., Kumar S. 2001 Toxicity, feeding deterrence and effect of activity of 1,8-cineole from *Artemisia annua* on progeny production of *Tribolium castaeum* (Coleoptera: Tenebrionidae). *Journal of Economic Entomology* 94(4): 979–983.

Ut B.M. 2011. Aligning *Artemisia* and ACT supplies: Vietnam production report 2011, *Artemisinin Conference*. Hanoi, November 2 [http://www.mmv.org/newsroom/events/pastevents/past-artemisinin-events].

Xiao S.H., Catto B.A. 1989. In vitro and in vivo studies of the effects of Artemether on Schistosoma mansoni. *Antimicrobs Agents Chemotheraphy* 33: 1557–1562.

Xiao S.H., Booth M., and Tanner, M. 2000. The prophylactic effects of artemether against Schistosoma japonicum infections. *Parasitology Today* 16: 122–126.

World Health Organization 2014. Factsheet on the World Malaria Report. Geneva, Switzerland, http://www.who.int/malaria/publications/world_malaria_report_2014/en/.

Wright C.W. 2002. Artemisia- Medicinal and Aromatic Plants-Industrial Profiles.

Yeboah S. 2010. Organic and inorganic fertilizers application on the growth, yield and artemisinin content of *Artemisia annua* L. in the humid tropics of Ghana. An unpublished MSC Thesis. 118.

Zobolo A.M. 2010. Effect of temperature, light intensity and growth regulators on propagation of *Ansellia africana* from cuttings. *African Journal of Biotechnology* 9(34): 5566–5574.

CHAPTER 14

Mode of Action of Artemisinin
An Update

Athar Ali, Abdul Qadir, Mather Ali Khan,
Parul Saxena, and Malik Zainul Abdin

CONTENTS

14.1 INTRODUCTION

Artemisinin (Qinghaosu) was first isolated from the aerial parts of the traditional Chinese medicinal herb *Artemisia annua* L., belonging to the family Asteraceae, by the Chinese pharmacologist Professor Youyou Tu on October 4, 1971 in the ongoing national project set up by the Chinese government under the leadership of Project 523 office China in the year 1967 with the aim of eliminating malaria worldwide (Klayman, 1985). For this work, she received half of the Nobel Prize given in the area of physiology or medicine on October 6, 2015. Today, it is a safe, potent, promising, and remarkable drug against both drug-resistant and cerebral malaria caused by strains of *Plasmodium* sp. (Newton and White, 1999), and still no serious side effects have been reported in the literature. The stereo-structure of artemisinin was first determined at the Institute of Biophysics, Chinese Academy of Sciences, in 1975 and published in 1977. It is a highly oxygenated sesquiterpene, containing a unique 1,2,4-trioxane ring structure, which is responsible for its antimalarial activity. Apart from its antimalarial activity, it is also found to be effective against other infectious diseases such as schistosomiasis

(Liu et al., 2011), HIV (Lubbe et al., 2012), hepatitis B and C (Romero et al., 2005; Obeid et al., 2013), leishmaniasis (Islamuddin et al., 2014), trypanosomiasis (Mishina et al., 2007), toxoplasmosis (Angelo et al., 2009), neosporosis (Müller et al., 2015), acanthamebiasis (Deng et al., 2015), naegleriasis (Gupta et al., 1995), cryptosporidiosis (Giacometti et al., 1996), giardiasis (Tian et al., 2010), and babesiosis (Iguchi et al., 2015). It is a natural herbicide (Duke et al., 1994). Artemisinin has shown effectiveness against a variety of cancer cell lines, including breast cancer, human leukemia, colon cancer, and small cell-lung carcinomas (Efferth et al., 2001; Singh and Lai, 2001). Due to the rampant development of drug resistance to traditional medicines such as chloroquine by malarial parasites in recent years, the World Health Organization (WHO) has strongly recommended the use of artemisinin-based combination therapy (ACT) as the first line of treatment of malaria due to its high efficacy, fast action, and lack of serious side effects (WHO, 2015).

Malaria is an infectious disease caused by five single cellular species of the genus *Plasmodium* parasite (mostly by *Plasmodium falciparum*). Four of these species, *P. falciparum*, *P. vivax*, *P. malariae*, and *P. ovale*, affect humans and are transmitted by female *Anopheles* mosquitoes. *P. falciparum* and *P. vivax* malaria causes the greatest public health challenge. *P. falciparum* is most prevalent in the African continent and is responsible for most deaths from malaria. *P. vivax* has a wider geographic distribution than *P. falciparum*, because it can develop in the *Anopheles* mosquito vector at lower temperatures and can survive at higher altitudes and in cooler climates. It also has a dormant liver stage that enables it to survive for long periods as a potential reservoir of infection. Although *P. vivax* can occur throughout Africa, the risk of infection with this species is quite low due to the absence in many African populations of the Duffy gene, which produces a protein necessary for *P. vivax* to invade red blood cells. In many areas outside Africa, however, infections due to *P. vivax* are more common than those due to *P. falciparum*.

Out of 97 endemic countries, malaria caused by *P. falciparum* is prevalent in 87, *P. vivax* in 55, and both in 47 countries. Forty percent of the world population living in these 97 countries are at risk of being infected, and half of these live in sub-Saharan Africa (WHO, 2015). In addition to causing untold suffering and disability, malaria ranks as one of the world's major killers, costing about 1.5 million human lives annually. It is estimated that 3.2 billion people are at risk and 1.2 million people are at high risk of malarial disease worldwide. According to the World Malaria Report 2015, 198 million malaria cases were reported globally in 2013, and the disease led to 438,000 deaths. According to WHO, 90% of all malaria deaths occurred in the African region alone, of which 80% were children under 5 years of age. This is the only disease from which most children in the poorest countries die after exposure to infection. Women in their first pregnancy are the next highest risk group for malaria in endemic areas. According to another estimation by WHO (2014), malaria is the sixth of the top ten causes of death in low-income countries (it accounts for 35 deaths/100,000 population).

Therefore, malaria remains a major global public health problem, even though it is both preventable and treatable.

To the poor sub-Saharan countries of Africa, WHO Roll Back Malaria has highlighted that

- The cost of malaria control and treatments drains African economies. Endemic countries have to use hard currency on drugs, nets, and insecticides.
- Africa's gross domestic product (GDP) today would be up to 32% greater if malaria had been eliminated 35 years ago. According to estimates from a Harvard study, malaria-endemic countries are among the world's most impoverished nations.
- A malaria-stricken family spends on an average over one quarter of its income on malaria treatment, as well as paying prevention costs and suffering loss of income.
- Workers suffering an attack of malaria can be incapacitated for at least 5 days.
- Malaria-afflicted families on average can only harvest 40% of the crops harvested by healthy families.

14.2 GLOBAL SCENARIO OF DEMAND, SUPPLY, AND COST OF ACTs

Due to an exponential increase in the number of countries adopting ACTs as a national health policy, the global demand for artemisinin and its derivatives (artemether, artesunate/lumefantrine) has increased up to approximately 180 metric tons (equivalent to 392 million courses of ACTs) against the global production of artemisinin, which is approximately 120 metric tons. Hence, there is a major gap between production of and demand for artemisinin worldwide (WHO, 2014). This gap has created instability in the supply and price fluctuation of ACTs. In addition, the cost of ACTs per dose is around USD 1.00 at government public health clinics and much higher at private pharmacies (Van Noorden, 2010). At present, artemisinin production at the commercial level is mainly based on plant extraction. In the year 2013, Paddon et al. successfully produced artemisinic acid through engineered yeast with a good yield (40%–45%). Artemisinic acid is further chemically converted into artemisinin (Paddon et al., 2013). Recently, a company based in Italy, Sanofi-Aventis, announced the commercial production of 50–60 tonnes artemisinin annually using this genetically engineered yeast, which is still not an adequate amount to meet the demand for ACTs and to reduce price fluctuations (www.path.org/news/pressroom/422/). Consequently, *A. annua* L. plants will remain an imperative source for artemisinin. Unfortunately, the low content of artemisinin in *A. annua* L. plants ($\simeq 0.5\%$–$0.9\ \%$ nationally and from $\simeq 0.9\%$ to 1.2% globally) is the main cause for low supply and price instability of ACTs in developing countries (Zhang et al., 2009).

14.3 TREATMENT OF MALARIA

Malaria is a disease that places a heavy burden in poor countries, particularly African countries, due to the lack of access to effective services for prevention, diagnosis, and treatment. In other words, malaria is retarding health system growth, infrastructure expansion, and poverty reduction. However, it is an absolutely curable and treatable disease. It was first treated with quinine worldwide until the 1940s but later replaced with chloroquine due to the horrible side effects of quinine. Quinine is still used in combination with doxycycline in the treatment of severe malaria, because it is cheaper than other antimalarial drugs. The development of *P. falciparum* resistance to chloroquine in African and Asian countries and South America was first documented in the 1950s, leading to a 40-fold decline in the therapeutic efficacy of the drug. Unfortunately, no other appropriate medicine was available for the treatment of malaria at that time. This problem of drug resistance was, however, sorted out through two decades' hard research work by Chinese pharmacologist Professor Yuyu Tu. She screened 2000 Chinese medicinal herb preparations from 200 medicinal herbs against the mouse model of malaria and exhaustively reviewed the literature. Finally, she was able to isolate an effective novel molecule called *artemisinin*. Since then, artemisinin has been effectively used alone for the treatment of malaria caused by all five species of *Plasmodium* and has saved the lives of millions of people globally. Hence, Professor Youyou Tu was awarded the Nobel Prize in Physiology or Medicine in 2015. Due to the short half-life of artemisinin in comparison with other antimalarial drugs, however, artemisinin leads to a high rate of recrudescence. Thus, for complete elimination of the parasite from the body, artemisinin should be given in prescribed doses with or without other antimalarial drugs. In addition, resistance against artemisinin has also been reported in some areas if sub-Saharan Africa and South-East Asia (Ariey et al., 2014). Therefore, WHO has strongly recommended artemisinin or its derivatives as a chief constituent in combination with other antimalarial drugs (ACTs) for the first line of cure of *P. falciparum* malaria (WHO, 2005). ACTs are effective (because the artemisinin component kills the majority of parasites at the start of the treatment, while the partner drug clears the remaining parasites), safe, and powerful medicines (White, 2004). Hence, they have been adopted in the national health policy of 79 out of 87 *P. falciparum*-endemic countries in 2013 for the first-line treatment of the disease (WHO, 2014). The number of doses of ACTs purchased from the manufacturers by both public and private sectors has significantly increased from 11 million in 2005 to 392 million in 2014. The percentage of children under the age of 5 years with *P. falciparum* malaria in Sub-Saharan Africa who were treated with ACTs also increased from 5% in 2005 to 26% in 2013 (WHO, 2015). ACTs are used in various combinations in both tablets as well as injection forms for the treatment of malaria (Figure 14.1). These are

- Amodiaquine plus artesunate
- Artemether plus lumefantrine (not recommended for the treatment of malaria of first trimester of pregnancy or children below 5 kg)
- Artesunate plus mefloquine or amodiaquine; or sulfadoxine plus pyrimethamine (used for the management of severe malaria)
- Artesunate plus mefloquine
- Sulfadoxine plus pyrimethamine plus artesunate

Of the abovementioned ACTs, artesunate and artemether are artemisinin derivates. Artemether plus lumefantrine shared 78% of the volume of total ACT procurement, followed by artesunate plus amodiaquine (26%) in the year 2013 (WHO, 2014). In addition, doxycycline, artemether, chloroquine, and primaquine are also used for prophylactic treatment. In all the abovementioned medicines, primaquine

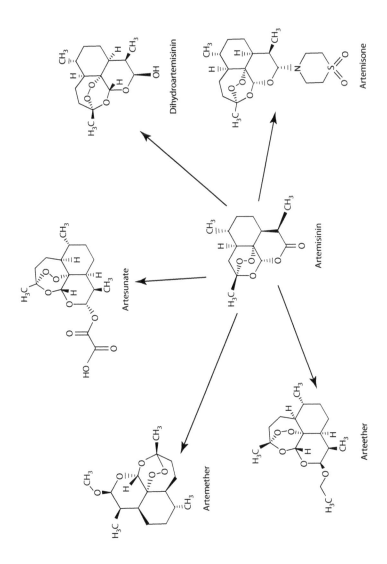

Figure 14.1 Artemisinin and its derivatives.

alone was used to achieve radical cure of *P. vivax* and *P. ovale* infection in 57 out of 97 countries. These medicines are categorized in both the "WHO Model List of Essential Medicine" and "WHO Model List of Essential Medicine for Children" (WHO, 2015). Recently, some research groups have reported for the first time that the direct use of dried whole plant (WP) of *A. annua* L. is more effective against the malaria parasite than a comparable dose of traditional medicine. This may be due to the synergistic effect of the presence of other antimalarial compounds in *A. annua* L. plants. This treatment is also known as *plant-based artemisinin combination therapy* (pACT) (Elfawal et al., 2012).

14.4 MODE OF ACTION OF ARTEMISININ

The mode of action of artemisinin as an antimalarial agent has been extensively studied by various research groups. However, they are still unable to propose a single mechanism for artemisinin's action. Therefore, the accurate mechanism of artemisinin's action is still an interesting research topic. Various theories, however, have been put forth to explain the antimalarial action of artemisinin. Among these is the "artemisinin activation theory." According to this theory, artemisinin is being activated to generate free radical species before its activation. This theory is supported by the fact that artemisinin interacts with hemoglobin-bound iron through a covalent reaction for its activation (Selmeczi et al., 2004; Kannan et al., 2005). This possibility, however, was over-ruled due to steric interference caused by bulky artemisinin when approaching the heme moiety of hemoglobin (Creek et al., 2009). Later on, it was assumed that ferrous heme (heme-Fe^{2+}) may be the most competent activator of artemisinin (Zhang and Gerhard, 2008), and artemisinin activation models were categorized into two main groups:

1. Reductive scission model
2. Open peroxide model

In reductive scission, ferrous-heme/non-heme exogenous Fe^{2+} is first attached to artemisinin by a covalent bond, causing the reductive scission of the endoperoxide bridge. This leads to the generation of oxygen-centered radicals, which subsequently self-arrange to give carbon-centered radicals. In addition, iron–peroxide interaction occurs in different ways to form either primary carbon-centered radicals (via C_3–C_4 bond scission) or secondary carbon-centered radicals (via 1,5 H-shifts). The open peroxide model, on the other hand, suggests that the endoperoxide group of artemisinin produces secondary carbon-centered radicals via a Fenton reaction involving the Fe^{2+} of hemoglobin (Haynes et al., 2007). It is reported, however, that these secondary carbon-centered radicals, which damage the parasite's essential biomolecules, are short-lived (Edikpo et al., 2013).

The "artemisinin activation theory" was further supported by research findings showing that artemisinin activity was significantly reduced in the presence of free radical scavengers and antioxidants (Hamacher-Brady et al., 2011), but enhanced in the presence of pro-oxidants (Krungkrai and Yuthawang, 1987). Recently, Gopalakrishnan and Kumar (2014) demonstrated artemisinin-induced parasite death due to an increase in both intercellular reactive oxygen species (ROS level) and DNA double-strand breaks (oxidative stress–mediated parasite damage) in *P. falciparum*. Asawamahasakda et al. (1994) demonstrated that artemisinin interacts covalently by alkylation of the PfTCTP molecule. This is an essential plasmodial biomolecule and a histamine-releasing factor that is involved in malaria treatment (Telerman and Amson, 2009; Bhisutthibhan et al., 1998; Asawamahasakda et al., 1994). Though it was not fully understood whether TCTP might function as a target for artemisinin, Eichhorn et al. (2013) have recently confirmed a role of TCTP in mediating artemisinin's effect due to its very high Kd values (77–120 μM), which are several orders of magnitude higher than the artemisinin IC_{50} for parasite growth inhibition.

A decade ago, another potential target for artemisinin action was identified as PfATP6, *P. falciparum*'s sarcoendoplasmic reticulum calcium ATPase (SERCA) (Eckstein-Ludwig et al., 2003). It reduces cytosolic free Ca^{2+} by pumping two Ca^{2+} into the endoplasmic reticulum in exchange for $<4H^+$, a mechanism critical for parasite survival (Brini and Carafoli, 2009). Eckstein-Ludwig et al. (2003) have also suggested that the mechanism of artemisinin-mediated PfATP6 inhibition is highly specific, not affecting any other malarial transporter, including the non-SERCA Ca^{2+} ATPase (PfATP4), and possibly functions through an allosteric mechanism (Shandilya et al., 2013). Further experiments dealing with the mutation of a single residue (L263) that directly modulates PfATP6 sensitivity to artemisinin suggest a potential resistance mechanism.

Artemisinin resistance has now been reported and is generally characterized by a delayed clearance of the malarial parasite. Based on research findings, it has been proposed to use PfATP6 as a molecular marker for artemisinin resistance. However, mutations in the Kelch 13 (K13) propeller domain have recently been demonstrated as important determinants for artemisinin resistance both *in vivo* and *in vitro* (Ariey et al., 2014; Straimer et al., 2015). Mok et al. (2015) conducted population transcriptomics of 1043 human malarial parasites and revealed that elevated expression of unfolded protein response (UPR) pathways, including the major PROSE and TRiC chaperon complexes, is the possible underlying mechanism behind artemisinin resistance. Further, they provided mechanistic evidence that artemisinin-resistant parasites upregulate the UPR pathway to overcome the protein damage caused by artemisinin. Hartwig et al. demonstrated in 2009 that artemisinin may cause parasite membrane damage by its accumulation within the neutral lipids of the parasite's membrane and suggested that the endoperoxide moiety of artemisinin may play a role in such effects. Experiments conducted on *Saccharomyces cerevisiae*, in which targeted deletion of genes encoding mitochondrial NADH dehydrogenases (NDE 1 or NDI 1) resulted in resistance to the inhibitory effects of artemisinin, whereas their overexpression increased sensitivity to artemisinin, suggested a mode of action associated with mitochondrial functions also (Li et al., 2005). This idea was supported by Wang et al. (2010), who demonstrated artemisinin-induced mitochondrial membrane depolarization via locally generated ROS that led to mitochondrial malfunctioning and ultimately parasite death.

Within the host erythrocytes, the parasite degrades host hemoglobin into "heme" and "globin" moieties through a series of proteases into its food vacuole. Further catabolism releases short peptides and amino acids, which are needed for parasite nutrition. It also releases a toxic by-product, hematin, which undergoes biomineralization to form non-toxic hemozoin (malaria pigment). It is proposed that Fe^{2+}-heme-activated artemisinin interferes in this catabolic pathway through heme alkylation at α-, β-, and δ-carbon atoms, which later create toxicity to the parasite (Meunier et al., 2010; Cazelles et al., 2001). Two proteins related to this catabolic pathway, heme detoxification protein (HDP) and histidine-rich protein II (HRP II), have been posited as targets of artemisinin action (Chugh et al., 2013).

An interesting demonstration regarding the interaction of heme and mitochondria with artemisinin has recently been revealed (Sun et al., 2015). It has been suggested that heme and mitochondria play distinct roles in mediating artemisinin-induced damage. The heme pathway has been shown to mediate artemisinin's killing of tumor cells, whereas the antimalarial action of artemisinin is driven by specific mitochondrial pathways.

To further uncover artemisinin's mode of action, gene expression analysis was undertaken and revealed that alterations in two genes (one encoding heat shock protein [HSP70] and the other encoding a hypoxanthine phosphoribosyl transferase [PF10-0121] related to purine biosynthesis) were associated with decreased sensitivity to artemisinin. Whether or not these proteins are directly part of artemisinin's mode of action is unclear, however.

14.5 CONCLUSION

Malaria is a disease that places a heavy burden in poor countries, particularly African countries, due to the lack of access to effective services for prevention, diagnosis, and treatment. In addition to causing untold suffering and disability, malaria ranks as one of the world's major killers, costing about 1.5 million human lives annually. This is the only disease from which most children in the poorest countries die after exposure to infection. Further, it is projected to remain at this level till 2030. A key setback in finding an effective cure for common malaria, cerebral malaria, and emerging multidrug-resistant malaria is the accessibility of artemisinin, the raw material used for manufacturing ACTs. The number of doses of ACTs purchased from the manufacturers by both public and private sectors has significantly increased from 11 million in 2005 to 392 million in 2013 (WHO, 2016). Therefore, tremendous research in different laboratories around the world has been conducted to meet the demand for artemisinin. Further, the exploration of the mode of action of artemisinin could open the door to the production of new ACTs. More studies, however, are needed to evaluate the mechanism of action of artemisinin. This will ultimately lead to combating drug-resistant malaria in developing countries.

ACKNOWLEDGMENT

The authors are thankful to University Grant Commission (UGC), Government of India for financial support to the Department of Biotechnology under the scheme of UGC-SAP (DRS-1) to develop the advanced research facilities used in this chapter. A. Ali is thankful to UGC for providing a Research Fellowship under the UGC-SAP (BSR) scheme. P. Saxena is thankful to DRDO for providing a Research Fellowship under the financial project scheme.

REFERENCES

Ariey, F., Witkowski, B., Amaratunga, C., Beghain, J., Langlois, A. C., Khim, N., and Lim, P., 2014. A molecular marker of artemisinin-resistant *Plasmodium falciparum* malaria. *Nature*. 505, 50–55.

Asawamahasakda, W., Ittarat, I., Chang, C. C., McElroy, P., and Meshnick, S. R., 1994. Effects of antimalarials and protease inhibitors on plasmodial hemozoin production. *Molecular and Biochemical Parasitology* 67, 183–191.

Bhisutthibhan, J., Pan, X. Q., Hossler, P. A., Walker, D. J., Yowell, C. A., and Carlton, J., 1998. The *Plasmodium falciparum* translationally controlled tumor protein homolog and its reaction with the antimalarial drug artemisinin. *Journal of Biological Chemistry* 273, 16192–16198.

Brini, M., and Carafoli, E., 2009. Calcium pumps in health and disease. *Physiological Reviews* 89, 1341–1378.

Cazelles, J., Robert, A., and Meunier, B., 2001. Alkylation of heme by artemisinin, an antimalarial drug. *Comptes Rendus de l'Académie des Sciences - Series IIC - Chemistry* 4, 85–89.

Chugh, M., Sundararaman, V., Kumar, S., Reddy, V. S., Siddiqui, W. A., Stuart, K. D., and Malhotra, P., 2013. Protein complex directs hemoglobin-to-hemozoin formation in *Plasmodium falciparum*. *Proceedings of the National Academy of Sciences* 110, 5392–5397.

Creek, D. J., Ryan, E., Charman, W. N., Chiu, F. C., Prankerd, R. J., Vennerstrom, J. L., and Charman, S. A., 2009. Stability of peroxide antimalarials in the presence of human hemoglobin. *Antimicrobial Agents and Chemotherapy* 53, 3496–3500.

D'Angelo, J. G., Bordón, C., Posner, G. H., Yolken, R., and Jones-Brando, L. 2009. Artemisinin derivatives inhibit *Toxoplasma gondii in vitro* at multiple steps in the lytic cycle. *Journal of Antimicrobial Chemotherapy* 63, 146–150.

Deng, Y., Ran, W., Man, S., Li, X., Gao, H., Tang, W., Tachibana, H., and Cheng, X. 2015. Artemether exhibits amoebicidal activity against *Acanthamoeba castellanii* through inhibition of the serine biosynthesis pathway. *Antimicrobial Agents and Chemotherapy* 59, 4680–4688.

Duke, M. V., Paul, R. N., Elsohly, H. N., Sturtz, G., and Duke, S. O., 1994. Localization of artemisinin and artemisitene in foliar tissues of glanded and glandless biotypes of *Artemisia annua*. *International Journal of Plant Sciences* 155, 365–372.

Eckstein-Ludwig, U., Webb, R. J., Van Goethem, I. D. A., East, J. M., Lee, A. G., Kimura, M., and Krishna, S., 2003. Artemisinins target the SERCA of *Plasmodium falciparum*. *Nature*. 424, 957–961.

Edikpo, N., Ghasi, S., Elias, A., and Oguanobi, N., 2013. Artemisinin and biomolecules: The continuing search for mechanism of action. *Molecular and Cellular Pharmacology* 5, 75–89.

Efferth, T., Dunstan, H., Sauerbrey, A., Miyachi, H., and Chitambar, C. R., 2001. The antimalarial artesunate is also active against cancer. *International Journal of Oncology* 18, 767–773.

Eichhorn, T., Winter, D., Buchele, B., Dirdjaja, N., Frank, M., Lehmann, W. D., and Efferth, T., 2013. Molecular interaction of artemisinin with translationally controlled tumor protein (TCTP) of *Plasmodium falciparum. Biochemical Pharmacology* 85, 38–45.

Elfawal, M. A., Towler, M. J., Reich, N. G., Golenbock, D., Weathers, P. J., and Rich, S. M., 2012. Dried whole plant *Artemisia annua* as an antimalarial therapy. *PLoS ONE* 7, e52746.

Giacometti, A., Cirioni, O., and Scalise, G. 1996. *In-vitro* activity of macrolides alone and in combination with artemisin, atovaquone, dapsone, minocycline or pyrimethamine against *Cryptosporidium parvum. Journal of Antimicrobial Chemotherapy* 38:399–408.

Gopalakrishnan, A. M., and Kumar, N., 2014. Anti-malarial action of artesunate involves DNA damage mediated by reactive oxygen species. *Antimicrobial Agents and Chemotherapy* 59, 317–325.

Gupta, S., Ghosh, P. K., Dutta, G., and Vishwakarma, R., 1995. *In vivo* study of artemisinin and its derivatives against primary amebic meningoencephalitis caused by *Naegleria fowleri. The Journal of Parasitology* 81, 1012–1013.

Hamacher-Brady, A., Stein, H. A., Turschner, S., Toegel, I., Mora, R., Jennewein, N., and Brady, N. R., 2011. Artesunate activates mitochondrial apoptosis in breast cancer cells via iron-catalyzed lysosomal reactive oxygen species production. *The Journal of Biological Chemistry* 286, 6587–6601.

Hartwig, C. L., Rosenthal, A. S., Dangelo, J., Griffin, C. E., Posner, G. H., and Cooper, R.A., 2009. Accumulation of artemisinin trioxane derivatives within neutral lipids of *Plasmodium falciparum* malaria parasites is endoperoxide-dependent. *Biochemical Pharmacology* 77(3), 322–336.

Haynes, R. K., Chan, W. C., Lung, C. M., Uhlemann, A. C., Eckstein, U., Taramelli, D., and Krishna, S., 2007. The Fe2+-mediated decomposition, PfATP6 binding, and antimalarial activities of artemisone and other artemisinins: The unlikelihood of C-centered radicals as bioactive intermediates. *ChemMedChem* 2, 1480–1497.

Iguchi, A., Matsuu, A., Matsuyama, K., and Hikasa, Y. 2015. The efficacy of artemisinin, artemether, and lumefantrine against *Babesia gibsoni in vitro. Parasitology International* 64:190–193.

Islamuddin, M., Chouhan, G., Tyagi, M., Abdin, M. Z., Sahal, D., and Afrin, F., 2014. Leishmanicidal activities of *Artemisia annua* leaf essential oil against visceral leishmaniasis. *Frontiers in Microbiology* 5: 626.

Kannan, R., Kumar, K., Sahal, D., Kukreti, S., and Chauhan, V. S., 2005. Reaction of artemisinin with haemoglobin: Implications for antimalarial activity. *Biochemical Journal* 385, 409–418.

Klayman, D. L., 1985. Qinghaosu (Artemisinin): An antimalarial drug from China. *Science* 228, 1049–1055.

Krungkrai, S. R., and Yuthawang, Y., 1987. The antimalarial action of *Plasmodium falciparum* of quinghaoso and artesunate in combination with agents which modulate oxidant stress. *Transactions of the Royal Society of Tropical Medicine and Hygiene* 81, 710–714.

Li, W., Mo, W., Shen, D., and Sun, L., 2005. Yeast model uncovers dual roles of mitochondria in the action of artemisinin. *PLoS Genetics* 1, 329–334.

Liu, R., Dong, H. F., Guo, Y., Zhao, Q. P., and Jiang, M. S., 2011. Efficacy of praziquantel and artemisinin derivatives for the treatment and prevention of human schistosomiasis: A systematic review and meta-analysis. *Parasites & Vectors* 4, 201.

Lubbe, A., Seibert, I., Klimkait, T., and Van der Kooy, F., 2012. Ethnopharmacology in overdrive: The remarkable anti-HIV activity of *Artemisia annua*. *Journal of Ethnopharmacology* 141, 854–859.

Meunier, F. A., Nguyen, T. H., Colasante, C., Luo, F., Sullivan, R. K., Lavidis, N. A., and Schiavo, G., 2010. Sustained synaptic-vesicle recycling by bulk endocytosis contributes to the maintenance of high-rate neurotransmitter release stimulated by glycerotoxin. *Journal of Cell Science* 123, 1131–1140.

Mishina, Y. V., Krishna, S., Haynes, R. K., and Meade, J.C. 2007. Artemisinin inhibits *Trypanosoma cruzi* and *Trypanosoma brucei rhodesiense in vitro* growth. *Antimicrobial Agents and Chemotherapy* 51, 1852–1854.

Müller, J., Balmer, V., Winzer, P., Rahman, M., Manser, V., Haynes, R. K., and Hemphill A. 2015. *In vitro* effects of new artemisinin derivatives in *Neospora caninum*-infected human fibroblasts. *International Journal of Antimicrobial Agents* 46, 88–93.

Mok, S., Ashley, E. A., Ferreira, P. E., Zhu, L., Lin, Z., Yeo, T., and Nguon, C., 2015. Population transcriptomics of human malaria parasites reveals the mechanism of artemisinin resistance. *Science* 347, 431–435.

Newton, P., and White, N., 1999. Malaria: New development in treatment and prevention. *Annual Review of Medicine* 50, 179–192.

Obeid, S., Alen, J., Pham, V. C., Meuleman, P., Pannecouque, C., Le, T. N., Neyts, J., Dehaen, W., and Paeshuyse, J., 2013. Artemisinin analogues as potent inhibitors of *in vitro* hepatitis C virus replication. *PLoS ONE* 8, e81783.

Paddon, C. J., Westfall, P., Pitera, D., Benjamin, K., Fisher, K., McPhee, D., Leavell, M., Tai, A., Main, A., and Eng, D., 2013. High-level semi-synthetic production of the potent antimalarial artemisinin. *Nature* 496, 528–532.

Romero, M. R., Efferth, T., Serrano, M. A., Castaño, B., Macias, R. I., Briz, O., and Marin, J. J., 2005. Effect of artemisinin/artesunate as inhibitors of hepatitis B virus production in an *in vitro* replicative system. *Antiviral Research* 68, 75–83.

Selmeczi, K., Robert, A., Claparols, C., and Meunier, B., 2004. Alkylation of human haemoglobin A0 by the antimalarial drug artemisinin. *FEBS Letters* 556, 245–248.

Shandilya, A., Chacko, S., Jayaram, B., and Ghosh, I., 2013. A plausible mechanism for the antimalarial activity of artemisinin: A computational approach. *Scientific Reports* 3, 251.

Singh, N. P., and Lai, H., 2001. Selective toxicity of dihydroartemisinin and holotransferrin on human breast cancer cells. *Life Sciences* 70, 49–56.

Straimer, J., Gnädig, N. F., Witkowski, B., Amaratunga, C., and Duru, V., 2015. Drug resistance K13-propeller mutations confer artemisinin resistance in *Plasmodium falciparum* clinical isolates. *Science* 347, 428–431.

Sun, C., Li, J., Cao, Y., Long, G., and Zhou, B., 2015. Two distinct and competitive pathways confer the cellcidal actions of artemisinins. *Microbial Cell* 2, 14–25.

Telerman, A., and Amson, R., 2009. The molecular programme of tumour reversion: The steps beyond malignant transformation. *Nature Reviews Cancer* 9, 206–216.

Tian, X. F., Shen, H. E., Li, J., Chen, Y., Yang, Z. H., and Lu, S. Q., 2010. The effects of dihydroartemisinin on *Giardia lamblia* morphology and cell cycle *in vitro*. *Parasitology Research* 107:369–375.

Van Noorden, R., 2010. Demand for malaria drug soars. *Nature* 466, 672–673.

Wang, J., Huang, L., Li, J., Fan, Q., Long, Y., Li, Y., and Zhou, B., 2010. Artemisinin directly targets malarial mitochondria through its specific mitochondrial activation. *PLoS ONE* 5, e9582.

World Health Organization, 2005. World malaria report 2005. WHO, Geneva.

World Health Organization, 2014. World malaria report 2014.

World Health Organization, 2015. World malaria report 2015. WHO, Geneva.

World Health Organisation, 2016. World malaria report. Geneva: World Health Organization. http://www.who.int/malaria/publications/world-malaria-report-2016/report/en/

Zhang, L., Jing, F., Li, F., Li, M., Wang, Y., Wang, G., Sun, X., and Tang, K., 2009. Development of transgenic *Artemisia annua* (Chinese wormwood) plants with an enhanced content of artemisinin, an effective anti-malarial drug, by hairpin-RNA mediated gene silencing. *Biotechnology and Applied Biochemistry* 52, 199–207.

Zhang, S., and Gerhard, G., 2008. Heme activates artemisinin more efficiently than hemin, inorganic iron or haemoglobin. *Bioorganic & Medicinal Chemistry* 16, 7853–7861.

Index